FIBER OPTICS
HANDBOOK

FIBER OPTICS HANDBOOK

Fiber, Devices, and Systems
for Optical Communications

Sponsored by the
OPTICAL SOCIETY OF AMERICA

Michael Bass   Editor in Chief

School of Optics / The Center for Research and Education in Optics and Lasers (CREOL)
University of Central Florida
Orlando, Florida

Eric W. Van Stryland   Associate Editor

School of Optics / The Center for Research and Education in Optics and Lasers (CREOL)
University of Central Florida
Orlando, Florida

McGRAW-HILL

New York   Chicago   San Francisco   Lisbon   London
Madrid   Mexico City   Milan   New Delhi   San Juan   Seoul
Singapore   Sydney   Toronto

Cataloging-in-Publication Data is on file with the Library of Congress

McGraw-Hill

A Division of The McGraw·Hill Companies

Copyright © 2002 by The McGraw-Hill Companies, Inc. All rights reserved. Printed in
the United States of America. Except as permitted under the United States Copyright
Act of 1976, no part of this publication may be reproduced or distributed in any form or
by any means, or stored in a data base or retrieval system, without the prior written permission of the publisher.

1 2 3 4 5 6 7 8 9 0   DOC/DOC   0 7 6 5 4 3 2 1

ISBN 0-07-138623-8

*The sponsoring editor for this book was Stephen S. Chapman and the production super-
visor was Pamela Pelton. It was set in Times Roman by North Market Street Graphics.*

Printed and bound by R. R. Donnelley & Sons Company.

 This book was printed on recycled, acid-free paper containing a minimum of
50% recycled, de-inked fiber.

*McGraw-Hill books are available at special quantity discounts to use as premiums and
sales promotions, or for use in corporate training programs. For more information, please
write to the Director of Special Sales, Professional Publishing, McGraw-Hill, Two Penn
Plaza, New York, NY 10121-2298. Or contact your local bookstore.*

CONTENTS

Chapter 9. Fiber Bragg Gratings   *Kenneth O. Hill*   **9.1**

Chapter 10. Micro-Optics-Based Components for Networking   *Joseph C. Palais*   **10.1**

Chapter 11. Semiconductor Optical Amplifiers and Wavelength Conversion
Ulf Österberg   **11.1**

Chapter 12. Optical Time-Division Multiplexed Communication Networks
Peter J. Delfyett   **12.1**

Chapter 13. Wavelength Domain Multiplexed (WDM) Fiber-Optic
Communication Networks   *Alan E. Willner and Yong Xie*   **13.1**

Chapter 14. Infrared Fibers   *James A. Harrington*   **14.1**

Chapter 15. Optical Fiber Sensors   *Richard O. Claus, Ignacio Matias, and Francisco Arregui*   **15.1**

Chapter 16. Fiber-Optic Communication Standards   *Casimer DeCusatis*   **16.1**

Index follows Chapter 16

CONTRIBUTORS

Francisco Arregui   *Public University Navarra, Pamplona, Spain* (CHAP. 15)

Tom G. Brown   *The Institute of Optics, University of Rochester, Rochester, New York* (CHAP. 1)

John A. Buck   *School of Electrical and Computer Engineering, Georgia Institute of Technology, Atlanta, Georgia* (CHAPS. 3 AND 5)

Richard O. Claus   *Virginia Tech, Blacksburg, Virginia* (CHAP. 15)

Casimer DeCusatis   *IBM Corporation, Poughkeepsie, New York* (CHAPS. 6 AND 16)

Peter J. Delfyett   *School of Optics/The Center for Research and Education in Optics and Lasers (CREOL), University of Central Florida, Orlando, Florida* (CHAP. 12)

Elsa Garmire   *Dartmouth College, Hanover, New Hampshire* (CHAP. 4)

James A. Harrington   *Rutgers University, Piscataway, New Jersey* (CHAP. 14)

Kenneth O. Hill   *New Wave Photonics, Ottawa, Ontario, Canada* (CHAP. 9)

Ira Jacobs   *Fiber and Electro-Optics Research Center, Virginia Polytechnic Institute and State University, Blacksburg, Virginia* (CHAP. 2)

Guifang Li   *School of Optics/The Center for Research and Education in Optics and Lasers (CREOL), University of Central Florida, Orlando, Florida* (CHAP. 6)

P. V. Mamyshev   *Bell Laboratories—Lucent Technologies, Holmdel, New Jersey* (CHAP. 7)

Ignacio Matias   *Public University Navarra Pamplona, Spain* (CHAP. 15)

Daniel Nolan   *Corning Inc., Corning, New York* (CHAP. 8)

Ulf Österberg   *Thayer School of Engineering, Dartmouth College, Hanover, New Hampshire* (CHAP. 11)

Joseph C. Palais   *Department of Electrical Engineering, College of Engineering and Applied Sciences, Arizona State University, Tempe, Arizona* (CHAP. 10)

Alan E. Willner   *Department of EE Systems, University of Southern California, Los Angeles, California* (CHAP. 13)

Yong Xie   *Department of EE Systems, University of Southern California, Los Angeles, California* (CHAP. 13)

PREFACE

Fiber optics has developed so rapidly during the last 30 years that it has become the backbone of our communications systems, critical to many medical procedures, the basis of many critical sensors, and utilized in many laser manufacturing applications. This book is part of the *Handbook of Optics*, Second Edition, Vol. IV, devoted to fiber optics and fiber optics communications.

The articles it contains cover both fiber optics and devices and systems for fiber optics communications. We thank Prof. Guifang Li of the School of Optics/CREOL and Dr. Casimir DeCusatis of IBM for organizing these articles and recruiting the authors. The result is a coherent and thorough presentation of the issues in fiber optics and in fiber optics communication systems. Some subjects covered in fiber optics overlap with the section in the *Handbook of Optics*, Second Edition, Vol. IV, on nonlinear and quantum optics. This is natural since the confinement of light in fibers produces high optical fields and long interaction lengths leading to important nonlinear effects.

This book contains 16 articles. The first is a general review of fiber optics and fiber optic communications that originally appeared in the *Handbook of Optics*, Second Edition, Vol. II. There are other articles from Vol. IV concerning fiber optic fundamentals and device issues. These include articles discussing nonlinear optical effects in fibers, sources, detectors, and modulators for communications, fiber amplifiers, fiber Bragg gratings, and infrared fibers. Fiber optics communications systems issues are treated in articles concerning telecommunication links, solitons, fiber couplers, MUX and deMUX, micro-optics for networking, semiconductor amplifiers and wavelength conversion, time and wavelength domain multiplexing, and fiber communications standards. An article on fiber optics sensors is also included.

The *Handbook of Optics*, Second Edition, and this topical volume are possible only through the support of the staff of the Optical Society of America and, in particular, Mr. Alan N. Tourtlotte and Ms. Laura Lee. We also thank Mr. Stephen Chapman of McGraw-Hill for his leadership in the production of this volume.

Michael Bass, *Editor-in-Chief*
Eric W. Van Stryland, *Associate Editor*

CHAPTER 1
OPTICAL FIBERS AND FIBER-OPTIC COMMUNICATIONS

Tom G. Brown
The Institute of Optics
University of Rochester
Rochester, New York

1.1 GLOSSARY

A	open loop gain of receiver amplifier
A	pulse amplitude
a	core radius
a_P	effective pump area
A_{eff}	effective (modal) area of fiber
A_i	cross-sectional area of ith layer
B	data rate
B_n	noise bandwidth of amplifier
c	vacuum velocity of light
D	fiber dispersion (total)
E_i	Young's modulus
e_{LO}, e_S	polarization unit vectors for signal and local oscillator fields
F	tensile loading
F_e	excess noise factor (for APD)
g_B	Brillouin gain
g_R	Raman gain
i_d	leakage current (dark)
I_m	current modulation
$I(r)$	power per unit area guided in single mode fiber
k	Boltzmann's constant
J_m	Bessel function of order m
K_m	modified Bessel function of order m

k_0	vacuum wave vector
l	fiber length
l_0	length normalization factor
L_D	dispersion length
m	Weibull exponent
M	modulation depth
N	order of soliton
n	actual number of detected photons
N_{eff}	effective refractive index
N_P	average number of detected photons per pulse
n_0	core index
n_1	cladding index
$n(r)$	radial dependence of the core refractive index for a gradient-index fiber
P	optical power guided by fiber
P_E	error probability
P_f	probability of fiber failure
P_s	received signal power
P_s	signal power
P_R	power in Raman-shifted mode
P_0	peak power
P_0	peak power of soliton
R	detector responsivity (A/W)
RIN	relative intensity noise
R_L	load resistor
$R(r)$	radial dependence of the electric field
S	failure stress
SNR	signal-to-noise ratio measured in a bandwidth B_n
S_0	location parameter
T	temperature (Kelvin)
t	time
T_0	pulse width
U	normalized pulse amplitude
z	longitudinal coordinate
$Z(z)$	longitudinal dependence of the electric field
α	profile exponent
α_f	frequency chirp
α_R	attenuation of Raman-shifted mode
$\tilde{\beta}$	complex propagation constant
β_1	propagation constant
β_2	dispersion (2d order)
Δ	peak index difference between core and cladding

Δf	frequency deviation
ΔL	change in length of fiber under load
$\Delta \phi$	phase difference between signal and local oscillator
$\Delta \nu$	source spectral width
$\Delta \tau$	time delay induced by strain
ε	strain
η_{HET}	heterodyne efficiency
θ_c	critical angle
$\Theta(\theta)$	azimuthal dependence of the electric field
λ	vacuum wavelength
λ_c	cut-off wavelength
ξ	normalized distance
$\Psi(r, \theta, z)$	scalar component of the electric field
Ψ_S, Ψ_{LO}	normalized amplitude distributions for signal and LO
σ_A^2	amplifier noise
$\sigma_d^2 = 2ei_d B_n$	shot noise due to leakage current
$\sigma_J^2 = \dfrac{4kT}{R_L} B_n$	Johnson noise power
$\sigma_R^2 = R^2 P_s^2 B_n \times 10^{-(\text{RIN}/10)}$	receiver noise due to source RIN
$\sigma_s^2 = 2eRP_s B_n Fe$	signal shot noise
τ	time normalized to moving frame
r, θ, z	cylindrical coordinates in the fiber

1.2   INTRODUCTION

Optical fibers were first envisioned as optical elements in the early 1960s. It was perhaps those scientists well-acquainted with the microscopic structure of the insect eye who realized that an appropriate bundle of optical waveguides could be made to transfer an image and the first application of optical fibers to imaging was conceived. It was Charles Kao[1] who first suggested the possibility that low-loss optical fibers could be competitive with coaxial cable and metal waveguides for telecommunications applications. It was not, however, until 1970 when Corning Glass Works announced an optical fiber loss less than the benchmark level of 10 dB/km[2,3] that commercial applications began to be realized. The revolutionary concept which Corning incorporated and which eventually drove the rapid development of optical fiber communications was primarily a materials one—it was the realization that low doping levels and very small index changes could successfully guide light for tens of kilometers before reaching the detection limit. The ensuing demand for optical fibers in engineering and research applications spurred further applications. Today we see a tremendous variety of commercial and laboratory applications of optical fiber technology. This chapter will discuss important fiber properties, describe fiber fabrication and chemistry, and discuss materials trends and a few commercial applications of optical fiber.

While it is important, for completeness, to include a treatment of optical fibers in any handbook of modern optics, an exhaustive treatment would fill up many volumes all by itself. Indeed, the topics covered in this chapter have been the subject of monographs, reference books, and textbooks; there is hardly a scientific publisher that has not published several books on fiber optics. The interested reader is referred to the "Further Reading" section at the end of this chapter for additional reference material.

Optical fiber science and technology relies heavily on both geometrical and physical optics, materials science, integrated and guided-wave optics, quantum optics and optical physics, communications engineering, and other disciplines. Interested readers are referred to other chapters within this collection for additional information on many of these topics.

The applications which are discussed in detail in this chapter are limited to information technology and telecommunications. Readers should, however, be aware of the tremendous activity and range of applications for optical fibers in metrology and medicine. The latter, which includes surgery, endoscopy, and sensing, is an area of tremendous technological importance and great recent interest. While the fiber design may be quite different when optimized for these applications, the general principles of operation remain much the same. A list of references which are entirely devoted to optical fibers in medicine is listed in "Further Reading."

1.3 PRINCIPLES OF OPERATION

The optical fiber falls into a subset (albeit the most commercially significant subset) of structures known as dielectric optical waveguides. The general principles of optical waveguides are discussed elsewhere in Chap. 6 of Vol. II, "Integrated Optics"; the optical fiber works on principles similar to other waveguides, with the important inclusion of a cylindrical axis of symmetry. For some specific applications, the fiber may deviate slightly from this symmetry; it is nevertheless fundamental to fiber design and fabrication. Figure 1 shows the generic optical fiber design, with a core of high refractive index surrounded by a low-index cladding. This index difference requires that light from inside the fiber which is incident at an angle greater than the critical angle

$$\theta_c = \sin^{-1}\left(\frac{n_1}{n_0}\right) \tag{1}$$

be totally internally reflected at the interface. A simple geometrical picture appears to allow a continuous range of internally reflected rays inside the structure; in fact, the light (being a wave) must satisfy a self-interference condition in order to be trapped in the waveguide. There are only a finite number of paths which satisfy this condition; these are analogous to the propagating electromagnetic modes of the structure. Fibers which support a large number of modes (these are fibers of large core and large numerical aperture) can be adequately analyzed by the tools of geometrical optics; fibers which support a small number of modes must be characterized by solving Maxwell's equations with the appropriate boundary conditions for the structure.

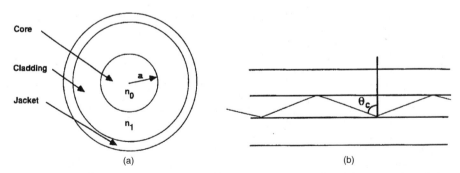

(a)   (b)

FIGURE 1   (a) Generic optical fiber design, (b) path of a ray propagating at the geometric angle for total internal reflection.

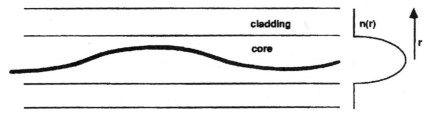

FIGURE 2   Ray path in a gradient-index fiber.

Fibers which exhibit a discontinuity in the index of refraction at the boundary between the core and cladding are termed *step-index fibers.* Those designs which incorporate a continuously changing index of refraction from the core to the cladding are termed *gradient-index fibers.* The geometrical ray path in such fibers does not follow a straight line—rather it curves with the index gradient as would a particle in a curved potential (Fig. 2). Such fibers will also exhibit a characteristic angle beyond which light will not internally propagate. A ray at this angle, when traced through the fiber endface, emerges at an angle in air which represents the maximum geometrical acceptance angle for rays *entering* the fiber; this angle is the numerical aperture of the fiber (Fig. 3). Both the core size and numerical aperture are very important when considering problems of fiber-fiber or laser-fiber coupling. A larger core and larger

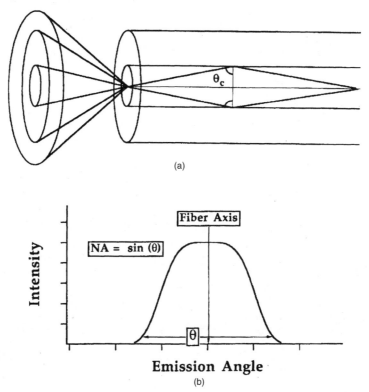

FIGURE 3   The numerical aperture of the fiber defines the range of external acceptance angles.

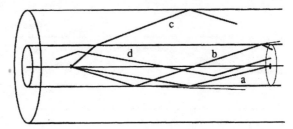

FIGURE 4 Classification of geometrical ray paths in an optical fiber. (*a*) Meridional ray; (*b*) leaky ray; (*c*) ray corresponding to a cladding mode; (*d*) skew ray.

numerical aperture will, in general, yield a higher coupling efficiency. Coupling between fibers which are mismatched either in core or numerical aperture is difficult and generally results in excess loss.

The final concept for which a geometrical construction is helpful is ray classification. Those geometrical paths which pass through the axis of symmetry and obey the self-interference condition are known as *meridional rays*. There are classes of rays which are nearly totally internally reflected and may still propagate some distance down the fiber. These are known as *leaky rays* (or modes). Other geometrical paths are not at all confined in the core, but internally reflect off of the cladding-air (or jacket) interface. These are known as *cladding modes*. Finally, there exists a class of geometrical paths which are bound, can be introduced outside of the normal numerical aperture of the fiber, and do not pass through the axis of symmetry. These are often called *skew rays*. Figure 4 illustrates the classification of geometrical paths.

Geometrical optics has a limited function in the description of optical fibers, and the actual propagation characteristics must be understood in the context of guided-wave optics. For waveguides such as optical fibers which exhibit a small change in refractive index at the boundaries, the electric field can be well described by a scalar wave equation,

$$\nabla^2 \Psi(r, \theta, z) + k_0^2 r^2(r) \Psi(r, \theta, z) = 0 \tag{2}$$

the solutions of which are the modes of the fiber. $\Psi(r, \theta, z)$ is generally assumed to be separable in the variables of the cylindrical coordinate system of the fiber:

$$\Psi(r, \theta, z) = R(r)\Theta(\theta)Z(z) \tag{3}$$

This separation results in the following eigenvalue equation for the radial part of the scalar field:

$$\frac{d^2R}{dr^2} + \frac{1}{r}\frac{dR}{dr} + \left(k_0^2 n^2(r) - \beta^2 - \frac{m^2}{r^2}\right)R = 0 \tag{4}$$

in which m denotes the azimuthal mode number, and β is the propagation constant. The solutions must obey the necessary continuity conditions at the core-cladding boundary. In addition, guided modes must decay to zero outside the core region. These solutions are readily found for fibers having uniform, cylindrically symmetric regions but require numerical methods for fibers lacking cylindrical symmetry or having an arbitrary index gradient. A common form of the latter is the so-called α-*profile* in which the refractive index exhibits the radial gradient.[4]

$$m(r) = \begin{cases} n_1\left[1 - \Delta\left(\dfrac{r}{a}\right)^\alpha\right] & r < a \\ n_1[1 - \Delta] = n_2 & r \geq a \end{cases} \tag{5}$$

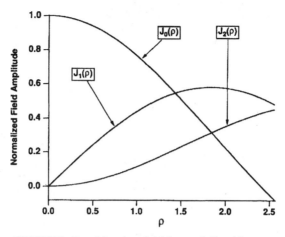

FIGURE 5   Bessel functions $J_m(\rho)$ for $m = 0, 1$, and 2.

The step-index fiber of circular symmetry is a particularly important case, because analytic field solutions are possible and the concept of the "order" of a mode can be illustrated. For this case, the radial dependence of the refractive index is the step function

$$n(r) = \begin{cases} n_1 \ r < a \\ n_2 \ r \geq a \end{cases} \qquad (6)$$

The solutions to this are Bessel functions[5] and are illustrated in Fig. 5. It can be seen that only the lowest-order mode ($m = 0$) has an amplitude maximum at the center. Its solution in the (core) propagating region ($r < a$) is

$$R(r) = \frac{J_0\left((n_1^2 k_0^2 - \beta^2)^{1/2}\left(\dfrac{r}{a}\right)\right)}{J_0((n_1^2 k_0^2 - \beta^2)^{1/2})} \qquad (7)$$

while the solution in the cladding ($r > a$) is the modified Bessel function

$$R(r) = \frac{K_0\left((\beta^2 - n_2^2 k_0^2)^{1/2}\left(\dfrac{r}{a}\right)\right)}{K_0((\beta^2 - n_2^2 k_0^2)^{1/2})} \qquad (8)$$

Higher-order modes will have an increasing number of zero crossings in the cross section of the field distribution.

Fibers which allow more than one bound solution for each polarization are termed *multimode* fibers. Each mode will propagate with its own velocity and have a unique field distribution. Fibers with large cores and high numerical apertures will typically allow many modes to propagate. This often allows a larger amount of light to be transmitted from incoherent sources such as light-emitting diodes (LEDs). It typically results in higher attenuation and dispersion, as discussed in the following section.

By far the most popular fibers for long distance telecommunications applications allow only a single mode of each polarization to propagate. Records for low dispersion and attenuation have been set using single-mode fibers, resulting in length-bandwidth products exceeding 10 Gb-km/s. In order to restrict the guide to single-mode operation, the core diameter must typically be 10 μm or less. This introduces stringent requirements for connectors and splices and increases the peak power density inside the guide. As will be discussed, this property of the single-mode fiber enhances optical nonlinearities which can act to either limit or increase the performance of an optical fiber system.

1.4 FIBER DISPERSION AND ATTENUATION

Attenuation

In most cases, the modes of interest exhibit a complex exponential behavior along the direction of propagation z.

$$Z(z) = \exp(i\tilde{\beta}z) \tag{9}$$

β is generally termed the propagation constant and may be a complex quantity. The real part of β is proportional to the phase velocity of the mode in question, and produces a phase shift on propagation which changes rather rapidly with optical wavelength. It is often expressed as an effective refractive index for the mode by normalizing to the vacuum wave vector:

$$N_{\text{eff}} = \frac{Re\{\tilde{\beta}\}}{k_0} \tag{10}$$

The imaginary part of β represents the loss (or gain) in the fiber and is a weak (but certainly not negligible) function of optical wavelength. Fiber attenuation occurs due to fundamental scattering processes (the most important contribution is Rayleigh scattering), absorption (both the OH-absorption and the long-wavelength vibrational absorption), and scattering due to inhomogeneities arising in the fabrication process. Attenuation limits both the short- and long-wavelength applications of optical fibers. Figure 6 illustrates the attenuation characteristics of a typical fiber.

The variation of the longitudinal propagation velocity with either optical frequency or path length introduces a fundamental limit to fiber communications. Since signaling necessarily requires a nonzero bandwidth, the dispersion in propagation velocity between different frequency components of the signal or between different modes of a multimode fiber produces a signal distortion and intersymbol interference (in digital systems) which is unacceptable. Fiber dispersion is commonly classified as follows.

Intermodal Dispersion

The earliest telecommunications links as well as many modern data communications systems have made use of multimode fiber. These modes (which we have noted have some connection to geometrical ray angles) will typically have a broad range of propagation velocities. An optical pulse which couples to this range of guided modes will tend to

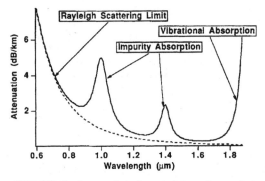

FIGURE 6 Attenuation characteristics of a typical fiber: (*a*) schematic, showing the important mechanisms of fiber attenuation.

broaden by an amount equal to the mean-squared difference in propagation time among the modes. This was the original purpose behind the gradient-index fiber; the geometrical illustrations of Figs. 1 and 2 show that, in the case of a step-index fiber, a higher-order mode (one with a steeper geometrical angle or a higher mode index m) will propagate by a longer path than an axial mode. A fiber with a suitable index gradient will support a wide range of modes with nearly the same phase velocity. Vassell was among the first to show this,[6] and demonstrated that a hyperbolic secant profile could very nearly equalize the velocity of all modes. The α-profile description eventually became the most popular due to the analytic expansions it allows (for certain values of α) and the fact that it requires the optimization of only a single parameter.

Multimode fibers are no longer used in long distance (>10 km) telecommunications due to the significant performance advantages offered by single-mode systems. Many short-link applications, for which intermodal dispersion is not a problem, still make use of multimode fibers.

Material Dispersion

The same physical processes which introduce fiber attenuation also produce a refractive index which varies with wavelength. This intrinsic, or material, dispersion is primarily a property of the glass used in the core, although the dispersion of the cladding will influence the fiber in proportion to the fraction of guided energy which actually resides outside the core. Material dispersion is particularly important if sources of broad spectral width are used, but narrow linewidth lasers which are spectrally broadened under modulation also incur penalties from material dispersion. For single-mode fibers, material dispersion must always be considered along with waveguide and profile dispersion.

Waveguide and Profile Dispersion

The energy distribution in a single-mode fiber is a consequence of the boundary conditions at the core-cladding interface, and is therefore a function of optical frequency. A change in frequency will therefore change the propagation constant independent of the dispersion of the core and cladding materials; this results in what is commonly termed *waveguide dispersion*. Since dispersion of the core and cladding materials differs, a change in frequency can result in a small but measurable change in index profile, resulting in *profile dispersion* (this contribution, being small, is often neglected). Material, waveguide, and profile dispersion act together, the waveguide dispersion being of opposite sign to that of the material dispersion. There exists, therefore, a wavelength at which the total dispersion will vanish. Beyond this, the fiber exhibits a region of anomalous dispersion in which the real part of the propagation constant increases with increasing wavelength. Anomalous dispersion has been used in the compression of pulses in optical fibers and to support long distance soliton propagation.

Dispersion, which results in a degradation of the signal with length, combines with attenuation to yield a length limit for a communications link operating at a fixed bandwidth. The bandwidth-length product is often cited as a practical figure of merit which can include the effects of either a dispersion or attenuation limit.

Normalized Variables in Fiber Description

The propagation constant and dispersion of guided modes in optical fibers can be conveniently expressed in the form of normalized variables. Two common engineering problems are the determination of mode content and the computation of total dispersion. For example, commonly available single-mode fibers are designed for a wavelength range of 1.3 to 1.55 μm.

TABLE 1 Normalized Variables in the Mathematical Description of Optical Fibers

Symbol	Description
$k_0 = \dfrac{2\pi}{\lambda}$	Vacuum wave vector
a	Core radius
n_0	Core index
n_1	Cladding index
$\tilde{\beta} = \beta' + i\beta''$	Mode propagation constant
$\alpha = 2\beta''$	Fiber attenuation
$N_{\text{eff}} = \beta'/k_0$	Effective index of mode
$\Delta = \dfrac{n_0^2 - n_1^2}{2n_1^2}$	Normalized core-cladding index differences
$V = \sqrt{2}k_0\, an_1\Delta$	Normalized frequency
$b = \left(\dfrac{N_{\text{eff}}}{n_1} - 1\right)\Big/\Delta$	Normalized effective index
$f(r)$	Gradient-index shape factor
$\Gamma = \dfrac{\displaystyle\int_0^a f(r)\Psi^2(r)r\,dr}{\displaystyle\int_0^a \Psi^2(r)r\,dr}$	Profile parameter ($\Gamma = 1$ for step-index)

Shorter wavelengths will typically support two or more modes, resulting in significant inter-modal interference at the output. In order to guarantee single-mode performance, it is important to determine the single-mode cut-off wavelength for a given fiber. Normalized variables allow one to readily determine the cut-off wavelength and dispersion limits of a fiber using universal curves.

The normalized variables are listed in Table 1 along with the usual designations for fiber parameters. The definitions here apply to the limit of the "weakly-guiding" fiber of Gloge,[7] for which $\Delta \ll 1$. The cutoff for single-mode performance appears at a normalized frequency of $V = 2.405$. For values of V greater than this, the fiber is multimode. The practical range of frequencies for good single-mode fiber operation lie in the range.

$$1.8 < V < 2.4 \tag{11}$$

An analytic approximation for the normalized propagation constant b which is valid for this range is given by

$$b(V) \approx \left(1 - 1.1428 - \frac{0.996}{V}\right)^2 \tag{12}$$

Operation close to the cutoff $V = 2.405$ risks introducing higher-order modes if the fiber parameters are not precisely targeted. A useful expression which applies to step-index fibers relates the core diameter and wavelength at the single-mode cutoff.[5]

$$\lambda_{\text{cutoff}} = \left(\frac{\pi}{2.405}\right)(2a)n_0\sqrt{2\Delta} \tag{13}$$

Evaluation of Fiber Dispersion

Evaluation of the fiber dispersion requires:

1. Detailed material dispersion curves such as may be obtained from a Sellmeier formula.[4] The Sellmeier constants for a range of silica-based materials used in fiber fabrication are contained in Chap. 33 of Vol. II, "Crystals and Glasses."

2. Complete information about the fiber profile, including compositional as well as refractive index information.

3. Numerical evaluation of the effective indices of the modes in question *and their first and second derivatives.* Several authors have noted the considerable numerical challenge involved in this,[8,9] particularly since measurements of the refractive index/composition possess intrinsic uncertainties.

Figure 7 shows an example of the dispersion exhibited by a step-index single-mode fiber. Different components of the dispersion are shown in order to illustrate the point of zero dispersion near 1.3 μm. The section devoted to fiber properties will describe how profile control can shift the minimum dispersion point to the very low-loss window near 1.55 μm.

1.5 POLARIZATION CHARACTERISTICS OF FIBERS

The cylindrical symmetry of an optical fiber leads to a natural decoupling of the radial and tangential components of the electric field vector. These polarizations are, however, so nearly degenerate that a fiber of circular symmetry is generally described in terms of orthogonal linear polarizations. This near-degeneracy is easily broken by any stresses or imperfections which break the cylindrical symmetry of the fiber. Any such symmetry breaking (which may arise accidentally or be introduced intentionally in the fabrication process) will result in two orthogonally polarized modes with slightly different propagation constants. These two modes need not be linearly polarized; in general, they are two elliptical polarizations. Such polarization splitting is referred to as *birefringence.*

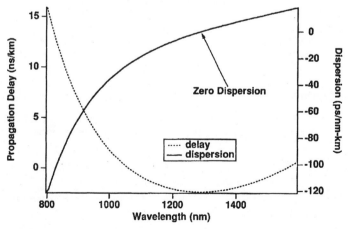

FIGURE 7   Dispersion of a typical single-mode fiber. The opposite contributions of the waveguide and material dispersion cancel near $\lambda = 1.3$ μm. (*Courtesy of Corning, Inc.*)

The difference in effective index between the two polarizations results in a state of polarization (SOP) which evolves through various states of ellipticity and orientation. After some propagation distance, the two modes will differ in phase by a multiple of 2π, resulting in a state of polarization identical to that at the input. This characteristic length is called the *beat length* between the two polarizations and is a measure of the intrinsic birefringence in the fiber. The time delay between polarizations is sometimes termed *polarization dispersion*, because it can have an effect on optical communication links which is similar to intermodal dispersion.

If this delay is much less than the coherence time of the source, coherence is maintained and the light in the fiber remains fully polarized. For sources of wide spectral width, however, the delay between the two polarizations may exceed the source coherence time and yield light which emerges from the fiber in a partially polarized or unpolarized state. The orthogonal polarizations then have little or no statistical correlation. The state of polarization of the output can have an important impact on systems with polarizing elements. For links producing an unpolarized output, a 3-dB power loss is experienced when passing through a polarizing element at the output.

The intentional introduction of birefringence can be used to provide polarization stability. An elliptical or double-core geometry will introduce a large birefringence, decoupling a pair of (approximately) linearly polarized modes.[10,11] It also will tend to introduce loss discrimination between modes. This combination of birefringence and loss discrimination is the primary principle behind polarization-maintaining fiber. As will be discussed in the description of optical fiber systems, there is a class of transmission techniques which requires control over the polarization of the transmitted light, and therefore requires polarization-maintaining fiber.

1.6   OPTICAL AND MECHANICAL PROPERTIES OF FIBERS

This section contains brief descriptions of fiber measurement methods and general information on fiber attenuation, dispersion, strength, and reliability. It should be emphasized that nearly all optical and mechanical properties of fibers are functions of chemistry, fabrication process, and transverse structure. Fibers are now well into the commercial arena and specific links between fiber structure, chemistry, and optical and mechanical properties are considered highly proprietary by fiber manufacturers. On the other hand, most fiber measurements now have established standards. We therefore give attention to the generic properties of fibers and the relevant evaluation techniques.

Attenuation Measurement

There are two general methods for the measurement of fiber attenuation. Source-to-fiber coupling must be taken into account in any scheme to measure attenuation, and destructive evaluation accomplishes this rather simply. The *cut-back method*[12,13] for attenuation measurement requires

1. Coupling light into a long length of fiber
2. Measuring the light output into a large area detector (so fiber-detector coupling remains constant)
3. Cutting the fiber back by a known distance and measuring the change in transmitted intensity

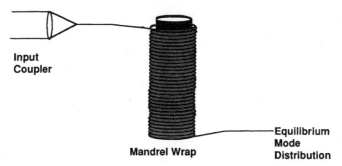

FIGURE 8   Mandrel wrap method of achieving an equilibrium mode distribution.

For single-mode fiber, the fiber can be cut back to a relatively short length provided that the cladding modes are effectively stripped. The concept of "mode stripping" is an important one for both attenuation and bandwidth measurements[14] (since modes near or just beyond cutoff can propagate some distance but with very high attenuation). If these modes are included in the measurement, the result yields an anomalously high attenuation. Lossy modes can be effectively stripped by a mandrel wrap or a sufficiently long length of fiber well-matched to the test fiber (see Fig. 8).

For multimode fiber (whether step-index or gradient-index) the excitation conditions are particularly important. This is because the propagating modes of a multimode fiber exhibit widely varying losses. If the laser used for performing the measurement is focused to a tight spot at the center of the core, a group of low-order modes may be initially excited. This group of lower-order modes will have lower loss and the first 10 to 1000 meters will show an anomalously low attenuation. As the propagation distance increases, lower-order modes gradually scatter into higher-order modes and the mode volume "fills up." The high-order modes are substantially lossier, so the actual power flow at equilibrium is that from the lower-order modes to the higher-order and out of the fiber. This process is illustrated in Fig. 9. It is easy to see that if the excitation conditions are set so that all modes guide approximately the same power at the input, the loss in the first hundred meters would be much higher than the equilibrium loss.

With modern single-mode splices, connectors, and couplers, it is sometimes possible to make nondestructive attenuation measurements simply by assuring that the connector loss is much less than the total loss of the fiber length being measured. With this method, care must be taken that the connector design exhibits no interference between fiber endfaces.

Connector loss measurements must have similar control over launch conditions. In addition, it is important to place a sufficiently long length of fiber (or short mandrel wrap) *after* the connector to strip the lossy modes. A slightly misaligned connector will often

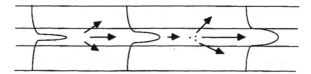

FIGURE 9   In a multimode fiber, low-order modes lose power to the high-order modes, and the high-order modes scatter into cladding and other lossy modes.

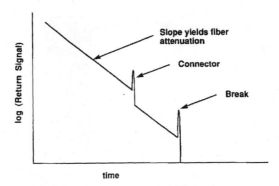

FIGURE 10   Typical OTDR signal. OTDR can be used for attenuation measurement, splice and connector evaluation, and fault location.

exhibit an extremely low loss prior to mode stripping. This is because power is coupled into modes which, while still guided, have high attenuation. It is important, in evaluation of fibers, to properly attribute this loss to the connector and not to the length of fiber which follows.

Another method of nondestructive evaluation of attenuation is optical time domain reflectometry (OTDR). The excitation of a fiber with a narrow laser pulse produces a continuous backscatter signal from the fiber. Assuming a linear and homogeneous scattering process, the reduction in backscattered light with time becomes a map of the round-trip attenuation versus distance. Sudden reductions in intensity typically indicate a splice loss, while a narrow peak will usually indicate a reflection. A typical OTDR signal is shown in Fig. 10. OTDR is extremely sensitive to excitation conditions—a fiber which is not properly excited will often exhibit anomalous behavior. Control of the launch conditions is therefore important for all methods of attenuation measurement.

A major theme of research and development in optical telecommunications has been the elimination of troublesome reflections from optical networks. In particular, high-return loss connectors have been developed which exhibit 30 to 40 dB of reflection suppression.[15–18] OTDR can be used to assess the reflection at network connections as well as perform on-line fault monitoring.

Dispersion and Bandwidth Measurement

The fiber has often been presented as the "multi-TeraHertz bandwidth transmission channel." While it is true that the total attenuation window of the fiber is extremely large by communications standards, the actual information bandwidth at any given wavelength is limited by the various sources of dispersion. The bandwidth of the fiber can be measured either in the time or frequency domain. Both measurements assume the fiber to be linear in its baseband (intensity) transfer characteristics. This assumption breaks down at very high powers and short pulses, but is nevertheless useful in most system applications.

The *time domain measurement*[19] measures the temporal broadening of a narrow input pulse. The ratio of the Fourier transform of the output intensity to that of the input yields a baseband transfer function for the fiber. If the laser and detector are linear, this transfer function relates the drive current of the laser to the photocurrent of the receiver and treats the fiber simply as a linear transmission channel of limited bandwidth. The use of the Fourier transform readily allows the phase to be extracted from the baseband transfer function. For intermodal pulse broadening in multimode fibers, this phase can be a nonlinear function of frequency, indicating a distortion as well as a broadening of the optical pulse.

Swept-frequency methods[20] have also been used for fiber evaluation. A pure sinusoidal modulation of the input laser is detected and compared in amplitude (and phase, if a network analyzer is available). In principle, this yields a transfer function similar to the pulse method. Both rely on the linearity of the laser for an accurate estimation, but since the swept-frequency method generally uses a single tone, the harmonics produced by laser nonlinearities can be rejected. Agreement between the two methods requires repeatable excitation conditions, a nontrivial requirement for multimode fibers.

The usual bandwidth specification of a multimode fiber is in the form of a 3-dB bandwidth (for a fixed length) or a length-bandwidth product. A single-mode fiber is typically specified simply in terms of the measured total dispersion. This dispersion can be measured either interferometrically, temporally, or using frequency domain techniques.

The *interferometric measurement*[21,22] is appropriate for short fiber lengths, and allows a detailed, direct comparison of the optical phase shifts between a test fiber and a reference arm with a suitable delay. This approach is illustrated in Fig. 11, which makes use of a Mach-Zehnder interferometer. This requires a source which is tunable, and one with sufficient coherence to tolerate small path differences between the two arms. The advantage of the approach is the fact that it allows measurements of extremely small absolute delays (a shift of one optical wavelength represents less than 10 fs time delay). It tends to be limited to rather short lengths of fiber; if a fiber is used in the reference arm to balance the interferometer, the properties of that fiber must be known with some accuracy.

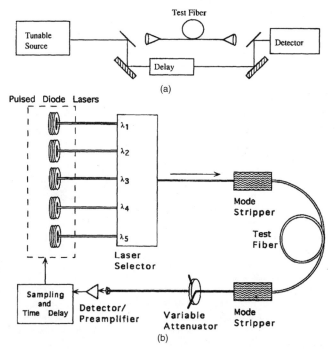

FIGURE 11 (*a*) Interferometric measurement of fiber dispersion; (*b*) time delay measurement of fiber dispersion.

Time-domain measurements[23] over a broad spectral range can be made provided a multiwavelength source is available with a sufficiently short optical pulse. One can make use of a series of pulsed diode lasers spaced at different wavelengths, use Raman scattering to generate many wavelengths from a single source, or make use of a tunable, mode-locked solid state laser. The relative delay between neighboring wavelengths yields the dispersion directly. This technique requires fibers long enough to adequately measure the delay, and the optical pulses must be weak enough not to incur additional phase shifts associated with fiber nonlinearities.

Frequency-domain or phase-shift measurements attempt to measure the effects of the dispersion on the baseband signal. A sinusoidally modulated signal will experience a phase shift with propagation delay; that phase shift can be readily measured electronically. This technique uses a filtered broadband source (such as an LED) or a CW, tunable, solid state source to measure the propagation delay as a function of wavelength.

Shifting and Flattening of Fiber Dispersion

A major dilemma facing system designers in the early 1980s was the choice between zero dispersion at 1.3 μm and the loss minimum at 1.55 μm. The loss minimum is an indelible consequence of the chemistry of silica fiber, as is the material dispersion. The waveguide dispersion can, however, be influenced by suitable profile designs.[24] Figure 12 illustrates a generic design which has been successfully used to shift the dispersion minimum to 1.55 μm.

The addition of several core and cladding layers to the fiber design allows for more complicated dispersion compensation to be accomplished. *Dispersion-flattened* fiber is designed for very low dispersion in an entire wavelength range; the spectral region from 1.3 to 1.6 μm is the usual range of interest. This is important for broadband WDM applications, for which the fiber dispersion must be as uniform as possible over a wide spectral region.

Reliability Assessment

The reliability of an optical fiber is of paramount importance in communications applications—long links represent large investments and require high reliability. There will, of course, always be unforeseen reliability problems. Perhaps the most famous such example was the fiber cable design on the first transatlantic link—the designers had not quite appreciated the Atlantic shark's need for a high-fiber diet. The sharks, apparently attracted by the scent of the cable materials, made short work of the initial cable installations. However, most of the stresses which an optical fiber will experience in the field can be replicated in the laboratory. A variety of accelerated aging models (usually relying on temperature as the accelerating factor) can be used to test for active and passive component reliability. In this section, we will review the reliability assessment of the fiber itself, referring interested readers to other sources for information on cable design.

Among the most important mechanical properties of the fiber in a wide range of applications is the tensile strength.[25] The strength is primarily measured destructively, by finding the maximum load just prior to fracture.[26] Full reliability information requires a knowledge of the maximum load, the relation between load and strain, a knowledge of the strain experienced by the fully packaged fiber, and some idea of how the maximum tolerable strain will change over long periods of time. One must finally determine the strain and associated failure probability for fibers with finite bends.

The tensile strength typically decreases slowly over time as the material exhibits fatigue, but in some cases can degrade rather rapidly after a long period of comparative strength. The former behavior is usually linked to fatigue associated with purely mechanical influences,

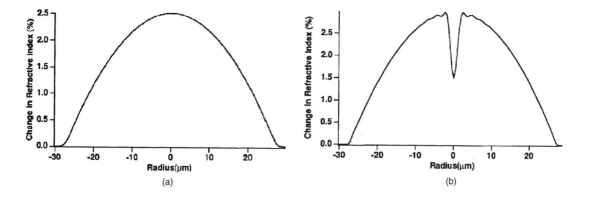

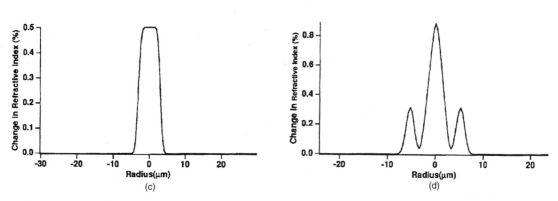

FIGURE 12   Typical index profiles for (*a*), (*b*) gradient-index multimode fiber; (*c*) step-index single-mode fiber; (*d*) dispersion-shifted fiber.

while the latter often indicates chemical damage to the glass matrix. The strain ε and tensile loading F are related through the fiber cross section and Young's modulus:[27]

$$\varepsilon = \frac{F}{\sum_i E_i A_i} \qquad (14)$$

E_i and A_i represent the Young's modulus and cross-sectional area of the i_{th} layer of the fiber-jacketing combination. Thus, if the Young's moduli are known, a measurement of the load yields the strain.

It is sometimes helpful to measure the fiber strain directly in cases where either the load or Young's moduli are not known. For example, a fiber does not necessarily have a uniform load after jacketing, cabling, and pulling; the load would (in any case) be a difficult quantity to measure. Using the relation between the strain and the optical properties of the fiber it is possible to infer the fiber strain from optical measurements. These techniques have been

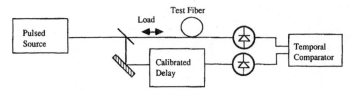

FIGURE 13 Single-pass technique for time-domain measurement of fiber strain.

successful enough to lead to the development of fiber strain gauges for use in mechanical systems.

Optical measurements of strain make use of the transit time of light through a medium of refractive index N_{eff}. (We will, for simplicity, assume single-mode propagation.) A change in length ΔL produced by a strain $\Delta L/L$ will yield a change in transit time

$$\frac{\Delta \tau}{\Delta L} = \frac{N_{eff}}{c}\left(1 + \frac{L}{N_{eff}} + \frac{dN_{eff}}{dL}\right) \tag{15}$$

For most cases of interest, the effective index is simply taken to be the value for that of the core. The ratio $\Delta \tau/\Delta L$ can be calculated (it is about 3.83 ns/m for a germania-silica fiber with $\Delta = 1\%$) or calibrated by using a control fiber and a measured load. It is important to note that this measurement yields only information on the *average* strain of a given fiber length.

There are three categories of optoelectronic techniques for measuring $\Delta \tau$; these are very similar to the approaches for dispersion measurement. A single-pass optical approach generally employs a short-pulse laser source passing through the fiber, with the delay of the transmitted pulse deduced by a comparison with a reference (which presumably is jitter-free). This is shown in Fig. 13. Figure 14a shows a multipass optoelectronic scheme, in which an optoelectronic oscillator circuit is set up with the fiber as a delay loop. The Q of the optoelectronic oscillator determines the effective number of passes in this measurement of optical delay. Finally, one can use an all-optical circuit in which the test fiber is placed in a fiber loop with weak optical taps to a laser and detector/signal processor (Fig. 14b). This "ring resonator" arrangement can also be set up with a fiber amplifier in the resonator to form the all-optical analog of the multipass optoelectronic scheme of Fig. 14a.

If the strain is being used to gain information about fiber reliability, it is necessary to understand how strain, load, and fiber failure are related. Fatigue, the delayed failure of the fiber, appears to be the primary model for fiber failure. One experimental evaluation of this process is to measure the mean time to failure as a function of the load on the fiber with the temperature, the chemical environment, and a host of other factors serving as control parameters.

Since the actual time to failure represents only the average of a performance distribution, the reliability of manufactured fibers is sometimes specified in terms of the two-parameter Weibull distribution[25,27-30]

$$P_f = 1 - \exp\left\{\left(\frac{l}{l_0}\right)\left(\frac{S}{S_0}\right)^m\right\} \tag{16}$$

where P_f denotes the cumulative failure probability and the parameters are as defined in Table 2. The Weibull exponent m is one of the primary descriptors of long-term fiber reliability. Figure 15 shows a series of Weibull plots associated with both bending and tensile strength measurements for low, intermediate, and high values of m.

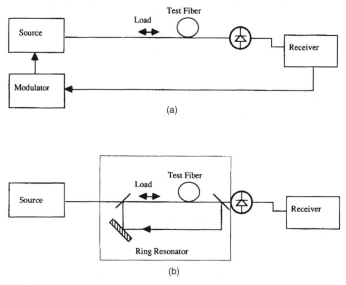

FIGURE 14 Multipass techniques for strain measurement. (*a*) Optoelectronic oscillator; (*b*) optical ring resonator.

One factor which has been shown to have a strong impact on reliability is the absolute humidity of the fiber environment and the ability of the protective coating to isolate the SiO_2 from the effects of H_2O. A recent review by Inniss, Brownlow, and Kurkjian[31] pointed out the correlation between a sudden change in slope, or "knee," in the time-to-failure curve and the H_2O content—a stark difference appeared between liquid and vapor environments. Before this knee, a combination of moisture and stress are required for fiber failure. In the case of fiber environments with a knee, a rather early mean time to failure will exist even for very low fiber stresses, indicating that chemistry rather than mechanical strain is responsible for the failure. The same authors investigated the effects of sodium solutions on the strength and aging of bare silica fibers.

1.7 OPTICAL FIBER COMMUNICATIONS

The optical fiber found its first large-scale application in telecommunications systems. Beginning with the first LED-based systems,[32,34,35] the technology progressed rapidly to longer

TABLE 2 Variables Used in the Weibull Distribution

l	Fiber length
l_0	Length normalization factor
S	Failure stress
S_0	Location parameter
m	Weibull exponent

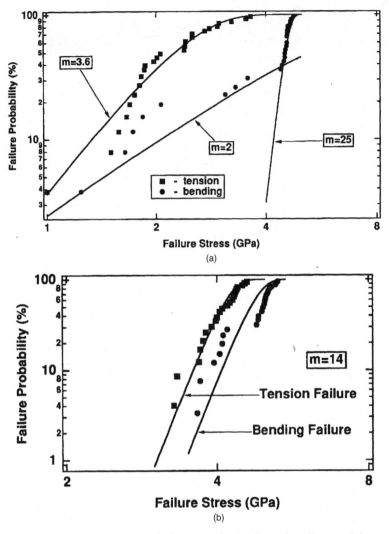

FIGURE 15   A series of Weibull plots comparing bending and tensile strength for (*a*) low, (*b*) intermediate, and (*c*) high values of the Weibull exponent *m*; (*d*) shows a typical mean time to failure plot. Actual fibers will often exhibit slope discontinuities, indicating a change in the dominant failure mechanism. (*Data Courtesy of Corning, Inc.*)

wavelengths and laser-based systems of repeater lengths over 30 km.[36] The first applications were primarily digital, since source nonlinearities precluded multichannel analog applications. Early links were designed for the 800- to 900-nm window of the optical fiber transmission spectrum, consistent with the emission wavelengths of the GaAs-AlGaAs materials system for semiconductor lasers and LEDs. The development of sources and detectors in the

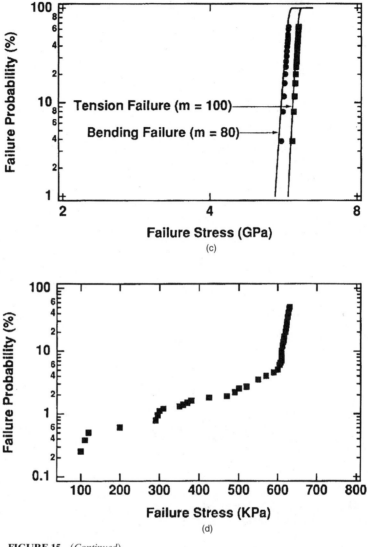

FIGURE 15   (*Continued*)

1.3- to 1.55-μm wavelength range and the further improvement in optical fiber loss over those ranges has directed most applications to either the 1.3-μm window (for low dispersion) or the 1.55-μm window (for minimum loss). The design of dispersion-shifted single-mode fiber along the availability of erbium-doped fiber amplifiers has solidified 1.55 μm as the wavelength of choice for high-speed communications.

The largest currently emerging application for optical fibers is in the local area network (LAN) environment for computer data communications, and the local subscriber loop for telephone, video, and data services for homes and small businesses. Both of these applications place a premium on reliability, connectivity, and economy. While existing systems still use point-to-point optical links as building blocks, there is a considerable range of networking components on the market which allow splitting, tapping, and multiplexing of optical components without the need for optical detection and retransmission.

Point-to-Point Links

The simplest optical communications system is the single-channel (no optical multiplexing) point-to-point digital link. As illustrated in Fig. 16, it consists of a diode laser (with associated driver circuitry and temperature control), optical fiber (with associated splices, connectors, and supporting material), and a detector (with appropriate electronics for signal processing and regeneration). The physics and principles of operation of the laser and detector are covered elsewhere in this collection (see Chap. 11 of Vol. I, "Lasers" Chap. 15 of Vol. I, "Photodetectors"), but the impact of certain device characteristics on the optical fiber communications link is of some importance.

Modulation and Source Characteristics. For information to be accurately transmitted, an appropriate modulation scheme is required. The most common modulation schemes employ direct modulation of the laser drive current, thereby achieving a modulation depth of 80 percent or better. The modulation depth is defined as

$$m = \frac{P_{max} - P_{min}}{P_{max} + P_{min}} \tag{17}$$

where P_{min} and P_{max} are the minimum and maximum laser power, respectively. The modulation depth is limited by the requirement that the laser always remain above threshold, since modulation near the lasing threshold results in a longer turn-on time, a broader spectrum, and higher source noise.

The transmitting laser contributes noise to the system in a fashion that is, generally speaking, proportional to the peak transmitted laser power. This noise is always evaluated as a fraction of the laser power and is therefore termed *relative intensity noise* (RIN). The RIN contribution from a laser is specified in dB/Hz, to reflect a spectral density which is approximately flat and at a fixed ratio (expressed in dB) to the laser power. Figure 17 shows a typical plot of the relative intensity noise of a source. The specification of RIN as a flat noise source is valid only at frequencies much less than the relaxation oscillation frequency and in situations where reflections are small.

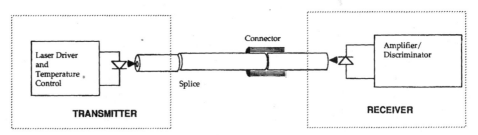

FIGURE 16   Typical point-to-point optical fiber communications link.

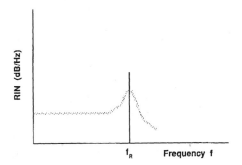

FIGURE 17 Typical RIN spectrum for a diode laser. The peak corresponds to the relaxation resonance frequency, f_R, of the laser.

The relative intensity noise is affected rather dramatically by the environment of the diode laser. A rather weak reflection back into the laser will both increase the magnitude of the relative intensity noise and modify its spectrum. As the reflection increases, it can produce self-pulsations and chaos in the output of the laser, rendering it useless for communications applications.[37] Thus, the laser cannot be thought of as an isolated component in the communications system. Just as RF and microwave systems require impedance matching for good performance, an optical communications system must minimize reflections. This is relatively easily accomplished for a long distance telecommunications link which makes use of low-reflection fusion splices. However, in a short link-network environment which must be modular, a small number of connectors can cause severe problems unless those connectors are designed to minimize reflections. It is now widely accepted that optical fiber connectors must be specified both in terms of insertion loss and reflection. A 1 percent reflection from a fiber connector can have far more serious implications for an optical fiber link than a 1 percent loss which is not reflected back to the laser. Optical isolators are available but only at considerable expense and are not generally considered economically realistic for network environments.

Impact of Fiber Properties on a Communications Link. For moderate power levels, the fiber is a passive, dispersive transmission channel. Dispersion can limit system performance in two ways. It results in a spreading of data pulses by an amount proportional to the spectral width of the source. This pulse spreading produces what is commonly termed "intersymbol interference." This should not be confused with an optical interference effect, but is simply the blurring of pulse energy into the neighboring time slot. In simple terms, it can be thought of as a reduction in the modulation depth of the signal as a function of link length. The effects of dispersion are often quantified in the form of a power penalty. This is simply a measure of the additional power required to overcome the effects of the dispersion, or bring the modulated power to what it would be in an identical link without dispersion. It is commonly expressed as a decibel ratio of the power required at the receiver compared to that of the ideal link.

Modulation-induced frequency chirp of the laser source will also result in pulse distortion. This is illustrated in Fig. 18, in which the drive current of the laser is modulated. The accompanying population relaxation produces a frequency modulation of the pulse. Upon transmission through a dispersive link, these portions of the pulse which are "chirped" will be advanced or retarded, resulting in both pulse distortion and intersymbol interference.

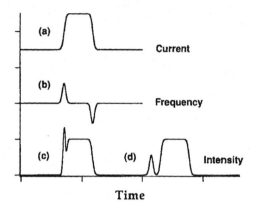

FIGURE 18 Modulation of the drive current in a semiconductor laser (*a*) results in both an intensity (*b*) and a frequency modulation (*c*). The pulse is distorted after transmission through the fiber (*d*).

System Design. The optical receiver must, within the signal bandwidth, establish an adequate signal-to-noise ratio (SNR) for accurate regeneration/retransmission of the signal. It must accomplish this within the constraints of the fiber dispersion and attenuation, the required system bandwidth, and the available source power. First-order system design normally requires the following steps:

1. Determine the maximum system bandwidth (or data rate for digital systems) and the appropriate transmission wavelength required for the system.

2. Find the maximum source RIN allowable for the system. For analog systems, in which a signal-to-noise ratio (SNR) must be specified in a bandwidth B_n, the RIN (which is usually specified in dB/Hz, indicating a measurement in a 1-Hz bandwidth) must obey the following inequality:

$$|RIN(dB/Hz)| \ll 10 \log (SNR \cdot B_n) \tag{18}$$

The SNR is specified here as an absolute ratio of carrier power to noise power. For an SNR specified as a decibel ratio,

$$|RIN(dB/Hz)| \ll SNR(dB) + 10 \log (B_n) \tag{19}$$

For digital systems, a Gaussian assumption allows a simple relationship between error probability (also termed bit error rate) and the signal-to-noise ratio:

$$P_E = 0.5 erfc[0.5(0.5SNR)^{1/2}] \tag{20}$$

Where *erfc* denotes the complementary error function and the decision threshold is assumed to be midway between the on and off states. The maximum error probability due to source noise should be considerably less than the eventual target error probability. For system targets of 10^{-9} to 10^{-12}, the error probability due to source RIN should be considerably less than 10^{-20}. This will allow at least a 3-dB margin to allow for increases in RIN due to device aging.

3. Establish a length limit associated with the source frequency chirp and spectral width. The frequency chirp α_f is specified in GHz per milliampere change in the drive current of the laser. A total current modulation I_m therefore yields a frequency deviation Δf of

$$\Delta f = I_m \alpha_f \tag{21}$$

This frequency deviation translates into a propagation delay via the fiber intramodal dispersion D. This delay must be kept less than the minimum pulse width (data rate). With D specified in ps/nm-km, the length in kilometers must obey the following inequality to avoid penalties due to frequency chirp:

$$L \ll \frac{c}{B\Delta f D \lambda_0^2} = \frac{c}{\alpha_f I_m B D \lambda_0^2} \tag{22}$$

where B denotes the data rate and is the reciprocal of the pulse width for data pulses that fill up an entire time slot. (These signals are designated non-return-to-zero, or NRZ.)

The length limit due to source spectral width Δv obeys a similar inequality—in this case, the delay associated with the spectral spread of the source must remain much less than one pulse width:

$$L \ll \frac{c}{\Delta v B D \lambda_0^2} \tag{23}$$

If the chirp is low and the unmodulated source bandwidth is less than the system bandwidth being considered, one must require that the delay distortion of the signal spectrum itself be small compared to a pulse width, requiring

$$L \ll \frac{c}{B^2 D \lambda_0^2} \tag{24}$$

For multimode fiber systems, the limiting length will generally be associated with the intermodal dispersion rather than the material and waveguide dispersion. A length-bandwidth product is generally quoted for such fibers. With the length and bandwidth limits established, it is now possible to design, within those limits, a receiver which meets the necessary specifications.

4. Determine the minimum power required at the receiver to achieve the target SNR or error probability. This minimum acceptable power (MAP) is first computed assuming an ideal source (no RIN contribution). A correction for the RIN can be carried out later. A computation of the MAP requires a knowledge of the noise sources and detector bandwidth. It is conventional to express the noise sources in terms of equivalent input noise current sources. The noise sources of importance for such systems are: the shot noise of the photocurrent, dark current, and drain current (in the case of a field effect transistor (FET) preamplifier); the Johnson noise associated with the load resistor or equivalent amplifier input impedance; 1/f noise from certain classes of FETs. The noise contributions from amplifiers other than the first stage are generally second-order corrections. Figure 19 shows a schematic of the receiver and relevant noise sources. Table 3 gives expressions for, and definitions of the important physical quantities which determine the receiver sensitivity.

Figure 20 illustrates two possible configurations for the detector/amplifier combination. Of these, the integrating front end is the simplest (particularly for high-frequency operation) but tends to be slower than a transimpedance amplifier with an equivalent load resistance. This is because the transimpedance amplifier reduces the effective input impedance of the circuit by $(A + 1)$, where A denotes the open loop gain of the amplifier.

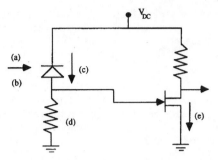

FIGURE 19 Schematic of the receiver, showing the introduction of noise into the system. Noise sources which may be relevant include (*a*) signal shot noise; (*b*) background noise (due to thermal background or channel crosstalk); (*c*) shot noise from the leakage current; (*d*) Johnson noise in the load resistor; (*e*) shot noise and 1/f noise in the drain current of the field effect transistor.

For equivalent bandwidth, the transimpedance amplifier exhibits a lower Johnson noise contribution since a higher feedback resistance is possible. It is worth mentioning that the transimpedance design tends to be much more sensitive to the parasitic capacitance which appears across the feedback resistor—small parasitics across the load resistor tend to be less important for the integrating front end.

The excess noise factor F_e is determined by the choice of detector. There are several choices over the wavelength range generally of interest for optical fiber transmission. (A detailed discussion of the principles of operation can be found in Chaps. 15–17 of Vol. I.)

TABLE 3 Symbols and Expressions for Receiver Noise

Symbol	Description
R_L	Load resistor
k	Boltzmann's constant
T	Temperature (Kelvin)
$\sigma_J^2 = \dfrac{4kT}{R_L} B_n$	Johnson noise power
R	Detector responsivity (A/W)
$P_g P$	Signal power
B_n	Noise bandwidth of amplifier
$\sigma_s^2 = 2eRP_s B_n Fe$	Signal shot noise
i_d	Leakage current (dark)
$\sigma_d^2 = 2ei_d B_n$	Shot noise due to leakage current
$\sigma_R^2 = R^2 P_s^2 B_n \times 10^{-(RIN/10)}$	Receiver noise due to source RIN
F_e	Excess noise factor (for APD)
σ_λ^2	Amplifier noise

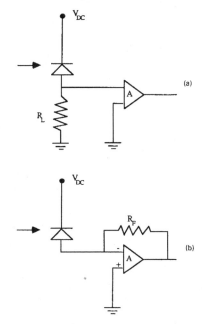

FIGURE 20 Two possible configurations for the detector/amplifier: (*a*) the integrating front end yields the simplest design for high speed operation; (*b*) the transimpedance amplifier provides an expansion of the receiver bandwidth by a factor of $A + 1$, where A is the open loop gain of the amplifier.

1. The p-i-n photodiode is described in some detail in Chap. 15 of Vol. I ("Photodetectors"). It can provide high quantum efficiencies and speeds in excess of 1 GHz. Dark currents range from less than 1 nA for silicon devices to 1 μA or more for Ge diodes. The dark current increases and the device slows down as the active area is increased.

2. The avalanche photodiode is a solid state device which exhibits internal multiplication of the photocurrent in a fashion that is sometimes compared with the gain in photomultiplier tubes. The multiplication does not come without a penalty, however, and that penalty is typically quantified in the form of an excess noise factor which multiplies the shot noise. The excess noise factor is a function both of the gain and the ratio of impact ionization rates between electrons and holes.

Figure 21 shows the excess noise factor for values of k ranging from 50 (large hole multiplication) to 0.03 (large electron multiplication). The former is claimed to be typical of certain III–V compounds while the latter is typical of silicon devices. Germanium, which would otherwise be the clear detector of choice for long wavelengths, has the unfortunate property of having k near unity. This results in maximum excess noise, and Ge avalanche photodiodes must typically be operated at low voltages and relatively small gains. The choice of a p-i-n detector, which exhibits no internal gain, yields $F_e = 1$.

3. The need for very high speed detectors combined with the fabrication challenges present in III-V detector technology has led to a renewed interest in Schottky barrier detectors for optical communications. A detector of considerable importance today is the metal-semiconductor-metal detector, which can operate at extremely high speed in an inter-

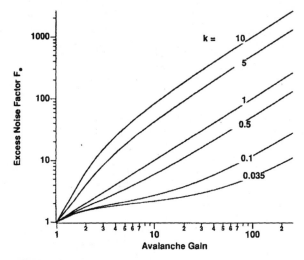

FIGURE 21 Excess noise factor for an avalanche photodiode with the electron/hole multiplication ratio k as a parameter. Small values of k indicate purely electron multiplication while large values of k indicate purely hole multiplication.

digitated electrode geometry. Chapter 17 of Vol. I provides further discussion of MSM detectors.

With all noise sources taken into account (see Table 3 for the relevant expressions), the signal-to-noise ratio of an optical receiver can be expressed as follows:

$$\text{SNR} = \frac{R^2 P_S^2}{\sigma_s^2 + \sigma_T^2} \tag{25}$$

where σ_T^2 denotes the total signal-independent receiver noise:

$$\sigma_T^2 = \sigma_D^2 + \sigma_J^2 + \sigma_A^2 \tag{26}$$

and σ_s^2 is the signal shot noise as in Table 3. If the effects of RIN are to be included, the following correction to the SNR may be made:

$$\text{SNR}^{-1} = \text{SNR}^{-1} + \sigma_R^2 \tag{27}$$

With the signal-to-noise ratio determined, the error probability may be expressed in terms of the signal-to-noise ratio

$$P_E = 0.5 erfc[0.5(0.5 \times \text{SNR})^{1/2}] \tag{28}$$

The just-noted expressions assume Gaussian distributed noise sources. This is a good assumption for nearly all cases of interest. The one situation in which the Gaussian assumption underestimates the effects of noise is for avalanche photodiodes with large excess noise. It was shown by McIntyre[38,39] and Personick[40] that the avalanche multiplication statistics are skewed and that the Gaussian assumption yields overly optimistic results.

5. Given the MAP of the receiver, the fiber attenuation and splice loss budget, and the available pigtailed laser power (the maximum power coupled into the first length of fiber by the laser), it is possible to calculate a link loss budget. The budget must include a substantial power margin to allow for device aging, imperfect splices, and a small measure of stupidity. The result will be a link length which, if shorter than the dispersion limit, will provide an adequate signal-to-noise ratio.

For further link modeling, a variety of approaches can be used to numerically simulate the link performance and fully include the effects of fiber dispersion, a realistic detector-preamplifier combination, and a variety of other factors which the first-order design does not include. Nevertheless, a first-order design is necessary to reduce the range of free parameters used in the simulation.

The ultimate goal of the point-to-point link is to transparently transmit the data (or the analog signal) in such a way that standard communications techniques may be used in the optical link. Examples include the use of block or error-correcting codes in digital systems, standard protocols for point-to-point links between nodes of a network, or frequency allocation in the case of a multichannel analog link.

Advanced Transmission Techniques

The optical bandwidth available in either of the low-loss transmission windows of the fiber exceeds 10^{13} Hz. Two ways of taking full advantage of this bandwidth are through the use of ultrashort pulse transmission combined with time-division multiplexing or the use of wavelength/frequency-division multiplexing. Either technique can overcome the limits imposed by the channel dispersion, but both techniques have their limitations. The first technique seeks to turn fiber dispersion to advantage; the second attempts to simply reduce the negative effects of dispersion on a broadband optical signal.

Ultrashort Pulse Transmission.   The most common form of multiplexing in digital communication systems is the combination of a number of low data rate signals into a single, high data rate signal by the use of time-division multiplexing. This requires much shorter optical pulses than are used in conventional transmission. As mentioned earlier, the normal (linear) limitation to the data rate is imposed by the fiber attenuation and dispersion. Both of these limits can be exceeded by the use of soliton transmission and optical amplification.

The physics of soliton formation[41-45] is discussed in "Nonlinear Optical Properties of Fibers," later in this chapter. Solitons, in conjunction with fiber amplifiers, have been shown to promise ultralong distance transmission without the need for optoelectronic repeaters/regenerators. Time-division multiplexing of optical solitons offers the possibility of extremely long distance repeaterless communications.

No communication technique is noise-free, and even solitons amplified by ideal amplifiers will exhibit phase fluctuations which broaden the spectrum and eventually cause the soliton to break up. This spontaneous-emission noise limit is known as the Gordon-Haus limit,[46] and had been thought to place a rather severe upper limit on the bit rate distance product for optical fiber systems. It has recently been noted,[47] that a unique series of linear filters can prevent the buildup of unwanted phase fluctuations in the soliton, thereby justifying amplified soliton transmission as a viable technology for undersea communications.

Such a communications system puts great demands on the signal processing both at the input and the output. For very high bit rates, one needs either all-optical demultiplexing or extremely fast electronic logic. Current limits on silicon logic are in the range of several Gb/s, which may be adequate for the first implementations of soliton transmission. It is anticipated that all-optical multiplexing and demultiplexing will be required in order to fully exploit the optical fiber bandwidth.

Solitons supported by an optical fiber bear a very specific relationship between pulse width T_0, peak power P_0, fiber dispersion D, effective area A_{eff}, and the intensity-dependent refractive index n_2. For a lowest-order ($N = 1$) soliton,

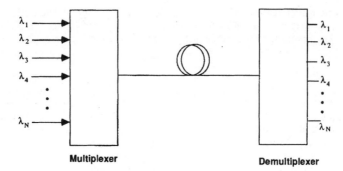

FIGURE 22 Schematic of a WDM transmission system. The main figures of merit are insertion loss (for both the multiplexer and demultiplexer) and channel crosstalk (for the demultiplexer).

$$T_0^2 = \frac{\lambda^3 D}{(2\pi)^2 n_2 (P_0/A_{\text{eff}})} \tag{29}$$

Under normal operation, a fiber will propagate lowest-order solitons of about 10 ps in duration. Even for a pulse train of comparatively high duty cycle, this represents less than 100 GHz of a much larger fiber bandwidth. To fully span the bandwidth requires wavelength-division multiplexing.

Wavelength-division Multiplexing (WDM). The troublesome delay between frequencies which is introduced by the fiber dispersion can also be overcome by division of the fiber transmission region into mutually incoherent (uncorrelated) wavelength channels. It is important for these channels to be uncorrelated in order to eliminate any worry about dispersion-induced delay between channels. Figure 22 shows a schematic picture of a WDM transmission system. The concept is quite simple, but reliable implementation can be a considerable challenge.

An attractive feature of WDM is the fact that the only active components of the system remain the optical sources and detectors. The multiplexers/demultiplexers are passive and are therefore intrinsically more reliable than active multiplexers. These schemes range from simple refractive/reflective beam combiners to diffractive approaches and are summarized in Fig. 23. For a multiplexing scheme, the key figure of merit is the insertion loss per channel. A simple 50-50 beam splitter for a two-channel combiner offers simple multiplexing with high insertion loss. If the beam splitter is coated to provide high reflectivity at one wavelength and high transmissivity at the other, the insertion loss is reduced, the coupler becomes wavelength-specific, and the element can act either as a multiplexer or demultiplexer.

Grating combiners offer an effective way to maximize the number of channels while still controlling the insertion loss. The grating shape must be appropriately designed—a problem which is easily solved for a single-wavelength, single-angle geometry. However, the diffraction efficiency is a function both of wavelength and angle of incidence. The optimum combination of a range of wavelengths over a wide angular range will typically require a tradeoff between insertion loss, wavelength range, and angular discrimination. Wavelength-division multiplexing technology has been greatly aided by the rapid advances in diffractive optics, synthetic holography, and binary optics in recent years. More on these subjects is included in Chap. 8 of Vol. II.

There have been considerable accomplishments in the past ten years in the fabrication of integrated optical components for WDM applications. Much of these involve the waveguide

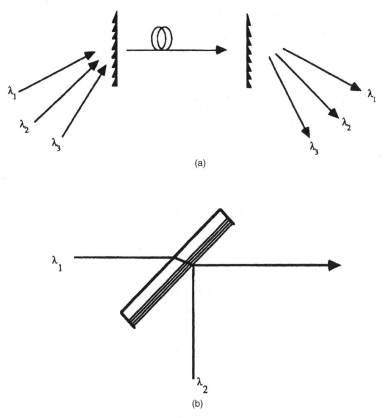

FIGURE 23   Multiplexing/demultiplexing schemes for WDM; (*a*) grating combiner (bulk optics); (*b*) wavelength selective beamsplitter (bulk optics); (*c*) directional coupler (integrated optics); (*d*) all-fiber multiplexer/demultiplexer.

equivalent of bulk diffractive optical elements. Since the optical elements are passive and efficient fiber coupling is required, glass waveguides have often been the medium of choice. A great variety of couplers, beam splitters, and multiplexer/demultiplexers have been successfully fabricated in ion-exchanged glass waveguides. Further details on the properties of these waveguides is contained in Chap. 36 of Vol. I. There has also been a major effort to fabricate low-cost polymer-based WDM components. These can be in the form of either waveguides or fibers.

From the point of view of connectivity and modular design, all-fiber WDM components are the most popular. Evanescent single-mode fiber couplers are inherently wavelength-sensitive and can be designed for minimum insertion loss. As with the bulk approaches, all-fiber components become more difficult to design and optimize as the number of channels increases. Most commercially available all-fiber components are designed for widely separated wavelength channels. For example, Corning, Inc. currently offers multiplexers designed for combining signals from 1.5-μm, 1.3-μm, and 0.8-μm sources.

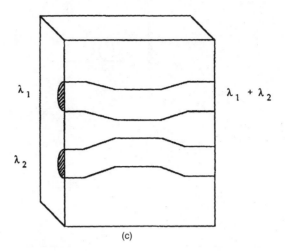

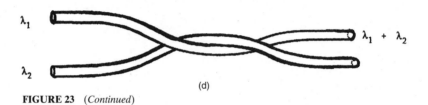

FIGURE 23 *(Continued)*

Advances in source fabrication technology in recent years have offered the possibility of fabricating diode laser arrays equipped with a controlled gradient in emission wavelength across the array. Such an array, equipped with appropriate beam-combining optics, could greatly reduce the packaging and alignment requirements in a large-scale WDM system. Minimizing crosstalk for closely spaced wavelength channels presents a significant challenge for demultiplexer design.

Coherent Optical Communications. Intensity modulation with direct detection remains the most popular scheme for optical communications systems. Under absolutely ideal transmission and detection conditions (no source RIN, perfect photon-counting detection, no background radiation), the probability of detecting n photons in a pulse train having an average of N_P photons per pulse would obey the Poisson distribution

$$p(n) = \frac{N_p^{\,n} e^{-N_P}}{n!} \tag{30}$$

The probability of an "error" P_E would be the detection of no photons during the pulse,

$$P_E = \exp{(-N_P)} \tag{31}$$

If we choose the benchmark error probability of 10^{-9}, we require an average of about 21 photons per pulse. This represents the quantum limit for the direct detection of optical signals. This limit can scarcely be reached, since it assumes no dark count and perfectly efficient photon counting.

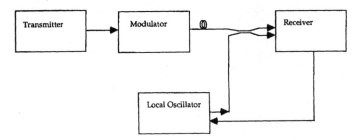

FIGURE 24   Generic coherent optical fiber communication link.

Current optical communication[48-54] offers a way to achieve quantum-limited receiver sensitivities even in the presence of receiver noise. By using either amplitude, phase, or frequency modulation combined with heterodyne or homodyne detection, it is possible to approach, and even exceed, the quantum limit for direct detection.

A generic coherent optical communication link is shown in Fig. 24. The crucial differences with direct detection lie in the role of the modulator in transmission and the presence of the local oscillator laser in reception. To understand the role of the modulator, we first consider the method of heterodyne detection. We will then discuss the component requirements for a coherent optical fiber communication link.

Heterodyne and Homodyne Detection.   We consider the receiver shown in Fig. 25, in which an optical signal arriving from some distant point is combined with an intense local oscillator laser by use of a 2×2 coupler. The power $I(r)$ guided in the single-mode fiber due to the interfering amplitudes can be expressed as

$$I(r) = P_S(t)|\Psi_S(r)|^2 + P_{LO}|\Psi_{LO}(r)|^2 + 2e_S(t) \cdot e_{LO}\Psi_S(r)\Psi_{LO}(r)\sqrt{P_S(t)P_{LO}}\cos(\omega_{IF}t + \Delta\phi) \quad (32)$$

in which $e_{LO}(t)$ and $e_S(t)$ denote the polarizations of the local oscillator and signal, P_{LO} and $P_S(t)$ denote the powers of the local oscillator and signal, $\Psi_S(r)$ and $\Psi_{LO}(r)$ are the spatial amplitude distributions, and $\Delta\phi(t)$ denotes the phase difference between the two sources. The two sources may oscillate at two nominally different frequencies, the difference being labeled the *intermediate frequency* ω_{IF} (from heterodyne radio nomenclature). If the intermediate fre-

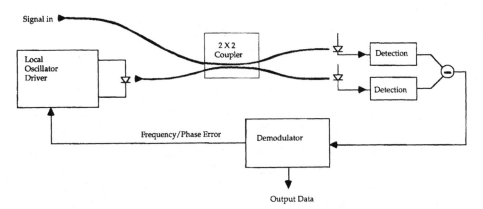

FIGURE 25   Heterodyne/homodyne receiver.

quency is zero, the detection process is termed *homodyne* detection; if a microwave or radio carrier frequency is chosen for postdetection processing, the detection process is referred to as *heterodyne* detection.

If the local oscillator power is much larger than the signal power, the first term is negligible. The second represents a large, continuous signal which carries no information but does provide a shot noise contribution. The third term represents the signal information. If the signal is coupled to a detector of responsivity R and ac-coupled to eliminate the local oscillator signal, the photocurrent $i(t)$ can be expressed as follows:

$$i(t) = 2R\eta_{\text{HET}}\sqrt{P_S(t)P_{LO}}\cos(\omega_{IF}t + \Delta\phi) \tag{33}$$

The heterodyne efficiency η_{HET} is determined by the spatial overlap of the fields and the inner product of the polarization components:

$$\eta_{\text{HET}} = (e_s(t) \cdot e_{LO})\int_{\substack{\text{Fiber} \\ \text{Area}}} \Psi_s(r)\Psi_{LO}(r)\,d^2r \tag{34}$$

These results illustrate four principles of coherent optical fiber communications:

1. The optical frequency and phase of the signal relative to those of the local oscillator are preserved, including the phase and frequency fluctuations.

2. The local oscillator "preamplifies" the signal, yielding a larger information-carrying component of the photocurrent than would be detected directly.

3. The local oscillator and signal fields must occupy the same spatial modes. Modes orthogonal to that of the local oscillator are rejected.

4. Only matching polarization components contribute to the detection process.

The first principle allows the detection of frequency or phase information, provided the local oscillator has sufficient stability. The second provides an improvement of the signal-to-noise ratio in the limit of large local oscillator power. Both the first and fourth lead to component requirements which are rather more stringent than those encountered with direct detection. The following sections will discuss the source, modulator, fiber, and receiver requirements in a coherent transmission system.

Receiver Sensitivity.   Let σ_T^2 represent the receiver noise described in Eq. (26). The signal-to-noise ratio for heterodyne detection may be expressed as

$$\text{SNR} = \frac{2\eta_{HET}R^2P_S P_{LO}}{2eRP_{LO}B_n + \sigma_T^2} \tag{35}$$

where B_n denotes the noise bandwidth of the receiver. (B_n is generally about half of the data rate for digital systems.) For homodyne detection, the signal envelope carries twice the energy, and

$$\text{SNR} = \frac{4\eta_{\text{HET}}R^2P_S P_{LO}}{2e\,RP_{LO}B_n + \sigma_T^2} \tag{36}$$

For a given modulation scheme, homodyne detection will therefore be twice as sensitive as heterodyne.

Modulation Formats.   The modulation formats appropriate for coherent optical communications can be summarized as follows:

1. *Amplitude-Shift Keying (ASK).*   This technique is simply on-off keying (similar to simple intensity modulation) but with the important constraint that the frequency and phase of the laser be kept constant. Direct modulation of ordinary semiconductor lasers produces a frequency chirp which is unacceptable for ASK modulation. An external modulator such as an electro-optic modulator, a Mach-Zehnder modulator, or an electroabsorption modulator would therefore be appropriate for ASK.

2. *Phase-Shift Keying (PSK).*   This technique requires switching the phase between two or more values. Any phase modulator can be suitable for phase-shift keying. Direct modulation of semiconductor lasers is not suitable for PSK for the same reasons mentioned for ASK.

3. *Frequency-Shift Keying (FSK).*   FSK has received a good deal of attention[55] because it can be achieved by direct modulation of the source. It is possible to make use of the natural frequency chirp of the semiconductor laser to frequency modulate the laser simply by a small modulation of the drive current.

All of the modulation techniques can operate between two states (binary) or extend to four or more levels. The only technique which benefits from an increase in the number of channels is FSK. The sensitivity of PSK to source phase noise generally precludes higher-level signaling. Multilevel FSK, being a bandwidth expansion technique, offers a receiver sensitivity improvement over binary FSK without placing severe constraints on the source.

Table 4 gives expressions for the receiver error probability as a function of received power for each modulation technique. The right-hand column gives, for comparison purposes, the number of photons required per pulse to assure an error rate of better than 10^{-9}. PSK modulation with homodyne detection is the most sensitive, requiring only nine photons per pulse, which is below the quantum limit for direct detection.

Source Requirements.   One of the ways coherent optical communications systems differ from their microwave counterparts is in the comparatively large phase noise of the source. Since the detection system is sensitive to the frequency and phase of the laser, the source linewidth is a major consideration. This is very different from intensity modulation/direct detection, in which the source spectral width limits the system only through the channel dispersion. When two sources are heterodyned to generate an intermediate frequency in the microwave region, the spectral spread of the heterodyned signal is the combined spectral spread of the signal and local oscillator. Thus, the rule of thumb for high-quality coherent detection is that the sum of the linewidths of the signal and local oscillator be much less than the receiver bandwidth.

TABLE 4   Receiver Sensitivities for a Variety of Modulation/Detection Schemes

Modulation/Detection Scheme	P_E	Photons per pulse @ $P_E = 10^{-9}$
ASK heterodyne	$0.5 erfc\left(\sqrt{\dfrac{\eta P_S}{4h\nu B}}\right)$	72
ASK homodyne	$0.5 erfc\left(\sqrt{\dfrac{\eta P_S}{2h\nu B}}\right)$	36
FSK heterodyne	$0.5 erfc\left(\sqrt{\dfrac{\eta P_S}{2h\nu B}}\right)$	36
PSK heterodyne	$0.5 erfc\left(\sqrt{\dfrac{\eta P_S}{h\nu B}}\right)$	18
PSK homodyne	$0.5 erfc\left(\sqrt{\dfrac{2\eta P_S}{h\nu B}}\right)$	9
Direction detection quantum limit	$0.5 \exp\left(\dfrac{-\eta P_S}{h\nu B}\right)$	21

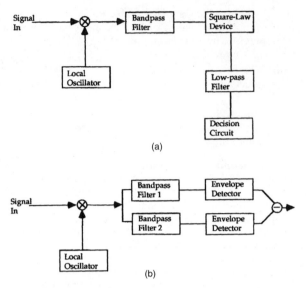

FIGURE 26   Noncoherent (asynchronous) demodulation schemes:
(*a*) ASK envelope detection; (*b*) FSK dual filter detection, in which
the signal is separated into complementary channels for ASK envelope detection.

Precisely how narrow the linewidth must be has been a topic of many papers.[49–52] The result
varies somewhat with modulation scheme and varies strongly with the demodulation process.
The general trends can be summarized as follows:

Incoherent Demodulation (Envelope Detection).   Either ASK or FSK can be demodulated
simply by using an appropriate combination of filters and nonlinear elements. The basic principle of incoherent ASK or dual-filter FSK detection is illustrated in Fig. 26. This type of
detection is, in general, least sensitive to the spectral width of the source. The primary effect
of a broad source is to broaden the *IF* signal spectrum, resulting in attenuation but not a catastrophic increase in bit error rate. Further, the receiver bandwidth can always be broadened to
accommodate the signal. This yields a penalty in excess receiver noise, but the source spectral
width can be a substantial fraction of the bit rate and still keep the receiver sensitivity within
tolerable limits.

There are two approaches to PSK detection which avoid the need for a phase-locked loop.
The first is differential phase-shift keying (DPSK), in which the information is transmitted in
the form of phase *differences* between neighboring time slots. The second is phase diversity
reception, in which a multiport interferometer is designed to yield signals proportional to the
power in different phase quadrants.

Coherent Demodulation with Electronic Phase-Locked Loop.   Some PSK signals cannot be
demodulated incoherently and require careful receiver design for proper carrier recovery.
Suppressed carrier communications schemes such as PSK require a nonlinear recovery circuit. The phase estimation required in proper carrier recovery is far more sensitive to phase
noise than is the case with envelope detection. In contrast to incoherent demodulation, source
spectral widths must generally be kept to less than 1 percent of the bit rate (10 percent of the
phase-locked loop bandwidth) to maintain reliable detection.

Coherent Demodulation with Optoelectronic Phase-Locked Loop.   Homodyne detection
requires that an error signal be fed back to the local oscillator; phase and frequency errors

must be corrected optically in order to maintain precise frequency and phase matching between the two signals. This generally results in a narrower phase-locked loop bandwidth and a much narrower spectral width requirement for the transmitter and local oscillator. Homodyne systems therefore require considerably narrower linewidths than their heterodyne counterparts.

Fiber Requirements.   Heterodyne or homodyne reception is inherently single-mode, and it is therefore necessary for coherent links to use single-mode fibers. Single-mode couplers can then be used to combine the signal and local oscillator lasers for efficient heterodyne detection.

As with other forms of fiber communications, fiber dispersion presents a degradation in the signal-to-noise ratio due to differential delay between different components of the signal spectrum. The power penalty associated with fiber dispersion is determined entirely by the dispersion, the fiber length, and the bit rate. Because of the stringent source linewidth requirements for coherent detection, the spectral broadening is entirely due to the signal itself. The coherent detection system is therefore inherently less sensitive to fiber dispersion.

One important requirement of any fiber that is to be used for coherent transmission is polarization control. As was discussed briefly under "Polarization Characteristics of Fibers" earlier in this chapter, the transmitted polarization of light from a single-mode fiber varies randomly with temperature, stress on the fiber, and other environmental influences. If heterodyning is attempted under these circumstances, the heat signal will fade in and out as the polarization of the signal changes.

Polarization fading can be controlled either by external compensation,[56] internal control,[11] or polarization diversity reception.[57] External compensation seeks to actively control the polarization of the output by sensing the error through a polarizer-analyzer combination and feeding back to correct the polarization. The latter can be accomplished through mechanical, electro-optical, or magneto-optical means.

There are classes of optical fiber sensors which have source and fiber requirements very similar to those of a coherent communication link. One of the most widely studied has been the optical fiber gyro, in which counterpropagating waves in a rotating fiber coil interfere with one another; the resulting beat frequency between the waves is proportional to the angular velocity of the coil. There are other interferometric sensors which make use of optical fibers. Most of them require polarization control and a high degree of frequency stability for the source. The relatively low frequencies and small bandwidths which are required for sensing represent the major difference between these applications and coherent data transmission.

1.8 NONLINEAR OPTICAL PROPERTIES OF FIBERS

Chapters and entire books have been devoted to the subject of optical nonlinearities in fibers. A selection of these are included in "Further Reading" at the end of this chapter. We will content ourselves with an overview of the subject, and consider nonlinear effects which are most important in either limiting or enhancing the performance of fibers. To date, most of the *applications* of nonlinear optics in fibers are in the area of ultralong distance telecommunications.[41,58–60] However, nonlinearities can limit the power-handling ability of fibers and can be an important limitation for certain medical/surgical applications.

Stimulated Scattering Processes

The low loss and long interaction length of an optical fiber makes it an ideal medium for stimulating even relatively weak scattering processes. Two important processes in fibers are: (1) stimulated Raman scattering, the interaction of the guided wave with high-frequency optical phonons in the material, and (2) stimulated Brillouin scattering, the emission, amplification, and scattering of low-frequency acoustic waves. These are distinguished by the size of the frequency shift and the dynamics of the process, but both can act to limit the power available for transmission.

Stimulated Raman scattering (SRS) produces a frequency shift of about 400 cm^{-1} from the incident laser line. The equation governing the power growth of the Raman-shifted mode is as follows

$$\frac{dP_R}{dz} = -\alpha_R P_R + \frac{g_R}{a_P} P_P P_R \tag{37}$$

where P_R denotes the power of the Stokes-shifted light, P_P is the pump power (this is the power in the initially excited mode), and a_P is the effective area of the pump. The Raman gain g_R ultimately determines the SRS-limited light intensity. For typical single-mode silica fibers, g_R is about 10^{-11} cm/W, and yields a power limit of

$$P_{CR} = \frac{16\alpha a_P}{g_R} \tag{38}$$

beyond which the guided wave power will be efficiently Raman-shifted and excess loss will begin to appear at the pump wavelength.

Stimulated Brillouin scattering (SBS) can yield an even lower stimulated scattering threshold. Acoustic waves in the fiber tend to form a Bragg index grating, and scattering occurs primarily in the backward direction. The Brillouin gain g_B is much higher than Raman gain in fibers ($g_B = 5 \times 10^{-9}$ cm/W) and leads to a stimulated scattering threshold of

$$P_{CR} = \frac{25\alpha a_P}{g_B} \tag{39}$$

for a narrowband, CW input.

Either type of stimulated scattering process can be used as a source of gain in the fiber. Injecting a signal within the frequency band of the stimulated scattering process will provide amplification of the input signal. Raman amplification tends to be the more useful of the two because of the relatively large frequency shift and the broader-gain bandwidth. SBS has been used in applications such as coherent optical communications[48] where amplification of a pilot carrier is desired.

The gain bandwidth for SBS is quite narrow—100 MHz for a typical fiber. SBS is therefore only important for sources whose spectra lie within this band. An unmodulated narrow-band laser source such as would be used as a local oscillator in a coherent system would be highly susceptible to SBS, but a directly modulated laser with a 1-GHz linewidth under modulation (modulated laser linewidths can extend well into the GHz range due to frequency chirp) would have an SBS threshold about ten times that of the narrow linewidth source.

Pulse Compression and Soliton Propagation

A major accomplishment in the push toward short pulse propagation in optical fibers was the prediction and observation of solitary wave propagation. In a nonlinear dispersive medium, solitary waves may exist provided the nonlinearity and dispersion act to balance one another. In the case of soliton propagation, the nonlinearity is a refractive index which follows the pulse intensity in a nearly instantaneous fashion:

$$n(t) = n_0 + n_2 I(t) \tag{40}$$

For silica fibers, $n_2 = 3 \times 10^{-16}$ cm^2/W.

TABLE 5   Normalized Variables of the Nonlinear Schrödinger Equation

A	Pulse amplitude
z	Longitudinal coordinate
t	Time
P_0	Peak power
T_0	Pulse width
U	$A/\sqrt{P_0}$ normalized pulse amplitude
β_1	Propagation constant
β_2	Dispersion (2d order)
L_D	$T_0^2/|\beta_2|$ dispersion length
n_2	Nonlinear refractive index
τ	$\dfrac{t - \beta_1 z}{T_0}$ time normalized to moving frame
ξ	$\dfrac{z}{L_D}$ normalized distance
N	$n_2\beta_1 P_0 T_0^2/|\beta_2|$ Order of soliton

The scalar equation governing pulse propagation in such a nonlinear dispersive medium is sometimes termed the *nonlinear Schrödinger equation*

$$i\frac{dU}{d\xi} + \frac{1}{2}\frac{d^2U}{d\tau^2} + N^2|U|^2U = 0 \qquad (41)$$

where the symbols are defined in Table 5. Certain solutions of this equation exist in which the pulse propagates without change in shape; these are the soliton solutions. Solitons can be excited in fibers and propagate great distances before breaking up. This is the basis for fiber-based soliton communication.

Figure 27 illustrates what happens to a pulse which propagates in such a medium. The local refractive index change produces what is commonly known as *self phase modulation*. Since n_2 is positive, the leading edge of the pulse produces a local increase in refractive index. This results in a red shift in the instantaneous frequency. On the trailing edge, the pulse experiences a blue shift. If the channel is one which exhibits normal dispersion, the red-shifted edge will advance while the blue-shifted edge will retard, resulting in pulse spreading If, however, the fiber exhibits anomalous dispersion (beyond 1.3 µm for most single-mode fibers), the red-shifted edge will retard and the pulse will be compressed. Fibers have been used in this way as pulse compressors for some time. In the normal dispersion regime, the fiber nonlinearity is used to chirp the pulse, and a grating pair supplies the dispersion necessary for compression.

FIGURE 27   A pulse propagating through a medium with an intensity-dependent refractive index will experience frequency shifts of the leading and trailing edges of the pulse (*left*). Upon transmission through a fiber having anomalous dispersion, the pulse compresses (*right*).

In the anomalous dispersion regime, the fiber can act both to chirp and compress the pulse. Near the dispersion minimum, higher-order dependence of the group delay on wavelength becomes important, and simple pulse compression does not take place.

Pulse compression cannot continue indefinitely, since the linear dispersion will always act to spread the pulse. At a critical shape, the pulse stabilizes and will propagate without change in shape. This is the point at which a soliton forms. The lowest-order soliton will propagate entirely without change in shape, higher order solitons (which also carry higher energy) experience a periodic evolution of pulse shape.

A soliton requires a certain power level in order to maintain the necessary index change. Distortion-free pulses will therefore propagate only until the fiber loss depletes the energy. Since solitons cannot persist in a lossy channel, they were long treated merely as laboratory curiosities. This was changed by several successful demonstrations of extremely long distance soliton transmission by the inclusion of gain to balance the loss. The gain sections, which initially made use of stimulated Raman scattering, now consist of rare-earth doped fiber amplifiers. The record for repeaterless soliton transmission is constantly being challenged. At the time of this writing, distance of well over 10,000 km have been demonstrated in recirculating loop experiments.

In the laboratory, solitons have most often been generated by mode-locked laser sources. Mode-locked solid state laser sources are generally limited to low duty-cycle pulses, with repetition rates in the 1-GHz range or less. The mode-locked pulse train must then be modulated to carry data, a process which must be carried out externally. There is a high level of current interest in Erbium-doped fiber lasers as mode-locked sources for ultralong distance data communications. Despite the capability of high duty cycle, directly modulated semiconductor lasers are generally rendered unsuitable for soliton communications by the spectral broadening that occurs under modulation.

Four-Wave Mixing

The nonlinear refractive index is simply a degenerate case of a third-order optical nonlinearity, in which the polarization of the medium responds to the cube of the applied electric field. It is possible for widely separated frequencies to phase modulate one another via the fiber nonlinearity, generating sidebands which interfere with neighboring channels in a multiplexed system. This represents an important limit to channel capacity in either WDM or FDM systems. The simplest picture of the four-wave mixing process in fibers can be illustrated by the transmission and cross-phase modulation of four equally spaced channels shown in Fig. 28. Channels 1 and 2 interfere, producing an index of refraction which oscillates at the difference frequency. This modulation in refractive index modulates channel 4, producing sidebands at channels 3 and 5. This is only the simplest combination of frequencies. Four-wave mixing allows any combination of three frequencies beating together to produce a fourth. If the fourth frequency lies within a communication band, that channel can be rendered unusable.

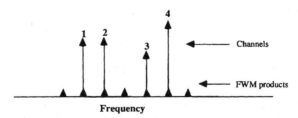

FIGURE 28 The effects of four-wave mixing on multichannel transmission through an optical fiber.

This channel interference can effect either closely spaced channels, as one encounters with coherent communications, or the rather widely separated channels of a WDM system. Efficient four-wave mixing requires phase matching of the interacting waves throughout the interaction length—widely separated channels will therefore be phase matched only in a region of low-fiber dispersion.

The communications engineer will recognize this as little more than the intermodulation products which must always be dealt with in a multichannel communications system with small nonlinearities. Four-wave mixing merely produces intermodulation products over an extremely wide bandwidth. Just as with baseband nonlinearities in analog communications systems, judicious allocation of channels can minimize the problem, but at the expense of bandwidth. The cumulative effect of the nonlinearities increases with interaction length and therefore imposes an important limit on frequency or wavelength-division multiplexed systems.

Photorefractive Nonlinearities in Fibers

There also exists a class of integrating, photorefractive nonlinearities in optical fibers which have been of some interest in recent years. We use the word photorefractive loosely here, simply to indicate a long-term change in either the first- or second-order susceptibility with light exposure. The effects appear strongest in fibers with a germania content, but the precise role of the glass constituents in these processes is still an area of active research.

Bragg Index Gratings.   Photons of energy near a UV absorption edge can often write permanent phase gratings by photoionizing certain constituents or impurities in the material. This is the case for $LiNbO_4$ and certain other ferroelectric materials, and such effects have also been observed in germania-silica fibers. The effects were first observed in the process of guiding relatively high power densities of green light—it was found that a high backscatter developed over a period of prolonged exposure. The fiber then exhibited the transmission characteristics of a Bragg grating, with extremely high resonant reflectivities.

The writing of permanent gratings in fibers using UV exposure is now relatively commonplace. Bragg gratings can be used as filters in WDM systems, reflectors on fiber lasers, and possibly optical switches. For short lengths, the gratings are most easily formed holographically, by using two interfering beams from a pulsed UV source such as an excimer laser. The fiber is exposed from the side; by controlling the angle of the two interfering beams, any grating period may be chosen.

Frequency Doubling in Germania-Silica Fibers.   While it is not surprising that UV exposure could produce refractive index changes, a rather unexpected discovery was the fact that strong optical fields inside the fiber could produce a second-order susceptibility, resulting in efficient frequency doubling. Electro-optic effects such as frequency doubling require that a crystalline material lack a center of symmetry while an amorphous material must lack a statistical center of symmetry. It has long been known that certain materials will develop an electro-optic effect under a suitable applied field. This process, known as *poling,* provides the necessary microscopic alignment of dipoles for development of the nonlinear susceptibility. In optical fibers, a type of self-poling occurs from the strong fundamental beam, resulting in a second-order susceptibility and efficient frequency doubling.

Efficient frequency doubling requires both a noncentrosymmetric material and adequate phase matching between the fundamental and second harmonic waves. The mechanism by which the fiber is both poled and phase matched is still not fully understood at the time of this writing, and it remains to be seen whether this represents an exciting, new application of germania-silica fibers or simply an internal damage mechanism which limits the ultimate power delivery of the fiber.

1.9 OPTICAL FIBER MATERIALS: CHEMISTRY AND FABRICATION

What is arguably the most important breakthrough in the history of optical fiber technology occurred in the materials development. Until 1970, many scientists felt that glasses of moderate softening points and smooth phase transitions would allow for easier drawing and better control. The choice of Corning Glass Works (now Corning, Inc.) to go to (what was then) the somewhat more difficult chemistry of the nearly pure silica fiber allowed both a dramatic reduction in fiber attenuation and a better understanding of the role of the chemical constituents in fiber loss. Researchers soon found that the best dopants for altering the refractive index were those which provided a weak index change without providing a large shift in the UV absorption edge. Conventional fiber chemistry consists of dopants such as GeO_2, P_2O_5 (for raising the refractive index) and B_2O_3 or SiF_4 (for lowering the refractive index).

Silica has both UV and mid-IR absorption bands; these two bands result in a fundamental limit to the attenuation which one can achieve in the silica system. This occurs despite the fact that the Rayleigh scattering contribution decreases as λ^{-4}, and the ultraviolet Urbach absorption edge decreases even faster with increasing λ. The infrared absorption increases with long wavelengths, and becomes dominant beyond wavelengths of about 1.6 μm, resulting in a fundamental loss minimum near 1.55 μm.

The promise of achieving a lower Rayleigh scattering limit in the mid-infrared (as well as the possible applications of fiber to the CO_2 laser wavelength range) have spurred a great deal of research in fiber materials which exhibit better infrared transparency. Two important representative materials are the heavy-metal fluoride glasses and the chalcogenide glasses. While both classes exhibit better infrared transparency, neither has yet improved to the point of serious competition with silica materials.

For a number of years, attenuation in optical fibers was limited by a strong absorption band near $\lambda = 1.4$ μm. (An examination of attenuation curves of early telecommunications-grade fiber shows it nearly unusable at what are now the wavelengths of prime interest—1.3 μm and 1.55 μm.) This absorption, which was linked to the presence of residual OH ions, grew steadily lower with the improvement of fiber fabrication techniques until the loss minimum at $\lambda = 1.55$ μm was eventually brought close to the Rayleigh scattering limit.

The low-cost, low-temperature processes by which polymers can be fabricated has led to continued research into the applications of plastic fiber to technologies which require low cost, easy connectivity, and that are not loss-limited. The additional flexibility of these materials makes them attractive for large-core, short-length applications in which one wishes to maximize the light insertion. Hybrid polymer cladding-silica core fibers have also received some attention in applications requiring additional flexibility.

The final triumph of fiber chemistry in recent years has been the introduction and successful demonstration of extremely long distance repeaterless fiber links using rare-earth doped fiber amplifiers. This represented the climax of a long period of research in rare-earth doped glasses which went largely unnoticed by the optics community. As a result, there has been an explosion of work in the materials science, materials engineering, and applications of rare-earth doped optical fibers.

Fabrication of Conventional Optical Fibers

Conventional fabrication of low-loss optical fibers requires two stages. The desired refractive index profile is first fabricated in macroscopic dimensions in a preform. A typical preform is several centimeters in width and a meter in length, maintaining the dimensions and dopant distribution in the core and cladding that will eventually form in the fiber.

Chemical vapor deposition (CVD) is the primary technology used in fiber manufacturing. The fabrication process must satisfy two requirements: (1) high purity, and (2) precise control over composition (hence, refractive index) profiles. Manufacturing considerations favor approaches which provide a fast deposition rate and comparatively large preforms. In CVD processes, submicron silica particles are produced through one (or both) of the following chemical reactions

$$SiCl_4 + O_2 \rightarrow SiO_2 + 2Cl_2$$

$$SiCl_4 + 2H_2O \rightarrow SiO_2 + HCl$$

The reactions are carried out at a temperature of about 1800°C. The deposition leads to a high-purity silica soot which must then be sintered in order to form optical quality glass.

Modern manufacturing techniques, generally speaking, use one of two processes.[61] In the so-called "inside process," a rotating silica substrate tube is subjected to an internal flow of reactive gases. The two inside processes which have received the most attention are modified chemical vapor deposition (MCVD) and plasma-assisted chemical vapor deposition (PCVD). Both techniques require a layer-by-layer deposition, controlling the composition at each step in order to reach the correct target refractive index. Oxygen, as a carrier gas, is bubbled through $SiCl_4$, which has a relatively high vapor pressure at room temperature.

The PCVD process provides the necessary energy for the chemical reaction by direct RF plasma excitation. The submicron-sized particles form on the inner layer of the substrate, and the composition of the layer is controlled strictly by the composition of the gas. PCVD does not require the careful thermal control of other methods, but requires a separate sintering step to provide a pore-free preform. A final heating to 2150°C collapses the preform into a state in which it is ready to be drawn.

The MCVD process (Fig. 29) accomplishes the deposition by an external, local application of a torch. The torch has the dual role of providing the necessary energy for oxidation and the heat necessary for sintering the deposited SiO_2. The submicron particles are deposited on the "leading edge" of the torch; as the torch moves over these particles, they are sintered into a vitreous, pore-free layer. Multiple passes result in a layered, glassy deposit which should approximate the target radial profile of the fiber. As with PCVD, a final pass is necessary for collapse of the preform before the fiber is ready to be drawn. MCVD requires rather precise control over the temperature gradients in the tube but has the advantage of accomplishing the deposition and sintering in a single step.

In the "outside process," a rotating, thin cylindrical target (or mandrel) is used as the substrate for a subsequent chemical vapor deposition, and requires removal before the boule is sintered. Much of the control in these deposition techniques lies in the construction of the

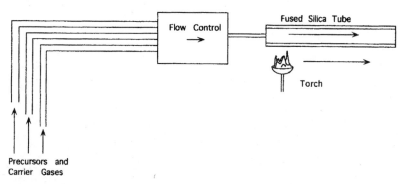

FIGURE 29 The modified chemical vapor deposition (MCVD) process for preform fabrication.

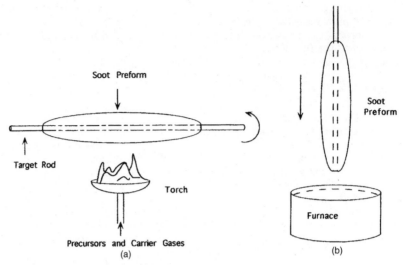

FIGURE 30 Outside method of preform fabrication. The soot deposition (*a*) is followed by sintering (*b*) to cast the preform.

torch. For an outside process, the torch supplies both the chemical constituents and the heat for the reaction.

Two outside processes which have been used a great deal are the outside vapor deposition (OVD) and the vapor axial deposition (VAD) techniques. Figure 30 illustrates a generic outside process. In the OVD process the torch consists of discrete holes formed in a pattern of concentric rings. The primary chemical stream is at the center, followed by O_2 (acting as a shield gas), premixed methane/oxygen, and another shield ring. The torch itself translates along the rotating boule and the dopants are dynamically controlled to achieve the necessary profiling.

The VAD torch is comprised of a set of concentric annular apertures, with a chemical sequence similar to the OVD. In contrast to the OVD method, the VAD torch is held stationary during the deposition; the rotating target is held vertically, and is lifted as the deposition continues.

Dopant Chemistry

Standard dopants for silica fiber include GeO_2, P_2O_5, B_2O_3, and SiF_4. The former two are used to increase the refractive index (and are therefore used in the core), while the latter decrease the index of refraction (and are therefore used in the cladding). The CVD processes will often use oxygen as a carrier gas with the high vapor pressure liquids $GeCl_4$, $POCl_3$, or SiF_4. The reaction which produces the dopant "soot" is then

$$GeCl_4 + O_2 \rightarrow GeO_2 + 2Cl_2$$

$$4POCl_3 + 3O_2 \rightarrow 2P_2O_5 + 6Cl_2$$

As noted in a recent article by Morse et al.,[62] "Nature has been kind in the creation of the high vapor pressure liquid precursors used in the fabrication of optical fibers for the transmission of telecommunication signals." This has been an extremely important factor in the success of CVD fiber fabrication techniques. The problem of introducing more exotic dopants, such as the rare-earth elements, is not quite so straightforward and there does not appear to exist, at this time, a single, widely used technique. The problem of control over the rare-earth dopant profile is compounded by the fact that research in laser and amplifier design is ongoing, and the optimum dopant profile for rare-earth doped fibers and amplifiers is, in many cases, still unknown. Despite these uncertainties, rare-earth doped fibers have already been introduced into commercial products and promise to spearhead the next generation of long distance telecommunications systems.

Other Fabrication Techniques

There are other preform fabrication and fiber drawing techniques. These are not generally used in telecommunications-grade silica fiber, but can be of advantage for glass chemistries which do not easily lend themselves to chemical vapor deposition. Several examples of this will be described in the following section on infrared fiber fabrication.

CVD materials, while the most popular, are not the only methods for preform fabrication. Alternative methods of preform fabrication include both bulk casting and a class of non-CVD tubular casting techniques. One such technique is the "rod-in-tube" method, in which the core and cladding materials are cast separately and combined in a final melting/collapsing step. This method assures a homogeneous, low-impurity content core but risks introducing defects and bubbles into the core/cladding interface.

The most well-known method of *preform-free drawing* is the double crucible method, in which the core and cladding melts are formed separately and combined in the drawing process itself. This eliminates the need for a very large preform in the case of long lengths of fiber. The index profile is established in the drawing process itself, and index gradients are therefore difficult to establish unless additional crucibles are added. Another difficulty of the crucible method is the sometimes inadequate control of the concentricity of the core and cladding.

Infrared Fiber Fabrication

The major applications of interest for infrared optical fibers are as follows:

1. Ultra-low-loss communication links
2. CO_2 laser transmission for medical applications
3. Thermal imaging and remote temperature monitoring
4. Gas sensing

These may differ rather dramatically in their attenuation requirements and spectral region of interest. For example, an ultra-low-loss communications link requires attenuation somewhat less than 0.1 dB/km in order to be competitive with silica fiber. Typical medical applications simply require high-power handling capabilities of a CO_2 laser over meter lengths. All of these applications require a departure from the silica-based chemistry which has been so successful for applications in the near infrared and visible. Much of the generic chemistry of these glasses is covered in Chap. 33 of Vol. II, "Crystals and glasses." Our intent here is to give an overview of the fiber types and the general state of the materials technology in each case.

Chalcogenide Fibers.   Sulfide, selenide, and telluride glasses have all been used for bulk infrared optics—particularly for applications involving CO_2 ($\lambda = 10.6 \, \mu m$) or CO laser transmission ($\lambda = 5.4 \, \mu m$). Infrared fibers have been drawn from these materials and yielded transmission losses of the order of 1 dB/meter in the 5- to 7-μm region.[63] The preform fabrication and drawing of chalcogenide fibers is much more difficult than that of silica due primarily to its sensitivity both to oxygen and moisture. Both oxidation and crystallization can occur at the temperatures necessary to draw the fiber. Either will result in catastrophically high losses and fiber weakness.

Fluoride Fibers. Fluoride fibers have received the most attention for low-loss telecommunications applications, because the theoretical limit for Rayleigh scattering is considerably lower. This is due both to a higher-energy UV absorption edge and better infrared transparency. The difficulty is that excess absorption has proven rather difficult to reduce, and the lowest published losses to date have been near 1 dB/km for long fiber lengths.[64,65] The state-of-the-art in fluoride fiber fabrication is still well above the Rayleigh scattering limit but does show the expected improvement over silica fiber in wavelengths beyond 1.6 μm. Fabrication of very short fiber lengths has been somewhat more successful, with reported losses as low as 0.025 dB/km at 2.55 μm.[66]

The residual loss for longer fibers has been largely linked to extrinsic impurity/defect content. Recent articles by Takahashi and Sanghera[64,65] have noted the role of transition metal and rare-earth ions, submicron platinum particles, oxyfluoride particles, fluoride microcrystals, and bubbles as both extrinsic absorbers and scatterers. The defects of interest originate from a variety of sources, and there has been much discussion on which defects dominate the scattering process. To date, the consensus appears to be that impurity absorption does *not* adequately account for the current loss limits, but does account for residual losses in the neighborhood of 0.2 dB/km.

The classes of defects which have been blamed for the current loss limits are as follows:

Platinum particles. These arise from the use of platinum crucibles. The use of vitreous carbon crucibles eases this contamination.

Core bubbles. This is clearly a problem in the preform fabrication and appears in some of the bulk casting techniques.

Interfacial bubbles. Bubbles appearing at the core-cladding interface have been named as being a major cause of excess scattering. These appear to be a particular problem for those techniques which employ separate core and cladding melts. This unfortunately negates some of the advantages offered by the crucible techniques in fluoride fiber fabrication.

Fluoride microcrystals. Crystals can nucleate at a variety of defect sites. Many of these sites appear at the core-cladding interface, producing interface roughness and scattering. Since, for step-index fibers, the integrity of the core-cladding interface is essential to the confinement of the optical wave, a small amount of interface roughness can produce rather high excess losses.

Chapter 33 of Vol. II, "Crystals and glasses," gives information on the composition and properties of a single-mode fiber grade fluoride glass. This class of compositions has received the designation ZBLAN, after its heavy-metal constituents. The large number of components makes it immediately obvious that both phase separation and crystallization are important issues in fabrication. Either can produce catastrophic increases in loss as well as mechanical weakening of the fiber, and it is clear that many materials science challenges remain in the area of fluoride fiber fabrication.

1.10 REFERENCES

1. K. C. Kao and G. A. Hockham, "Dielectric Fibre Surface Waveguides for Optical Frequencies," *Proc. IEE,* **113**:1151–1158 (1966).

2. D. B. Keck, P. C. Schultz, and F. W. Zimar, U.S. Patent 3,737,393.

3. F. P. Kapron, D. B. Keck, and R. D. Maurer, "Radiation Losses in Glass Optical Waveguides" *Appl. Phys. Lett.* **17**:423 (1970).

4. M. J. Adams, *An Introduction to Optical Waveguides,* John Wiley and Sons, Chichester, 1981.

5. D. Davidson, "Single-Mode Wave Propagation in Cylindrical Optical Fibers," in E. E. Basch (ed.), *Optical Fiber Transmission,* Howard W. Sams, Indianapolis, 1987, 27–64.

6. M. O. Vassell, "Calculation of Propagating Modes in a Graded-Index Optical Fiber," *Optoelectronics* **6**:271–286 (1974).

7. D. Gloge, "Weakly Guiding Fibers," *Appl. Opt.* **10**:2252–2258 (1971).

8. R. W. Davies, D. Davidson, and M. P. Singh, "Single Mode Optical Fiber with Arbitrary Refractive Index Profile: Propagation Solution by the Numerov Method," *J. Lightwave Tech.* **LT-3**:619–627 (1985).

9. P. C. Chow, "Computer Solutions to the Schroedinger Problem," *Am. J. of Physics* **40**:730–734 (1972).

10. A. Kumar, R. K. Varshney, and K. Thyagarajan, "Birefringence Calculations in Elliptical-core Optical Fibers," *Electron. Lett.* **20**:112–113 (1984).

11. K. Sano and Y. Fuji, "Polarization Transmission Characteristics of Optical Fibers With Elliptical Cross Section," *Electron. Commun. Japan* **63**:87 (1980).

12. EIA-FOTP-46, *Spectral Attenuation Measurement for Long-Length, Graded-Index Optical Fibers, Procedure B,* Electronic Industries Association (Washington, D.C.).

13. EIA-FOTP-78, *Spectral Attenuation Cutback Measurement for Single Mode Optical Fibers,* Electronic Industries Association (Washington, D.C.).

14. EIA-FOTP-50, *Light Launch Conditions for Long-Length, Graded-Index Optical Fiber Spectral Attenuation Measurements, Procedure B,* Electronic Industries Association (Washington, D.C.).

15. A. W. Carlisle, "Small Size High-performance Lightguide Connector for LAN's," *Proc. Opt. Fiber Comm.* **TUQ 18**:74–75 (1985).

16. E. Sugita et al., "SC-Type Single-Mode Optical Fiber Connectors," *J. Lightwave Tech* **LT-7**:1689–1696 (1989).

17. N. Suzuki, M. Saruwatari, and M. Okuyama, "Low Insertion- and High Return-loss Optical Connectors with Spherically Convex-polished Ends," *Electron. Lett.* **22(2)**:110–112 (1986).

18. W. C. Young et al., "Design and Performance of the Biconic Connector Used in the FT3 Lightwave System," in *30th IWCS,* 1981.

19. EIA-FOTP-168, *Chromatic Dispersion Measurement of Multimode Graded-Index and Single-Mode Optical Fibers by Spectral Group Delay Measurement in the Time Domain,* Electronic Industries Association (Washington, D.C.).

20. R. Rao, "Field Dispersion Measurement—A Swept-Frequency Technique," in *NBS Special Publication 683,* Boulder, 1984, p. 135.

21. L. G. Cohen and J. Stone, "Interferometric Measurements of Minimum Dispersion Spectra in Short Lengths of Single-Mode Fiber," *Electron. Lett.* **18**:564 (1982).

22. L. G. Cohen, et al., "Experimental Technique for Evaluation of Fiber Transmission Loss and Dispersion," *Proc. IEEE* **68**:1203 (1980).

23. R. A. Modavis and W. F. Love, "Multiple-Wavelength System for Characterization of Dispersion in Single-Mode Optical Fibers," in *NBS Special Publication 683,* Boulder, 1984, p. 115.

24. T. Miya et al., "Fabrication of Low-dispersion Single-Mode Fiber Over a Wide Spectral Range," *IEEE J. Quantum Electronics,* **QE-17**:858 (1981).

25. R. Olshansky and R. D. Maurer, "Tensile Strength and Fatigue of Optical Fibers," *J. Applied Physics* **47**:4497–4499 (1976).

26. EIA-FOTP-28, *Method for Measuring Dynamic Tensile Strength of Optical Fibers,* Electronic Industries Association (Washington, D.C.).

27. M. R..Brininstool, "Measuring Longitudinal Strain in Optical Fibers," *Optical Engineering* **26**:1113 (1987).

28. M. J. Matthewson, C. R. Kurkjian, and S. T. Gulati, "Strength Measurement of Optical Fibers by Bending," *J. Am. Ceramic. Soc.* **69**:815 (1986).

29. W. Weibull, "A Statistical Distribution Function of Wide Applicability," *J. Appl. Mech.* **24**:293–297 (1951).

30. J. D. Helfinstine, "Adding Static and Dynamic Fatigue Effects Directly to the Weibull Distribution," *J. Am. Ceramic Soc.* **63**:113 (1980).

31. D. Inniss, D. L. Brownlow, and C. R. Kurkjian, "Effect of Sodium Chloride Solutions on the Strength and Fatigue of Bare Silica Fibers," *J. Am. Ceramic Soc.* **75**:364 (1992).

32. J. S. Cook and O. I. Scentesi, "North American Field Trials and Early Applications in Telephony," *IEEE J. Selected Areas in Communications* **SAC-1**:393–397 (1983).

33. H. Ishio "Japanese Field Trials and Early Applications in Telephony," *IEEE J. Selected Areas in Communications* **SAC-1**:398–403 (1983).

34. A. Moncolvo and F. Tosco, "European Field Trials and Early Applications in Telephony," *IEEE J. Selected Areas in Communications* **SAC-1**:398–403 (1983).

35. E. E. Basch, R. A. Beaudette, and H. A. Carnes, "Optical Transmission for Interoffice Trunks," *IEEE Trans. on Communications* **COM-26**:1007–1014 (1978).

36. G. P. Agrawal, *Fiber-Optic Communication Systems,* Wiley Series in Microwave and Optical Engineering, K. Chang (ed.), John Wiley and Sons, New York, 1992.

37. K. Petermann, *Laser Diode Modulation and Noise,* Kluwer Academic, Dordrecht, The Netherlands, 1991.

38. R. J. McIntyre, "Multiplication Noise in Uniform Avalanche Diodes," *IEEE Trans. Electron Devices* **ED-13**:164 (1966).

39. R. J. McIntyre, "The Distribution of Gains in Uniformly Multiplying Avalanche Photodiodes: Theory," *IEEE Trans. Electron Devices* **ED-19**:703–713 (1972).

40. S. D. Personick, "Statistics of a General Class of Avalanche Detectors with Applications to Optical Communications," *Bell System Technical Journal* **50**:167–189 (1971).

41. R. K. Dodd, et al., *Solitons and Nonlinear Wave Equations,* Academic Press, Orlando, Florida, 1984.

42. A. Hasegawa, *Solitons in Optical Fibers,* Springer-Verlag, Berlin, 1989.

43. L. F. Mollenauer, R. H. Stolen, and J. P. Gordon, "Experimental Observation of Picosecond Pulse Narrowing and Solitons in Optical Fibers," *Phys. Rev. Letters* **45**:1095 (1980).

44. L. F. Mollenauer et al., "Extreme Picosecond Pulse Narrowing by Means of Soliton Effect in Single-Mode Optical Fibers," *Opt. Lett.* **8**:289 (1983).

45. L. F. Mollenauer, R. H. Stolen, and M. N. Islam, "Experimental Demonstration of Soliton Propagation in Long Fibers: Loss Compensated by Raman Gain," *Opt. Lett.* **10**:229 (1985).

46. J. P. Gordon and H. A. Haus, *Opt. Lett.* **11**:665 (1986).

47. L. F. Mollenauer, J. P. Gordon, and S. G. Evangelides, "The Sliding Frequency Guiding Filter—an Improved Form of Soliton Jitter Control," *Opt. Lett.* **17**:1575 (1992).

48. A. R. Chraplyvy and R. W. Tkach, *Electron. Lett.* **22**:1084 (1986).

49. I. Garrett and G. Jacobsen, "Theoretical Analysis of Heterodyne Optical Receivers using Semiconductor Lasers of Non-negligible Linewidth," *J. Lightwave Technology* **4**:323 (1986).

50. B. Glance, "Performance of Homodyne Detection of Binary PSK Optical Signals," *J. Lightwave Technology* **4**:228 (1986).

51. L. G. Kazovsky, "Performance Analysis and Laser Linewidth Requirements for Optical PSK Heterodyne Communications," *J. Lightwave Technology* **4**:415 (1986).

52. K. Kikuchi, et al., *J. Lightwave Technology* **2**:1024 (1984).

53. T. Okoshi and K. Kikuchi, *Coherent Optical Fiber Communications,* Kluwer, Boston, 1988.

54. N. A. Olsson et al., "400 Mbit/s 372 = Km Coherent Transmission Experiment," *Electron. Lett.* **24**:36 (1988).

55. E. G. Bryant et al., "A 1.2 Gbit/s Optical FSK Field Trial Demonstration," *British Telecom Technology Journal* **8**:18 (1990).

56. T. Okoshi, "Polarization-State Control Schemes for Heterodyne of Homodyne Optical Fiber Communications," *J. Lightwave Technology* **3**:1232–1237 (1985).

57. B. Glance, "Polarization Independent Coherent Optical Receiver," *J. Lightwave Technology* **5**:274 (1987).

58. R. H. Stolen, "Nonlinear Properties of Optical Fibers," in S. E. Miller and A. G. Chynowth, eds., *Optical Fiber Telecommunications,* Academic Press, New York.

59. G. P. Agrawal, "Nonlinear Interactions in Optical Fibers," in G. P. Agrawal and R. W. Boyd, eds., *Contemporary Nonlinear Optics,* Academic Press, San Diego, California, 1992.

60. G. P. Agrawal, *Nonlinear Fiber Optics,* Academic Press, San Diego, California, 1989.

61. J. R. Bautista and R. M. Atkins, "The Formation and Deposition of SiO_2 Aerosols in Optical Fiber Manufacturing Torches," *J. Aerosol Science* **22**:667–675 (1991).

62. T. F. Morse et al., "Aerosol Transport for Optical Fiber Core Doping: A New Technique for Glass Formation," *J. Aerosol Science* **22**:657–666 (1991).

63. J. Nishii et al., "Recent Advances and Trends in Chalcogenide Glass Fiber Technology: A Review," *J. Noncrystalline Solids* **140**:199–208 (1992).

64. J. S. Sanghera, B. B. Harbison, and I. D. Aggarwal, "Challenges in Obtaining Low Loss Fluoride Glass Fibers," *J. Non-Crystalline Solids* **140**:146–149 (1992).

65. S. Takahashi, "Prospects for Ultra-low Loss Using Fluoride Glass Optical Fiber," *J. Non-Crystalline Solids* **140**:172–178 (1992).

66. I. Aggarwal, G. Lu, and L. Busse, *Materials Science Forum* **32 & 33**: Plenum, New York, 1988, p. 495.

1.11 FURTHER READING

Agrawal, G. P., *Fiber-Optic Communication Systems,* John Wiley and Sons, New York, 1992.

Baack, C. (ed.), *Optical Wideband Transmission Systems,* CRC Press, Boca Raton, Florida, 1986.

Baker, D. G., *Fiber Optic Design and Applications,* Reston Publishing, Reston, Virginia, 1985.

Barnoski, M. K. (ed.), *Fundamentals of Optical Fiber Communications,* Academic Press, New York, 1981.

Basch, E. E. (ed.), *Optical Fiber Transmission,* Howard W. Sams, Indianapolis, Indiana, 1987.

Chaffee, C. D., *The Rewiring of America: The Fiber Optics Revolution,* Academic Press, Boston, 1988.

Chaimowitz, J. C. A., *Lightwave Technology,* Butterworths, Boston, 1989.

Cheo, P. K., *Fiber Optics: Devices and Systems,* Prentice-Hall, Englewood Cliffs, New Jersey, 1985.

Cheo, P. K., *Fiber Optics and Optoelectronics,* Prentice-Hall, Englewood Cliffs, New Jersey, 1990.

Cherin, A. H., *An Introduction to Optical Fibers,* McGraw-Hill, New York, 1983.

Culshaw, B., *Optical Fibre Sensing and Signal Processing,* Peter Peregrinus, London, 1984.

Daly, J. C. (ed.), *Fiber Optics,* CRC Press, Boca Raton, Florida, 1984.

Day, G. W., *Measurement of Optical-Fiber Bandwidth in the Frequency Domain,* NBS Special Publication, No. 637, National Bureau of Standards, Boulder, 1983.

Edwards, T. C., *Fiber-Optic Systems: Network Applications,* John Wiley and Sons, New York, 1989.

Geckeler, S., *Optical Fiber Transmission Systems,* Artech House, Norwood, Massachusetts, 1987.

Gowar, J., *Optical Communications Systems,* Prentice-Hall, London, 1984.

Howes, M. J. and D. V. Morgan (eds.), *Optical Fiber Communications,* John Wiley and Sons, New York, 1980.

Jones, W. B., Jr., *Introduction to Optical Fiber Communications,* Holt, Rinehart and Winston, New York, 1988.

Kaiser, G. E., *Optical Fiber Communications,* McGraw-Hill, New York, 1991.

Kao, C. K., *Optical Fibre,* Peter Pereginus, London, 1988.

Karp, S., R. Gagliardi et al., *Optical Channels: Fibers, Clouds, Water, and the Atmosphere,* Plenum Press, New York, 1988.

Killen, H. B., *Fiber Optic Communications,* Prentice-Hall, Englewood Cliffs, New Jersey, 1991.

Li, T., (ed.), *Optical Fiber Data Transmission,* Academic Press, Boston, 1991.

Lin, C. (ed.), *Optoelectronic Technology and Lightwave Communications Systems,* Van Nostrand Reinhold, New York, 1989.

Mahlke, G. and P. Gossing, *Fiber Optic Cables,* John Wiley and Sons, New York, 1987.

Miller, S. E. and J. P. Kaminow (eds.), *Optical Fiber Telecommunications II,* Academic Press, San Diego, California, 1988.

Okoshi, T. and K. Kikuchi, *Coherent Optical Fiber Communications,* Kluwer, Boston, 1988.

Palais, J. C., *Fiber-Optic Communications,* Prentice-Hall, Englewood Cliffs, New Jersey, 1988.

Personick, S. D., *Optical Fiber Transmission Systems,* Plenum Press, New York, 1981.

Personick, S. D., *Fiber Optics: Technology and Applications,* Plenum Press, New York, 1985.

Runge, P. K. and P. R. Trischitta (eds.), *Undersea Lightwave Communications,* IEEE Press, New York, 1986.

Senior, J. M., *Optical Fiber Communications,* Prentice-Hall, London, 1985.

Sharma, A. B., S. J. Halme, et al., *Optical Fiber Systems and Their Components,* Springer-Verlag, Berlin, 1981.

Sibley, M. J. N., *Optical Communications,* Macmillan, London, 1990.

Taylor, H. F. (ed.), *Fiber Optics Communications,* Artech House, Norwood, Massachusetts, 1983.

Taylor, H. F. (ed.), *Advances in Fiber Optics Communications,* Artech House, Norwood, Massachusetts, 1988.

Tsang, W. T. (ed.), *Lightwave Communications Technology,* Semiconductors and Semimetals. Academic Press, Orlando, Florida, 1985.

Fibers in Medicine

Harrington, J. A. (ed.), *Infrared Fiber Optics III,* SPIE, Bellingham, Washington, 1991.

Joffe, S. N., *Lasers in General Surgery,* Williams & Wilkins, Baltimore, 1989.

Katzir, A. (ed.), *Selected papers on optical fibers in medicine,* SPIE, Bellingham, Washington, 1990.

Katzir, A. (ed.), *Proc. Optical Fibers in Medicine VII,* Bellingham, Washington, SPIE, 1992.

Nonlinear Properties of Fibers

Agrawal, G. P., *Nonlinear Fiber Optics,* Academic Press, San Diego, California, 1989.

Agrawal, G. P., "*Nonlinear Interactions in Optical Fibers,*" in *Contemporary Nonlinear Optics,* Academic Press, San Diego, California, 1992.

Hasegawa, A., *Solitons in Optical Fibers,* Springer-Verlag, Berlin, 1989.

CHAPTER 2
OPTICAL FIBER COMMUNICATION TECHNOLOGY AND SYSTEM OVERVIEW

Ira Jacobs
Fiber and Electro-Optics Research Center
Virginia Polytechnic Institute and State University
Blacksburg, Virginia

2.1 INTRODUCTION

Basic elements of an optical fiber communication system include the transmitter [laser or light-emitting diode (LED)], fiber (multimode, single-mode, or dispersion-shifted), and the receiver [positive-intrinsic-negative (PIN) diode and avalanche photodetector (APD) detectors, coherent detectors, optical preamplifiers, receiver electronics]. Receiver sensitivities of digital systems are compared on the basis of the number of photons per bit required to achieve a given bit error probability, and eye degradation and error floor phenomena are described. Laser relative intensity noise and nonlinearities are shown to limit the performance of analog systems. Networking applications of optical amplifiers and wavelength-division multiplexing are considered, and future directions are discussed.

Although the light-guiding property of optical fibers has been known and used for many years, it is only relatively recently that optical fiber communications has become both a possibility and a reality.[1] Following the first prediction in 1966[2] that fibers might have sufficiently low attenuation for telecommunications, the first low-loss fiber (20 dB/km) was achieved in 1970.[3] The first semiconductor laser diode to radiate continuously at room temperature was also achieved in 1970.[4] The 1970s were a period of intense technology and system development, with the first systems coming into service at the end of the decade. The 1980s saw both the growth of applications (service on the first transatlantic cable in 1988) and continued advances in technology. This evolution continued in the 1990s with the advent of optical amplifiers and with the applications emphasis turning from point-to-point links to optical networks.

This chapter provides an overview of the basic technology, systems, and applications of optical fiber communication. It is an update and compression of material presented at a 1994 North Atlantic Treaty Organization (NATO) Summer School.[5]

2.2   *BASIC TECHNOLOGY*

This section considers the basic technology components of an optical fiber communications link, namely the fiber, the transmitter, and the receiver, and discusses the principal parameters that determine communications performance.

Fiber

An optical fiber is a thin filament of glass with a central core having a slightly higher index of refraction than the surrounding cladding. From a physical optics standpoint, light is guided by total internal reflection at the core-cladding boundary. More precisely, the fiber is a dielectric waveguide in which there are a discrete number of propagating modes.[6] If the core diameter and the index difference are sufficiently small, only a single mode will propagate. The condition for single-mode propagation is that the normalized frequency V be less than 2.405, where

$$V = \frac{2\pi a}{\lambda}\sqrt{n_1^2 - n_2^2} \tag{1}$$

and a is the core radius, λ is the free space wavelength, and n_1 and n_2 are the indexes of refraction of the core and cladding, respectively. Multimode fibers typically have a fractional index difference (Δ) between core and cladding of between 1 and 1.5 percent and a core diameter of between 50 and 100 μm. Single-mode fibers typically have $\Delta \approx 0.3\%$ and a core diameter of between 8 and 10 μm.

The fiber numerical aperture (NA), which is the sine of the half-angle of the cone of acceptance, is given by

$$\text{NA} = \sqrt{n_1^2 - n_2^2} = n_1\sqrt{2\Delta} \tag{2}$$

Single-mode fibers typically have an NA of about 0.1, whereas the NA of multimode fibers is in the range of 0.2 to 0.3.

From a transmission system standpoint, the two most important fiber parameters are attenuation and bandwidth.

Attenuation.   There are three principal attenuation mechanisms in fiber: absorption, scattering, and radiative loss. Silicon dioxide has resonance absorption peaks in the ultraviolet (electronic transitions) and in the infrared beyond 1.6 μm (atomic vibrational transitions), but is highly transparent in the visible and near-infrared.

Radiative losses are generally kept small by using a sufficiently thick cladding (communication fibers have an outer diameter of 125 μm), a compressible coating to buffer the fiber from external forces, and a cable structure that prevents sharp bends.

In the absence of impurities and radiation losses, the fundamental attenuation mechanism is Rayleigh scattering from the irregular glass structure, which results in index of refraction fluctuations over distances that are small compared to the wavelength. This leads to a scattering loss

$$\alpha = \frac{B}{\lambda^4}, \text{ with } B \approx 0.9 \frac{dB}{km}\mu m^4 \tag{3}$$

for "best" fibers. Attenuation as a function of wavelength is shown in Fig. 1. The attenuation peak at $\lambda = 1.4$ μm is a resonance absorption due to small amounts of water in the fiber, although fibers now may be made in which this peak is absent. Initial systems operated at a wavelength around 0.85 μm owing to the availability of sources and detectors at this wave-

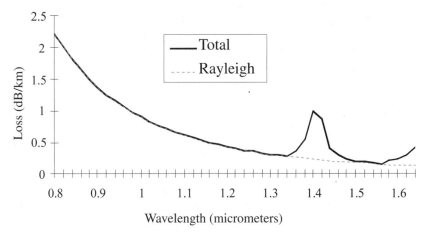

FIGURE 1   Fiber attenuation as a function of wavelength. Dashed curve shows Rayleigh scattering. Solid curve indicates total attenuation including resonance absorption at 1.38 μm from water and tail of infrared atomic resonances above 1.6 μm.

length. Present systems (other than some short-distance data links) generally operate at wavelengths of 1.3 or 1.55 μm. The former, in addition to being low in attenuation (about 0.32 dB/km for best fibers), is the wavelength of minimum intramodal dispersion (see the next section) for standard single-mode fiber. Operation at 1.55 μm allows even lower attenuation (minimum is about 0.16 dB/km) and the use of erbium-doped-fiber amplifiers (see Sec. 2.5), which operate at this wavelength.

Dispersion.   Pulse spreading (dispersion) limits the maximum modulation bandwidth (or maximum pulse rate) that may be used with fibers. There are two principal forms of dispersion: intermodal dispersion and intramodal dispersion. In multimode fiber, the different modes experience different propagation delays resulting in pulse spreading. For graded-index fiber, the lowest dispersion per unit length is given approximately by[7]

$$\frac{\delta\tau}{L} = \frac{n_1 \Delta^2}{10c} \text{ (intermodal)} \tag{4}$$

[Grading of the index of refraction of the core in a nearly parabolic function results in an approximate equalization of the propagation delays. For a step-index fiber, the dispersion per unit length is $\delta\tau/L = n_1\Delta/c$, which for $\Delta=0.01$ is 1000 times larger than that given by Eq. (4).]

Bandwidth is inversely proportional to dispersion, with the proportionality constant dependent on pulse shape and how bandwidth is defined. If the dispersed pulse is approximated by a Gaussian pulse with $\delta\tau$ being the full width at the half-power point, then the −3-dB bandwidth B is given by

$$B = 0.44/\delta\tau \tag{5}$$

Multimode fibers are generally specified by their bandwidth in a 1-km length. Typical specifications are in the range from 200 MHz to 1 GHz. Fiber bandwidth is a sensitive function of the index profile and is wavelength dependent, and the scaling with length depends on whether there is mode mixing.[8] Also, for short-distance links, the bandwidth is dependent on the launch conditions. Multimode fibers are generally used only when the bit rates and distances are sufficiently small that accurate characterization of dispersion is not of concern, although this may be changing with the advent of graded-index plastic optical fiber for high-bit-rate short-distance data links.

Although there is no intermodal dispersion in single-mode fibers,* there is still dispersion within the single mode (intramodal dispersion) resulting from the finite spectral width of the source and the dependence of group velocity on wavelength. The intramodal dispersion per unit length is given by

$$\delta\tau/L = D\,\delta\lambda \qquad \text{for } D \neq 0$$

$$= 0.2S_o\,(\delta\lambda)^2 \qquad \text{for } D = 0 \tag{6}$$

where D is the dispersion coefficient of the fiber, $\delta\lambda$ is the spectral width of the source, and S_o is the dispersion slope

$$S_o = \frac{dD}{d\lambda} \text{ at } \lambda = \lambda_0, \text{ where } D(\lambda_0) = 0 \tag{7}$$

If both intermodal and intramodal dispersion are present, the square of the total dispersion is the sum of the squares of the intermodal and intramodal dispersions. For typical digital systems, the total dispersion should be less than half the interpulse period T. From Eq. (5) this corresponds to an effective fiber bandwidth that is at least $0.88/T$.

There are two sources of intramodal dispersion: material dispersion, which is a consequence of the index of refraction being a function of wavelength, and waveguide dispersion, which is a consequence of the propagation constant of the fiber waveguide being a function of wavelength.

For a material with index of refraction $n(\lambda)$, the material dispersion coefficient is given by

$$D_{\text{mat}} = -\frac{\lambda}{c}\frac{d^2n}{d\lambda^2} \tag{8}$$

For silica-based glasses, D_{mat} has the general characteristics shown in Fig. 2. It is about -100 ps/km·nm at a wavelength of 820 nm, goes through zero at a wavelength near 1300 nm, and is about 20 ps/km·nm at 1550 nm.

For step-index single-mode fibers, waveguide dispersion is given approximately by[10]

$$D_{\text{wg}} \approx -\frac{0.025\lambda}{a^2cn_2} \tag{9}$$

For conventional single-mode fiber, waveguide dispersion is small (about -5 ps/km·nm at 1300 nm). The resultant $D(\lambda)$ is then slightly shifted (relative to the material dispersion curve) to longer wavelengths, but the zero-dispersion wavelength (λ_0) remains in the vicinity of 1300 nm. However, if the waveguide dispersion is made larger negative by decreasing a or equivalently by tapering the index of refraction in the core the zero-dispersion wavelength may be shifted to the vicinity of 1550 nm (see Fig. 2). Such fibers are called *dispersion-shifted fibers* and are advantageous because of the lower fiber attenuation at this wavelength and the advent of erbium-doped-fiber amplifiers (see Sec. 8.5). Note that dispersion-shifted fibers have a smaller slope at the dispersion minimum ($S_0 \approx 0.06$ ps/km·nm^2 compared to $S_0 \approx 0.09$ ps/km·nm^2 for conventional single-mode fiber).

---

* A single-mode fiber actually has two degenerate modes corresponding to the two principal polarizations. Any asymmetry in the transmission path removes this degeneracy and results in polarization dispersion. This is typically very small (in the range of 0.1 to 1 ps/km$^{1/2}$), but is of concern in long-distance systems using linear repeaters.[9]

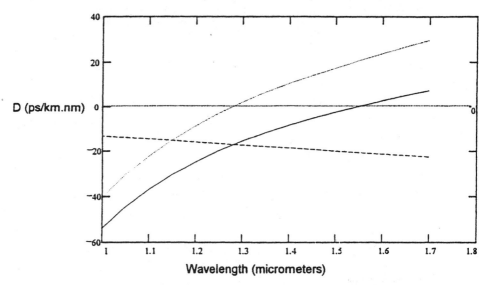

FIGURE 2   Intramodal dispersion coefficient as a function of wavelength. Dotted curve shows D_{mat}; dashed curve shows D_{wg} to achieve D (solid curve) with dispersion 0 at 1.55 μm.

With more complicated index of refraction profiles, it is possible, at least theoretically, to control the shape of the waveguide dispersion such that the total dispersion is small in both the 1300- and 1550-nm bands, leading to dispersion-flattened fibers.[11]

Transmitting Sources

Semiconductor light-emitting diodes (LEDs) or lasers are the primary light sources used in fiber-optic transmission systems. The principal parameters of concern are the power coupled into the fiber, the modulation bandwidth, and (because of intramodal dispersion) the spectral width.

Light-Emitting Diodes (LEDs).   LEDs are forward-biased positive-negative (PN) junctions in which carrier recombination results in spontaneous emission at a wavelength corresponding to the energy gap. Although several milliwatts may be radiated from high-radiance LEDs, the radiation is over a wide angular range, and consequently there is a large coupling loss from an LED to a fiber. Coupling efficiency (η = ratio of power coupled to power radiated) from an LED to a fiber is given approximately by[12]

$$\eta \approx (NA)^2 \text{ for } r_s < a$$

$$\eta \approx (a/r_s)^2 \, (NA)^2 \text{ for } r_s > a \tag{10}$$

where r_s is the radius of the LED. Use of large-diameter, high-NA multimode fiber improves the coupling from LEDs to fiber. Typical coupling losses are 10 to 20 dB for multimode fibers and more than 30 dB for single-mode fibers.

In addition to radiating over a large angle, LED radiation has a large spectral width (about 50 nm at λ = 850 nm and 100 nm at λ = 1300 nm) determined by thermal effects. Systems employing LEDs at 850 nm tend to be intramodal-dispersion-limited, whereas those at 1300 nm are intermodal-dispersion-limited.

Owing to the relatively long time constant for spontaneous emission (typically several nanoseconds), the modulation bandwidths of LEDs are generally limited to several hundred MHz. Thus, LEDs are generally limited to relatively short-distance, low-bit-rate applications.

Lasers.   In a laser, population inversion between the ground and excited states results in stimulated emission. In edge-emitting semiconductor lasers, this radiation is guided within the active region of the laser and is reflected at the end faces.* The combination of feedback and gain results in oscillation when the gain exceeds a threshold value. The spectral range over which the gain exceeds threshold (typically a few nanometers) is much narrower than the spectral width of an LED. Discrete wavelengths within this range, for which the optical length of the laser is an integer number of half-wavelengths, are radiated. Such a laser is termed a *multilongitudinal mode Fabry-Perot laser.* Radiation is confined to a much narrower angular range than for an LED, and consequently may be efficiently coupled into a small-NA fiber. Coupled power is typically about 1 mW.

The modulation bandwidth of lasers is determined by a resonance frequency caused by the interaction of the photon and electron concentrations.[14] Although this resonance frequency was less than 1 GHz in early semiconductor lasers, improvements in materials have led to semiconductor lasers with resonance frequencies (and consequently modulation bandwidths) in excess of 10 GHz. This not only is important for very high-speed digital systems, but now also allows semiconductor lasers to be directly modulated with microwave signals. Such applications are considered in Sec. 2.7.

Although multilongitudinal-mode Fabry-Perot lasers have a narrower spectral spread than LEDs, this spread still limits the high-speed and long-distance capability of such lasers. For such applications, single-longitudinal-mode (SLM) lasers are used. SLM lasers may be achieved by having a sufficiently short laser (less than 50 μm), by using coupled cavities (either external mirrors or cleaved coupled cavities[15]), or by incorporating a diffraction grating within the laser structure to select a specific wavelength. The latter has proven to be most practical for commercial application, and includes the distributed feedback (DFB) laser, in which the grating is within the laser active region, and the distributed Bragg reflector (DBR) laser, where the grating is external to the active region.[16]

There is still a finite line width for SLM lasers. For lasers without special stabilization, the line width is on the order of 0.1 nm. Expressed in terms of frequency, this corresponds to a frequency width of 12.5 GHz at a wavelength of 1550 nm. (Wavelength and frequency spread are related by $\delta f/f = -\delta\lambda/\lambda$, from which it follows that $\delta f = -c\delta\lambda/\lambda^2$.) Thus, unlike electrical communication systems, optical systems generally use sources with spectral widths that are large compared to the modulation bandwidth.

The finite line width (phase noise) of a laser is due to fluctuations of the phase of the optical field resulting from spontaneous emission. In addition to the phase noise contributed directly by the spontaneous emission, the interaction between the photon and electron concentrations in semiconductor lasers leads to a conversion of amplitude fluctuations to phase fluctuations, which increases the line width.[17] If the intensity of a laser is changed, this same phenomenon gives rise to a change in the frequency of the laser (*chirp*). Uncontrolled, this causes a substantial increase in line width when the laser is modulated, which may cause difficulties in some system applications, possibly necessitating external modulation. However, the phenomenon can also be used to advantage. For appropriate lasers under small signal modulation, a change in frequency proportional to the input signal can be used to frequency-modulate and/or to tune the laser. Tunable lasers are of particular importance in networking applications employing wavelength-division multiplexing (WDM).

---

* In vertical cavity surface-emitting lasers (VCSELs), reflection is from internal "mirrors" grown within the semiconductor structure.[13]

Photodetectors

Fiber-optic systems generally use PIN or APD photodetectors. In a reverse-biased PIN diode, absorption of light in the intrinsic region generates carriers that are swept out by the reverse-bias field. This results in a photocurrent (I_p) that is proportional to the incident optical power (P_R), where the proportionality constant is the responsivity ($\Re$) of the photodetector; that is, $\Re = I_P/P_R$. Since the number of photons per second incident on the detector is power divided by the photon energy, and the number of electrons per second flowing in the external circuit is the photocurrent divided by the charge of the electron, it follows that the quantum efficiency (η = electrons/photons) is related to the responsivity by

$$\eta = \frac{hc}{q\lambda}\frac{I_p}{P_R} = \frac{1.24\ (\mu m \cdot V)}{\lambda}\Re \tag{11}$$

For wavelengths shorter than 900 nm, silicon is an excellent photodetector, with quantum efficiencies of about 90 percent. For longer wavelengths, InGaAs is generally used, with quantum efficiencies typically around 70 percent. Very high bandwidths may be achieved with PIN photodetectors. Consequently, the photodetector does not generally limit the overall system bandwidth.

In an avalanche photodetector (APD), a larger reverse voltage accelerates carriers, causing additional carriers by impact ionization resulting in a current $I_{\text{APD}} = MI_p$, where M is the current gain of the APD. As noted in Sec. 2.3, this can result in an improvement in receiver sensitivity.

2.3 RECEIVER SENSITIVITY

The receiver in a direct-detection fiber-optic communication system consists of a photodetector followed by electrical amplification and signal-processing circuits intended to recover the communications signal. Receiver sensitivity is defined as the average received optical power needed to achieve a given communication rate and performance. For analog communications, the communication rate is measured by the bandwidth of the electrical signal to be transmitted (B), and performance is given by the signal-to-noise ratio (SNR) of the recovered signal. For digital systems, the communication rate is measured by the bit rate (R_b) and performance is measured by the bit error probability (P_e).

For a constant optical power transmitted, there are fluctuations of the received photocurrent about the average given by Eq. (11). The principal sources of these fluctuations are signal shot noise (quantum noise resulting from random arrival times of photons at the detector), receiver thermal noise, APD excess noise, and relative intensity noise (RIN) associated with fluctuations in intensity of the source and/or multiple reflections in the fiber medium.

Digital On-Off-Keying Receiver

It is instructive to define a normalized sensitivity as the average number of photons per bit ($\overline{N}_p$) to achieve a given error probability, which we take here to be $P_e = 10^{-9}$. Given $\overline{N}_p$, the received power when a 1 is transmitted is obtained from

$$P_R = 2\overline{N}_p R_b \frac{hc}{\lambda} \tag{12}$$

where the factor of 2 in Eq. (12) is because P_R is the peak power, and $\overline{N}_p$ is the average number of photons per bit.

Ideal Receiver. In an ideal receiver individual photons may be counted, and the only source of noise is the fluctuation of the number of photons counted when a 1 is transmitted. This is a Poisson random variable with mean $2\overline{N}_p$. No photons are received when a 0 is transmitted. Consequently, an error is made only when a 1 is transmitted and no photons are received. This leads to the following expression for the error probability

$$P_e = \tfrac{1}{2} \exp(-2\overline{N}_p) \tag{13}$$

from which it follows that $\overline{N}_p = 10$ for $P_e = 10^{-9}$. This is termed the *quantum limit.*

PIN Receiver. In a PIN receiver, the photodetector output is amplified, filtered, and sampled, and the sample is compared with a threshold to decide whether a 1 or 0 was transmitted. Let I be the sampled current at the input to the decision circuit scaled back to the corresponding value at the output of the photodetector. (It is convenient to refer all signal and noise levels to their equivalent values at the output of the photodetector.) I is then a random variable with means and variances given by

$$\mu_1 = I_p \qquad\qquad \mu_0 = 0 \tag{14a}$$
$$\sigma_1^2 = 2qI_pB + \frac{4kTB}{R_e} \qquad \sigma_0^2 = \frac{4kTB}{R_e} \tag{14b}$$

where the subscripts 1 and 0 refer to the bit transmitted, kT is the thermal noise energy, and R_e is the effective input noise resistance of the amplifier. Note that the noise values in the 1 and 0 states are different owing to the shot noise in the 1 state.

Calculation of error probability requires knowledge of the distribution of I under the two hypotheses. Under the assumption that these distributions may be approximated by gaussian distributions with means and variances given by Eq. (14), the error probability may be shown to be given by (Chap. 4 in Ref. 18)

$$P_e = K\left(\frac{\mu_1 - \mu_0}{\sigma_1 + \sigma_0}\right) \tag{15}$$

where

$$K(Q) = \frac{1}{\sqrt{2\pi}} \int_Q^\infty dx\ \exp(-x^2/2) = \frac{1}{2}\mathrm{erfc}(Q/\sqrt{2}) \tag{16}$$

It can be shown from Eqs. (11), (12), (14), and (15) that

$$\overline{N}_p = \frac{B}{\eta R_b} Q^2 \left[1 + \frac{1}{Q}\sqrt{\frac{8\pi kTC_e}{q^2}}\right] \tag{17}$$

where

$$C_e = \frac{1}{2\pi R_e B} \tag{18}$$

is the effective noise capacitance of the receiver, and from Eq. (16), $Q = 6$ for $P_e = 10^{-9}$. The minimum bandwidth of the receiver is half the bit rate, but in practice B/R_b is generally about 0.7.

The gaussian approximation is expected to be good when the thermal noise is large compared to the shot noise. It is interesting, however, to note that Eq. (17) gives $\overline{N}_p = 18$ when $C_e = 0$, $B/R_b = 0.5$, $\eta = 1$, and $Q = 6$. Thus, even in the shot noise limit, the gaussian approximation gives a surprisingly close result to the value calculated from the correct Poisson distribution. It must be pointed out, however, that the location of the threshold calculated by the gaussian approximation is far from correct in this case. In general, the gaussian approximation is much better in estimating receiver sensitivity than in establishing where to set receiver thresholds.

Low-input-impedance amplifiers are generally required to achieve the high bandwidths required for high-bit-rate systems. However, a low input impedance results in high thermal noise and poor sensitivity. High-input-impedance amplifiers may be used, but this narrows the bandwidth, which must be compensated for by equalization following the first-stage amplifier. Although this may result in a highly sensitive receiver, the receiver will have a poor dynamic range owing to the high gains required in the equalizer.[19] Receivers for digital systems are generally implemented with transimpedance amplifiers having a large feedback resistance. This reduces the effective input noise capacitance to below the capacitance of the photodiode, and practical receivers can be built with $C_e \approx 0.1$ pF. Using this value of capacitance and $B/R_b = 0.7$, $\eta = 0.7$, and $Q = 6$, Eq. (17) gives $\overline{N}_p \approx 2600$. Note that this is about 34 dB greater than the value given by the quantum limit.

APD Receiver.   In an APD receiver, there is additional shot noise owing to the excess noise factor F of the avalanche gain process. However, thermal noise is reduced because of the current multiplication gain M before thermal noise is introduced. This results in a receiver sensitivity given approximately by*

$$\overline{N}_p = \frac{B}{\eta R_b} Q^2 \left[F + \frac{1}{Q} \sqrt{\frac{8\pi k T C_e}{q^2 M^2}} \right] \tag{19}$$

The excess noise factor is an increasing function of M, which results in an optimum M to minimize $\overline{N}_p$.[19] Good APD receivers at 1300 and 1550 nm typically have sensitivities of the order of 1000 photons per bit. Owing to the lower excess noise of silicon APDs, sensitivity of about 500 photons per bit can be achieved at 850 nm.

Impairments.   There are several sources of impairment that may degrade the sensitivity of receivers from the values given by Eqs. (17) and (19). These may be grouped into two general classes: eye degradations and signal-dependent noise.

An eye diagram is the superposition of all possible received sequences. At the sampling point, there is a spread of the values of a received 1 and a received 0. The difference between the minimum value of a received 1 and the maximum value of the received 0 is known as the *eye opening*. This is given by $(1 - \varepsilon)I_p$ where ε is the eye degradation. The two major sources of eye degradation are intersymbol interference and finite laser extinction ratio. Intersymbol interference results from dispersion, deviations from ideal shaping of the receiver filter, and low-frequency cutoff effects that result in direct current (DC) offsets.

Signal-dependent noises are phenomena that give a variance of the received photocurrent that is proportional to I_p^2 and consequently lead to a maximum signal-to-noise ratio at the output of the receiver. Principal sources of signal-dependent noise are laser relative intensity noise (RIN), reflection-induced noise, mode partition noise, and modal noise. RIN is a consequence of inherent fluctuations in laser intensity resulting from spontaneous emission (Ref. 17; Chap. 4 in Ref. 18). This is generally sufficiently small that it is not of concern in digital systems, but is an important limitation in analog systems requiring high signal-to-noise ratios (see Sec. 2.7). Reflection-induced noise is the conversion of laser phase noise to intensity noise by multiple reflections from discontinuities (such as at imperfect connectors.) This may result in a substantial RIN enhancement that can seriously affect digital as well as analog systems.[20] Mode partition noise occurs when Fabry-Perot lasers are used with dispersive fiber.

---

* The gaussian approximation is not as good for an APD as for a PIN receiver owing to the nongaussian nature of the excess APD noise.

Fiber dispersion results in changing phase relation between the various laser modes, which results in intensity fluctuations. The effect of mode partition noise is more serious than that of dispersion alone.[21] Modal noise is a similar phenomenon that occurs in multimode fiber when relatively few modes are excited and these interfere.

Eye degradations are accounted for by replacing Eq. (14a) by

$$\mu_1 - \mu_0 = (1 - \varepsilon)I_p \tag{20a}$$

and signal-dependent noise by replacing Eq. (14b) by

$$\sigma_1^2 = 2qI_pB + \frac{4kTB}{R_e} + \alpha^2 I_p^2 B \qquad \sigma_0^2 = \frac{4kTB}{R_e} + \alpha^2 I_p^2 B \tag{20b}$$

and α^2 is the relative spectral density of the signal-dependent noise. (It is assumed that the signal-dependent noise has a large bandwidth compared to the signal bandwidth B.) With these modifications, the sensitivity of an APD receiver becomes

$$\overline{N}_p = \frac{\dfrac{B}{\eta R_b}\left(\dfrac{Q}{1-\varepsilon}\right)^2\left[F + \left(\dfrac{1-\varepsilon}{Q}\right)\sqrt{\dfrac{8\pi kTC_e}{q^2M^2}}\right]}{1 - \alpha^2 B\left(\dfrac{Q}{1-\varepsilon}\right)^2} \tag{21}$$

where the PIN expression is obtained by setting $F = 1$ and $M = 1$. It follows from Eq. (21) that there is a minimum error probability (*error floor*) given by

$$P_{e,\min} = K(Q_{\max}) \text{ where } Q_{\max} = \frac{1-\varepsilon}{\alpha\sqrt{B}} \tag{22}$$

The existence of eye degradations and signal-dependent noise causes an increase in the receiver power (called *power penalty*) required to achieve a given error probability.

2.4   BIT RATE AND DISTANCE LIMITS

Bit rate and distance limitations of digital links are determined by loss and dispersion limitations. The following example is used to illustrate the calculation of the maximum distance for a given bit rate. Consider a 2.5-Gbit/s system at a wavelength of 1550 nm. Assume an average transmitter power of 0 dBm coupled into the fiber. Receiver sensitivity is taken to be 3000 photons per bit, which from Eq. (12) corresponds to an average receiver power of −30.2 dBm. Allowing a total of 8 dB for margin and for connector and cabling losses at the two ends gives a loss allowance of 22.2 dB. If the cabled fiber loss, including splices, is 0.25 dB/km, this leads to a loss-limited transmission distance of 89 km.

Assuming that the fiber dispersion is $D = 15$ ps/km·nm and source spectral width is 0.1 nm, this gives a dispersion per unit length of 1.5 ps/km. Taking the maximum allowed dispersion to be half the interpulse period, this gives a maximum dispersion of 200 ps, which then yields a maximum dispersion-limited distance of 133 km. Thus, the loss-limited distance is controlling.

Consider what happens if the bit rate is increased to 10 Gbit/s. For the same number of photons per bit at the receiver, the receiver power must be 6 dB greater than that in the preceding example. This reduces the loss allowance by 6 dB, corresponding to a reduction of 24 km in the loss-limited distance. The loss-limited distance is now 65 km (assuming all other parameters are unchanged). However, dispersion-limited distance scales inversely with bit rate, and is now 22 km. The system is now dispersion-limited. Dispersion-shifted fiber would be required to be able to operate at the loss limit.

Increasing Bit Rate

There are two general approaches for increasing the bit rate transmitted on a fiber: time-division multiplexing (TDM), in which the serial transmission rate is increased, and wavelength-division multiplexing (WDM), in which separate wavelengths are used to transmit independent serial bit streams in parallel. TDM has the advantage of minimizing the quantity of active devices but requires higher-speed electronics as the bit rate is increased. Also, as indicated by the preceding example, dispersion limitations will be more severe.

WDM allows use of existing lower-speed electronics, but requires multiple lasers and detectors as well as optical filters for combining and separating the wavelengths. Technology advances, including tunable lasers, transmitter and detector arrays, high-resolution optical filters, and optical amplifiers (Sec. 2.5) are making WDM more attractive, particularly for networking applications (Sec. 2.6).

Longer Repeater Spacing

In principal, there are three approaches for achieving longer repeater spacing than that calculated in the preceding text: lower fiber loss, higher transmitter powers, and improved receiver sensitivity (smaller $\overline{N}_p$). Silica-based fiber is already essentially at the theoretical Rayleigh scattering loss limit. There has been research on new fiber materials that would allow operation at wavelengths longer than 1.6 μm, with consequent lower theoretical loss values.[22] There are many reasons, however, why achieving such losses will be difficult, and progress in this area has been slow.

Higher transmitter powers are possible, but there are both nonlinearity and reliability issues that limit transmitter power. Since present receivers are more than 30 dB above the quantum limit, improved receiver sensitivity would appear to offer the greatest possibility. To improve the receiver sensitivity, it is necessary to increase the photocurrent at the output of the detector without introducing significant excess loss. There are two main approaches for doing so: optical amplification and optical mixing. Optical preamplifiers result in a theoretical sensitivity of 38 photons per bit[23] (6dB above the quantum limit), and experimental systems have been constructed with sensitivities of about 100 photons per bit.[24] This will be discussed further in Sec. 2.5. Optical mixing (coherent receivers) will be discussed briefly in the following text.

Coherent Systems. A photodetector provides an output current proportional to the magnitude square of the electric field that is incident on the detector. If a strong optical signal (*local oscillator*) coherent in phase with the incoming optical signal is added prior to the photodetector, then the photocurrent will contain a component at the difference frequency between the incoming and local oscillator signals. The magnitude of this photocurrent, relative to the direct detection case, is increased by the ratio of the local oscillator to the incoming field strengths. Such a coherent receiver offers considerable improvement in receiver sensitivity. With on-off keying, a heterodyne receiver (signal and local oscillator frequencies different) has a theoretical sensitivity of 36 photons per bit, and a homodyne receiver (signal and local oscillator frequencies the same) has a sensitivity of 18 photons per bit. Phase-shift keying (possible with coherent systems) provides a further 3-dB improvement. Coherent systems, however, require very stable signal and local oscillator sources (spectral linewidths need to be small compared to the modulation bandwidth) and matching of the polarization of the signal and local oscillator fields.[25]

An advantage of coherent systems, more so than improved receiver sensitivity, is that because the output of the photodetector is linear in the signal field, filtering for WDM demultiplexing may be done at the difference frequency (typically in the microwave range). This allows considerably greater selectivity than is obtainable with optical filtering techniques. The advent of optical amplifiers has slowed the interest in coherent systems.

2.5   *OPTICAL AMPLIFIERS*

There are two types of optical amplifiers: laser amplifiers based on stimulated emission and parametric amplifiers based on nonlinear effects (Chap. 8 in Ref. 18). The former are currently of most interest in fiber-optic communications. A laser without reflecting end faces is an amplifier, but it is more difficult to obtain sufficient gain for amplification than it is (with feedback) to obtain oscillation. Thus, laser oscillators were available much earlier than laser amplifiers.

Laser amplifiers are now available with gains in excess of 30 dB over a spectral range of more than 30 nm. Output saturation powers in excess of 10 dBm are achievable. The amplified spontaneous emission (ASE) noise power at the output of the amplifier, in each of two orthogonal polarizations, is given by

$$P_{ASE} = n_{sp} \frac{hc}{\lambda} B_o (G - 1) \qquad (23)$$

where G is the amplifier gain, B_o is the bandwidth, and the spontaneous emission factor n_{sp} is equal to 1 for ideal amplifiers with complete population inversion.

Comparison of Semiconductor and Fiber Amplifiers

There are two principal types of laser amplifiers: semiconductor laser amplifiers (SLAs) and doped-fiber amplifiers. The erbium-doped-fiber amplifier (EDFA), which operates at a wavelength of 1.55 μm, is of most current interest.

The advantages of the SLA, similar to laser oscillators, are that it is pumped by a DC current, it may be designed for any wavelength of interest, and it can be integrated with electrooptic semiconductor components.

The advantages of the EDFA are that there is no coupling loss to the transmission fiber, it is polarization-insensitive, it has lower noise than SLAs, it can be operated at saturation with no intermodulation owing to the long time constant of the gain dynamics, and it can be integrated with fiber devices. However, it does require optical pumping, with the principal pump wavelengths being either 980 or 1480 nm.

Communications Application of Optical Amplifiers

There are four principal applications of optical amplifiers in communication systems:[26,27]

1. Transmitter power amplifiers
2. Compensation for splitting loss in distribution networks
3. Receiver preamplifiers
4. Linear repeaters in long-distance systems

The last application is of particular importance for long-distance networks (particularly undersea systems), where a bit-rate-independent linear repeater allows subsequent upgrading of system capacity (either TDM or WDM) with changes only at the system terminals. Although amplifier noise accumulates in such long-distance linear systems, transoceanic lengths are achievable with amplifier spacings of about 60 km corresponding to about 15-dB fiber attenuation between amplifiers.

However, in addition to the accumulation of ASE, there are other factors limiting the distance of linearly amplified systems, namely dispersion and the interaction of dispersion and nonlinearity.[28] There are two alternatives for achieving very long-distance, very high-bit-rate systems with linear repeaters: *solitons,* which are pulses that maintain their shape in a dispersive medium,[29] and dispersion compensation.[30]

2.6 FIBER-OPTIC NETWORKS

Networks are communication systems used to interconnect a number of terminals within a defined geographic area—for example, local area networks (LANs), metropolitan area networks (MANs), and wide area networks (WANs). In addition to the transmission function discussed throughout the earlier portions of this chapter, networks also deal with the routing and switching aspects of communications.

Passive optical networks utilize couplers to distribute signals to users. In an $N \times N$ ideal star coupler, the signal on each input port is uniformly distributed among all output ports. If an average power P_T is transmitted at a transmitting port, the power received at a receiving port (neglecting transmission losses) is

$$P_R = \frac{P_T}{N}(1 - \delta_N) \tag{24}$$

where δ_N is the excess loss of the coupler. If N is a power of 2, an $N \times N$ star may be implemented by $\log_2 N$ stages of 2×2 couplers. Thus, it may be conservatively assumed that

$$1 - \delta_N = (1 - \delta_2)^{\log_2 N} = N^{\log_2(1 - \delta_2)} \tag{25}$$

The maximum bit rate per user is given by the average received power divided by the product of the photon energy and the required number of photons per bit (N_p). The throughput Y is the product of the number of users and the bit rate per user, and from Eqs. (24) and (25) is therefore given by

$$Y = \frac{P_T}{N_p} \frac{\lambda}{hc} N^{\log_2(1 - \delta_2)} \tag{26}$$

Thus, the throughput (based on power considerations) is independent of N for ideal couplers ($\delta_2 = 0$) and decreases slowly with N ($\sim N^{-0.17}$) for $10 \log (1 - \delta_2) = 0.5$ dB. It follows from Eq. (26) that for a power of 1 mW at $\lambda = 1.55$ μm and with $N_p = 3000$, the maximum throughput is 2.6 Tbit/s.

This may be contrasted with a tapped bus, where it may be shown that optimum tap weight to maximize throughput is given by $1/N$, leading to a throughput given by[31]

$$Y = \frac{P_T}{N_p} \frac{\lambda}{hc} \frac{1}{Ne^2} \exp(-2N\delta) \tag{27}$$

Thus, even for ideal ($\delta = 0$) couplers, the throughput decreases inversely with the number of users. If there is excess coupler loss, the throughput decreases exponentially with the number of users and is considerably less than that given by Eq. (26). Consequently, for a power-limited transmission medium, the star architecture is much more suitable than the tapped bus. The same conclusion does not apply to metallic media, where bandwidth rather than power limits the maximum throughput.

Although the preceding text indicates the large throughput that may be achieved in principle with a passive star network, it doesn't indicate how this can be realized. Most interest is in WDM networks.[32] The simplest protocols are those for which fixed-wavelength receivers and tunable transmitters are used. However, the technology is simpler when fixed-wavelength transmitters and tunable receivers are used, since a tunable receiver may be implemented with a tunable optical filter preceding a wideband photodetector. Fixed-wavelength transmitters and receivers involving multiple passes through the network are also possible, but this requires utilization of terminals as relay points. Protocol, technology, and application considerations for gigabit networks (networks having access at gigabit rates and throughputs at terabit rates) is an extensive area of current research.[32,33]

2.7   ANALOG TRANSMISSION ON FIBER

Most interest in fiber-optic communications is centered around digital transmission, since fiber is generally a power-limited rather than a bandwidth-limited medium. There are applications, however, where it is desirable to transmit analog signals directly on fiber without converting them to digital signals. Examples are cable television (CATV) distribution and microwave links such as entrance links to antennas and interconnection of base stations in mobile radio systems.

Carrier-to-Noise Ratio (CNR)

Optical intensity modulation is generally the only practical modulation technique for incoherent-detection fiber-optic systems. Let $f(t)$ be the carrier signal that intensity modulates the optical source. For convenience, assume that the average value of $f(t)$ is equal to 0, and that the magnitude of $f(t)$ is normalized to be less than or equal to 1. The received optical power may then be expressed as

$$P(t) = P_o[1 + mf(t)] \tag{28}$$

where m is the optical modulation index

$$m = \frac{P_{\max} - P_{\min}}{P_{\max} + P_{\min}} \tag{29}$$

The carrier-to-noise ratio is then given by

$$\text{CNR} = \frac{\frac{1}{2}m^2 \Re^2 P_o^2}{\text{RIN } \Re^2 P_o^2 B + 2q\Re P_o B + <i_{th}^2>B} \tag{30}$$

where $\Re$ is the photodetector responsivity, RIN is the relative intensity noise spectral density (denoted by α^2 in Sec. 2.3), and $<i_{th}^2>$ is the thermal noise spectral density (expressed as $4kT/R_e$ in Sec. 2.3). CNR is plotted in Fig. 3 as a function of received optical power for a bandwidth of $B = 4$ MHz (single video channel), optical modulation index $m = 0.05$, $\Re = 0.8$ A/W, RIN = -155 dB/Hz, and $\sqrt{<i_{th}^2>} = 7$ pA/$\sqrt{\text{Hz}}$. At low received powers (typical of digital systems) the CNR is limited by thermal noise. However, to obtain the higher CNR generally needed by analog systems, shot noise and then ultimately laser RIN become limiting.

Analog Video Transmission on Fiber[34]

It is helpful to distinguish between single-channel and multiple-channel applications. For the single-channel case, the video signal may directly modulate the laser intensity [amplitude-modulated (AM) system], or the video signal may be used to frequency-modulate an electrical subcarrier, with this subcarrier then intensity-modulating the optical source [frequency-modulated (FM) system]. Equation (30) gives the CNR of the recovered subcarrier. Subsequent demodulation of the FM signal gives an additional increase in signal-to-noise ratio. In addition to this FM improvement factor, larger optical modulation indexes may be used than in AM systems. Thus FM systems allow higher signal-to-noise ratios and longer transmission spans than AM systems.

Two approaches have been used to transmit multichannel video signals on fiber. In the first (AM systems), the video signals undergo electrical frequency-division multiplexing (FDM),

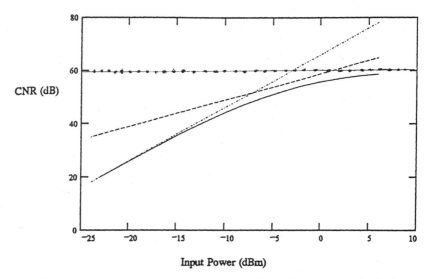

FIGURE 3   CNR as a function of input power. Straight lines indicate thermal noise (-.-.-), shot noise (–), and RIN (.....) limits.

and this combined FDM signal intensity modulates the optical source. This is conceptually the simplest system, since existing CATV multiplexing formats may be used.

In FM systems, the individual video channels frequency-modulate separate microwave carriers (as in satellite systems). These carriers are linearly combined and the combined signal intensity modulates a laser. Although FM systems are more tolerant than AM systems to intermodulation distortion and noise, the added electronics costs have made such systems less attractive than AM systems for CATV application.

Multichannel AM systems are of interest not only for CATV application but also for mobile radio applications to connect signals from a microcellular base station to a central processing station. Relative to CATV applications, the mobile radio application has the additional complication of being required to accommodate signals over a wide dynamic power range.

Nonlinear Distortion

In addition to CNR requirements, multichannel analog communication systems are subject to intermodulation distortion. If the input to the system consists of a number of tones at frequencies ω_i, then nonlinearities result in intermodulation products at frequencies given by all sums and differences of the input frequencies. Second-order intermodulation gives intermodulation products at frequencies $\omega_i \pm \omega_j$, whereas third-order intermodulation gives frequencies $\omega_i \pm \omega_j \pm \omega_k$. If the signal frequency band is such that the maximum frequency is less than twice the minimum frequency, then all second-order intermodulation products fall outside the signal band, and third-order intermodulation is the dominant nonlinearity. This condition is satisfied for the transport of microwave signals (e.g., mobile radio signals) on fiber, but is not satisfied for wideband CATV systems, where there are requirements on composite second-order (CSO) and composite triple-beat (CTB) distortion.

The principal causes of intermodulation in multichannel fiber-optic systems are laser threshold nonlinearity,[35] inherent laser gain nonlinearity, and the interaction of chirp and dispersion.

2.8 TECHNOLOGY AND APPLICATIONS DIRECTIONS

Fiber-optic communication application in the United States began with metropolitan and short-distance intercity trunking at a bit rate of 45 Mbit/s, corresponding to the DS-3 rate of the North American digital hierarchy. Technological advances, primarily higher-capacity transmission and longer repeater spacings, extended the application to long-distance intercity transmission, both terrestrial and undersea. Also, transmission formats are now based on the synchronous digital hierarchy (SDH), termed synchronous optical network (SONET) in the U.S. OC-48 systems* operating at 2.5 Gbit/s are widely deployed, with OC-192 10-Gbit/s systems also available as of 1999. All of the signal processing in these systems (multiplexing, switching, performance monitoring) is done electrically, with optics serving solely to provide point-to-point links.

For long-distance applications, dense wavelength-division multiplexing (DWDM), with channel spacings of 100 GHz and with upward of 80 wavelength channels, has extended the bit rate capability of fiber to greater than 400 Gbit/s in commercial systems and up to 3 Tbit/s in laboratory trials.[36] For local access, there is extensive interest in hybrid combinations of optical and electronic technologies and transmission media.[37,38] Owing to the criticality of communications, network survivability has achieved growing importance, with SONET rings being implemented so that no single cable cut will result in system failure.[39]

The huge bandwidth capability of fiber optics (measured in tens of terahertz) is not likely to be utilized by time-division techniques alone, and DWDM technology and systems are receiving considerable emphasis, although work is also under way on optical time-division multiplexing (OTDM) and optical code-division multiplexing (OCDM).

Nonlinear phenomena, when uncontrolled, generally lead to system impairments. However, controlled nonlinearities are the basis of devices such as parametric amplifiers and switching and logic elements. Nonlinear optics will consequently continue to receive increased emphasis.

2.9 REFERENCES

1. Jeff Hecht, *City of Light: The Story of Fiber Optics,* Oxford University Press, 1999.

2. C. K. Kao and G. A. Hockham, "Dielectric-Fiber Surface Waveguides for Optical Frequencies," *Proc. IEE* **113**:1151–1158 (July 1966).

3. F. P. Kapron et al., "Radiation Losses in Glass Optical Waveguides," *Appl. Phys. Lett.* **17**:423 (November 15, 1970).

4. I. Hayashi, M. B. Panish, and P. W. Foy, "Junction Lasers which Operate Continuously at Room Temperature," *Appl. Phys. Lett.* **17**:109 (1970).

5. Ira Jacobs, "Optical Fiber Communication Technology and System Overview," in O. D. D. Soares (ed.), *Trends in Optical Fibre Metrology and Standards,* NATO ASI Series, vol. 285, pp. 567–591, Kluwer Academic Publishers, 1995.

6. D. Gloge, "Weakly Guiding Fibers," *Appl. Opt.* **10**:2252–2258 (October 1971).

7. R. Olshansky and D. Keck, "Pulse Broadening in Graded Index Fibers," *Appl. Opt.* **15**:483–491 (February 1976).

8. D. Gloge, E. A. J. Marcatili, D. Marcuse, and S. D. Personick, "Dispersion Properties of Fibers," in S. E. Miller and A. G. Chynoweth (eds.), *Optical Fiber Telecommunications,* chap. 4, Academic Press, 1979.

9. Y. Namihira and H. Wakabayashi, "Fiber Length Dependence of Polarization Mode Dispersion Measurements in Long-Length Optical Fibers and Installed Optical Submarine Cables," *J. Opt. Commun.* **2**:2 (1991).

10. W. B. Jones Jr., *Introduction to Optical Fiber Communication Systems,* pp. 90–92, Holt, Rinehart and Winston, 1988.

---

* OC-*n* systems indicate optical channel at a bit rate of (51.84)*n* Mbit/s.

11. L. G. Cohen, W. L. Mammel, and S. J. Jang, "Low-Loss Quadruple-Clad Single-Mode Lightguides with Dispersion Below 2 ps/km·nm over the 1.28 μm–1.65 μm Wavelength Range," *Electron. Lett.* **18**:1023–1024 (1982).

12. G. Keiser, *Optical Fiber Communications*, 3d ed., chap. 5, McGraw-Hill, 2000.

13. N. M. Margalit, S. Z. Zhang, and J. E. Bowers, "Vertical Cavity Lasers for Telecom Applications," *IEEE Commun. Mag.* **35**:164–170 (May 1997).

14. J. E. Bowers and M. A. Pollack, "Semiconductor Lasers for Telecommunications," in S. E. Miller and I. P. Kaminow (eds.), *Optical Fiber Telecommunications II,* chap. 13, Academic Press, 1988.

15. W. T. Tsang, "The Cleaved-Coupled-Cavity (C^3) Laser," in *Semiconductors and Semimetals,* vol. 22, part B, chap. 5, pp. 257–373, 1985.

16. K. Kobayashi and I. Mito, "Single Frequency and Tunable Laser Diodes," *J. Lightwave Technol.* **6**:1623–1633 (November 1988).

17. T. Mukai and Y. Yamamoto, "AM Quantum Noise in 1.3 μm InGaAsP Lasers," *Electron. Lett.* **20**:29–30 (January 5, 1984).

18. G. P. Agrawal, *Fiber-Optic Communication Systems*, 2d ed., Wiley Interscience, 1997.

19. S. D. Personick, "Receiver Design for Digital Fiber Optic Communication Systems I," *Bell Syst. Tech. J.* **52**: 843–874 (July–August 1973).

20. J. L. Gimlett and N. K. Cheung, "Effects of Phase-to-Intensity Noise Conversion by Multiple Reflections on Gigabit-per-Second DFB Laser Transmission Systems," *J. Lightwave Technol.* **LT-7**:888–895 (June 1989).

21. K. Ogawa, "Analysis of Mode Partition Noise in Laser Transmission Systems," *IEEE J. Quantum Electron.* **QE-18**:849–855 (May 1982).

22. D. C. Tran, G. H. Sigel, and B. Bendow, "Heavy Metal Fluoride Fibers: A Review," *J. Lightwave Technol.* **LT-2**:566–586 (October 1984).

23. P. S. Henry, "Error-Rate Performance of Optical Amplifiers," *Optical Fiber Communications Conference (OFC'89 Technical Digest)*, THK3, Houston, Texas, February 9, 1989.

24. O. Gautheron, G. Grandpierre, L. Pierre, J.-P. Thiery, and P. Kretzmeyer, "252 km Repeaterless 10 Gbits/s Transmission Demonstration," *Optical Fiber Communications Conference (OFC'93) Post-deadline Papers*, PD11, San Jose, California, February 21–26, 1993.

25. I. W. Stanley, "A Tutorial Review of Techniques for Coherent Optical Fiber Transmission Systems," *IEEE Commun. Mag.* **23**:37–53 (August 1985).

26. Bellcore, "Generic Requirements for Optical Fiber Amplifier Performance," Technical Advisory TA-NWT-001312, Issue 1, December 1992.

27. T. Li, "The Impact of Optical Amplifiers on Long-Distance Lightwave Telecommunications," *Proc. IEEE* **81**:1568–1579 (November 1993).

28. A. Naka and S. Saito, "In-Line Amplifier Transmission Distance Determined by Self-Phase Modulation and Group-Velocity Dispersion," *J. Lightwave Technol.* **12**:280–287 (February 1994).

29. G. P. Agrawal, *Nonlinear Fiber Optics,* chap. 5, Academic Press, 1989.

30. Bob Jopson and Alan Gnauck, "Dispersion Compensation for Optical Fiber Systems," *IEEE Commun. Mag.* **33**:96–102 (June 1995).

31. P. E. Green, Jr., *Fiber Optic Networks,* chap. 11, Prentice Hall, 1993.

32. M. Fujiwara, M. S. Goodman, M. J. O'Mahony, O. K. Tonguz, and A. E. Willner (eds.), Special Issue on Multiwavelength Optical Technology and Networks, *J. Lightwave Technology* **14**(6):932–1454 (June 1996).

33. P. J. Smith, D. W. Faulkner, and G. R. Hill, "Evolution Scenarios for Optical Telecommunication Networks Using Multiwavelength Transmission," *Proc. IEEE,* **81**:1580–1587 (November 1993).

34. T. E. Darcie, K. Nawata, and J. B. Glabb, Special Issue on Broad-Band Lightwave Video Transmission, *J. Lightwave Technol.* **11**(1) (January 1993).

35. A. A. M. Saleh, "Fundamental Limit on Number of Channels in SCM Lightwave CATV System," *Electron. Lett.* **25**(12):776–777 (1989).

36. A. R. Chraplyvy, "High-Capacity Lightwave Transmission Experiments," *Bell Labs Tech. J.* **4**:230–245 (January–March 1999).

37. C. Baack and G. Walf, "Photonics in Future Telecommunications," *Proc. IEEE* **81**:1624–1632 (November 1993).

38. Gordon C. Wilson et al., "FiberVista: An FTTH or FTTC System Delivering Broadband Data and CATV Services," *Bell Labs Tech. J.* **4**:300–322 (January–March 1999).

39. Bellcore (Telcordia), "SONET Bidirectional Line Switched Ring Equipment Generic Criteria," Bellcore GR-1230, December 1996.

CHAPTER 3
NONLINEAR EFFECTS IN OPTICAL FIBERS

John A. Buck
Georgia Institute of Technology,
School of Electrical and Computer Engineering
Atlanta, Georgia

Fiber nonlinearities are important in optical communications, both as useful attributes and as characteristics to be avoided. They must be considered when designing long-range high-data-rate systems that involve high optical power levels and in which signals at multiple wavelengths are transmitted. The consequences of nonlinear transmission can include (1) the generation of additional signal bandwidth within a given channel, (2) modifications of the phase and shape of pulses, (3) the generation of light at other wavelengths at the expense of power in the original signal, and (4) crosstalk between signals at different wavelengths and polarizations. The first two, arising from self-phase modulation, can be used to advantage in the generation of *solitons*—pulses whose nonlinear phase modulation compensates for linear group dispersion in the fiber channel[1] or in fiber gratings,[2] leading to pulses that propagate without changing shape or width (see Chap. 7). The third and fourth effects arise from stimulated Raman or Brillouin scattering or four-wave mixing. These can be used to advantage when it is desired to generate or amplify additional wavelengths, but they must usually be avoided in systems.

3.1 KEY ISSUES IN NONLINEAR OPTICS IN FIBERS

Optical fiber waveguides, being of glass compositions, do not possess large nonlinear coefficients. Nonlinear processes can nevertheless occur with high efficiencies since intensities are high and propagation distances are long. Even though power levels are usually modest (a few tens of milliwatts), intensities within the fiber are high due to the small cross-sectional areas involved. This is particularly true in single-mode fiber, where the LP_{01} mode typically presents an effective cross-sectional area of between 10^{-7} and 10^{-8} cm^2, thus leading to intensities on the order of MW/cm^2. Despite this, long interaction distances are usually necessary to achieve nonlinear mixing of any significance, so processes must be phase matched, or nearly so. Strategies to avoid unwanted nonlinear effects usually involve placing upper limits on optical power levels, and if possible, choosing other parameters such that phase mismatching occurs. Such choices may include wavelengths or wavelength spacing in wavelength-division multiplexed systems, or may be involved in special fiber waveguide designs.[3]

The generation of light through nonlinear mixing arises through polarization of the medium, which occurs through its interaction with intense light. The polarization consists of an array of phased dipoles in which the dipole moment is a nonlinear function of the applied field strength. In the classical picture, the dipoles, once formed, reradiate light to form the nonlinear output. The medium polarization is conveniently expressed through a power series expansion involving products of real electric fields:

$$\mathcal{P} = \varepsilon_0[\chi^{(1)} \cdot \mathcal{E} + \chi^{(2)} \cdot \mathcal{E}\mathcal{E} + \chi^{(3)} \cdot \mathcal{E}\mathcal{E}\mathcal{E} + ...] = \mathcal{P}_L + \mathcal{P}_{NL} \tag{1}$$

in which the χ terms are the linear, second-, and third-order susceptibilities. Nonlinear processes are described through the product of two or more optical fields to form the nonlinear polarization $\mathcal{P}_{NL}$, consisting of all terms of second order and higher in Eq. (1).

The second-order term in Eq. (1) [involving $\chi^{(2)}$] describes three-wave mixing phenomena, such as second-harmonic generation. The third-order term describes four-wave mixing (FWM) processes and stimulated scattering phenomena. In the case of optical fibers, second-order processes are generally not possible, since these effects require noncentrosymmetric media.[4] In amorphous fiber waveguides, third-order effects [involving $\chi^{(3)}$] are usually seen exclusively, although second-harmonic generation can be observed in special instances.[5]

The interactions between fields and polarizations are described by the nonlinear wave equation:

$$\nabla^2 \mathcal{E} + n_0^2 \mu_0 \varepsilon_0 \frac{\partial^2 \mathcal{E}}{\partial t^2} = \mu_0 \frac{\partial^2 \mathcal{P}_{NL}}{\partial t^2} \tag{2}$$

where $\mathcal{E}$ and $\mathcal{P}$ are the sums of all electric fields and nonlinear polarizations that are present, and where n_0 is the refractive index of the medium. The second-order differential equation is usually reduced to first order through the slowly varying envelope approximation (SVEA):

$$\left| \frac{\partial^2 E}{\partial z^2} \right| \ll \left| \frac{2\pi}{\lambda} \frac{\partial E}{\partial z} \right| \tag{3}$$

where E is the complex field amplitude. The interpretation of the SVEA is that the changes in field amplitude that occur over distances on the order of a wavelength are very large compared to variations in the rate of change over the same distance. The wave equation will separate according to frequencies or propagation directions, yielding sets of coupled differential equations that, under the SVEA, are first order. These describe the growth or decay of fields involved in the mixing process.

The requirement for phase matching is that the nonlinear polarization wave and the electric field associated with the generated wave propagate with the same phase constant; that is, their phase velocities are equal. Phase-matched processes in fiber include those that involve (1) interacting waves at the same wavelength and polarization, such as self- and cross-phase modulation, as well as other degenerate Kerr-type interactions; and (2) stimulated scattering processes (Raman and Brillouin), in addition to cross-phase modulation involving two wavelengths. Four-wave mixing processes involving light at different wavelengths can occur that are not precisely phase matched but that can nevertheless yield high efficiencies. Matters are further complicated by the fact that different nonlinear processes can occur simultaneously, with each affecting the performance of the other. Nonlinear effects are usually favored to occur under pulsed operation, since high peak powers can be achieved with comparatively modest average powers. Consequently, group velocity matching is desirable (although not always required) to achieve efficient mixing between pulses.

3.2 SELF- AND CROSS-PHASE MODULATION

Self-phase modulation (SPM) can occur whenever a signal having a time-varying amplitude is propagated in a nonlinear material. The origin of the effect is the refractive index of the medium, which will change with the instantaneous signal intensity. The complex nonlinear polarization for the process is:

$$P_{NL} = \tfrac{3}{4}\varepsilon_0\chi^{(3)}|E_0(z, t)|^2 E_0(z, t)\exp[i(\omega t - \beta z)] \tag{4}$$

where $E_0(t)$ is the time-varying electric field amplitude that describes the pulse or signal envelope, and where the frequency ω is the same as that of the input light. Incorporating this polarization and the field into the wave equation leads to a modified refractive index over the original zero-field value n_0. The net index becomes:[6]

$$n = n_0 + n_2'|E_0(z, t)|^2 \tag{5}$$

where the nonlinear refractive index is given by $n_2' = Re\{3\chi^{(3)}/8n_0\}$. In fused silica it has the value $n_2' = 6.1 \times 10^{-23}\, m^2/V^{2}$.[7] Equation (5) can also be expressed in terms of light intensity through $n(I) = n_0 + n_2 I(z, t)$, where $n_2 = 3.2 \times 10^{-20}\, m^2/W$. In optical fibers the index is modified from the effective mode index of the single-mode fiber n_{eff} (which assumes the role of n_0).

The complex field as it propagates through the medium can be expressed as:

$$E = E_0(z, t)\exp\{i[\omega_0 t - [n_0 + n_2 I(z, t)]k_0 z]\} \tag{6}$$

which exhibits phase modulation that follows the shape of the intensity envelope. The instantaneous frequency is found through the time derivative of the phase:

$$\omega' = \omega_0 - n_2 k_0 z \frac{\partial I}{\partial t} \tag{7}$$

The effects of self-phase modulation on pulse propagation can be qualitatively observed from Eqs. (6) and (7). First, additional frequency components are placed on the pulse, thus increasing its spectral width. Second, a frequency sweep (chirp) imposed on the pulse, the direction of which depends on the sign of $\partial I/\partial t$. The latter feature is particularly important in optical fibers, since the imposed frequency sweep from SPM will either add to or subtract from the chirp imposed by linear group dispersion. If the chirp directions for self-phase modulation and group dispersion are opposite, an effective cancellation may occur, leading to the formation of an optical soliton. In more conventional systems in which solitons are not employed, SPM must be considered as a possible benefit or detriment to performance, as some pulse shaping (which could include broadening or compression) can occur;[8,9] however, such systems can in theory yield excellent performance.[10] Furthermore, in systems employing fiber amplifiers, the change in refractive index associated with the signal-induced upper state population in erbium has been shown to be an important performance factor.[11] An additional effect can occur when pulse spectra lie within the anomalous group dispersion regime of the fiber; pulse breakup can occur as a result of *modulation instability,* in which the interplay between dispersive and nonlinear contributions to pulse shaping becomes unstable.[12]

Cross-phase modulation (XPM) is similar to SPM, except that two overlapping but distinguishable pulses (having, for example, different frequencies or polarizations) are involved. One pulse will modulate the index of the medium, which then leads to phase modulation of an overlapping pulse. XPM thus becomes a cross-talk mechanism between two channels if phase encoding is employed or if intensity modulation is used in dispersive systems.[13,14] No transfer of energy occurs between channels, however, which distinguishes the process from other crosstalk mechanisms in which growth of signal power in one channel occurs at the expense of power in another. The strength of the effect is enhanced by a factor of 2 over that which can be obtained by a single field acting on itself (the nonlinear refractive index n_2 is effectively doubled in XPM).[15] The XPM process, while twice as strong as SPM, is effectively weakened by the fact that pulses of differing frequencies or polarizations are generally not group velocity matched, and so cannot maintain overlap indefinitely. The efficiency is further reduced if the interaction occurs between cross-polarized waves; in this case the nonlinear tensor element (and thus the effective nonlinear index) is a factor of ⅓ less than the tensor element that describes copolarized waves (pp. 164–165 in Ref. 6).

Self- and cross-phase modulation are analyzed by way of coupled equations of the *nonlinear Schrödinger* form,[16] which describes the evolution over time and position of the electric field envelopes of two pulses, E_{0a} and E_{0b}, where SVEA is used and where pulse widths are on the order of 1 ps or greater:

$$\frac{\partial E_{0a}}{\partial z} + \beta_{1a}\frac{\partial E_{0a}}{\partial t} = -\frac{i}{2}\,\beta_{2a}\frac{\partial^2 E_{0a}}{\partial t^2} + i\gamma_a|E_{0a}|^2 E_{0a} + i\delta\gamma_a|E_{0b}|^2 E_{0a} - \frac{\alpha_a}{2}\,E_{0a} \tag{8}$$

$$\frac{\partial E_{0b}}{\partial z} + \beta_{1b}\frac{\partial E_{0b}}{\partial t} = -\frac{i}{2}\,\beta_{2b}\frac{\partial^2 E_{0b}}{\partial t^2} + i\gamma_b|E_{0b}|^2 E_{0b} + i\delta\gamma_b|E_{0a}|^2 E_{0b} - \frac{\alpha_b}{2}\,E_{0b} \tag{9}$$

In these equations, β_{1j} $(j = a, b)$ are the group delays of the pulses at the two frequencies or polarizations over a unit distance; β_{2j} are the group dispersion parameters associated with the two pulses; and $\gamma_j = n_2'\omega_j/(cA_{\text{eff}})$, where A_{eff} is the effective cross-sectional area of the fiber mode. The coefficient δ is equal to 2 for copolarized pulses of different frequencies and is ⅔ if the pulses are cross-polarized. Propagation loss characterized by coefficients α_j is assumed. The equation form that describes the propagation with SPM of a single pulse—E_{0a}, for example—is found from Eq. (8) by setting $E_{0b} = 0$. The terms on the right sides of Eqs. (8) and (9) describe in order the effects of group dispersion, SPM, XPM, and loss. The equations can be solved using numerical techniques that are described on pages 50–55 of Ref. 16.

For subpicosecond pulses, the accuracy of Eqs. (8) and (9) begins to degrade as pulse bandwidths increase with decreasing temporal width. Additional terms are usually incorporated in the equations as pulse widths are reduced to the vicinity of 100 fs. These embody (1) cubic dispersion, which becomes important as bandwidth increases, and (2) changes in group velocity with intensity. This latter effect can result in *self-steepening,* in which the pulse trailing edge shortens to the point of forming an optical shock front (pp. 113–120 of Ref. 16) under appropriate conditions. An additional consequence of broad pulse spectra is that power conversion from high-frequency components within a pulse to those at lower frequencies can occur via stimulated Raman scattering, provided the interacting components are sufficiently separated in wavelength. The effect is an overall red shift of the spectrum. At sufficiently high intensities, cross-coupling between pulses having different center wavelengths can also occur through Raman scattering, regardless of pulse width.

3.3   STIMULATED RAMAN SCATTERING

In *stimulated Raman scattering* (SRS), coupling occurs between copropagating light waves whose frequency difference is in the vicinity of resonances of certain molecular oscillation modes. In silica-based fibers, stretch vibrational resonances occur between Si and O atoms in several possible modes within the glass matrix (see Ref. 17 for illustrations of the important modes in pure silica). In the Stokes process, light at frequency ω_2 (pump wave) is downshifted to light at ω_1 (Stokes wave), with the excess energy being absorbed by the lattice vibrational modes (manifested in the generation of optical phonons). The process is either spontaneous, in which the Stokes wave builds up from noise, or is stimulated, in which both waves are present in sufficient strength to generate a beat frequency that excites the oscillators and promotes coupling. A fiber Raman amplifier works on this principle, in which an input signal at ω_1 experiences gain in the presence of pump light at ω_2. Figure 1 shows the beam geometry in which an input wave at ω_1 and intensity I_{10} can emerge at the far end with amplified value I_{1L}. This occurs in the presence of the pump wave at ω_2 that has initial intensity I_{20} and that emerges with depleted intensity I_{2L}.

Back-conversion from ω_1 to ω_2 (the inverse Raman effect) will also occur once the Stokes wave reaches sufficient intensity, but gain will only occur for the Stokes wave. Both processes are phase matched, and so occur with high efficiency in long fibers. The back-conversion process is to be distinguished from anti-Stokes scattering, in which pump light at ω_2 is upshifted to frequency ω_3, with the additional energy being supplied by optical phonons associated with the previously excited medium. The anti-Stokes process is rarely seen in fiber transmission because (1) it is phase mismatched and (2) it requires a substantial population of excited oscillators, which is not the case at thermal equilibrium.

Figure 2 shows the measured Raman gain for the Stokes wave in fused silica. The gain is plotted as a function of difference frequency between the interacting waves measured in cm^{-1} (to convert this to wavelength shift, use the formula $\Delta\lambda = \lambda_p^2 \Delta f(cm^{-1})$, where λ_p is the pump wavelength). Other fiber constituents such as GeO_2, P_2O_5, and B_2O_3 exhibit their own Raman resonances, which occur at successively greater wavelength shifts;[19] the effects of these will be weak, since their concentration in the fiber is generally small. Thus the dominant Raman shifts in optical fiber are associated with SiO_2, and occur within the range of 440 to 490 cm^{-1}, as is evident in Fig. 2.

Nonlinear polarizations at frequencies ω_1 and ω_2 can be constructed that are proportional to products of the Stokes and pump fields, $E_1^{\omega_1}$ and $E_2^{\omega_2}$. These are of the form $P_{NL}^{\omega_1} \propto |E_2^{\omega_2}|^2 E_1^{\omega_1}$ (Stokes generation) and $P_{NL}^{\omega_2} \propto |E_1^{\omega_1}|^2 E_2^{\omega_2}$ (the inverse Raman effect). Substituting these polarizations and the two fields into the wave equation, using the SVEA, and assuming copolarized fields leads to the following coupled equations involving the Stokes and pump wave intensities I_1 and I_2:[18]

$$\frac{dI_1}{dz} = g_r I_1 I_2 - \alpha I_1 \tag{10}$$

$$\frac{dI_2}{dz} = -\frac{\omega_2}{\omega_1} g_r I_1 I_2 - \alpha I_2 \tag{11}$$

where the loss terms involving α (the fiber loss per unit distance) are added phenomenologically. The Raman gain function g_r is expressed in a general way as

$$g_r = \frac{A}{\lambda_2} f(\lambda_1 - \lambda_2) \tag{12}$$

where A is a function of the material parameters and $f(\lambda_1 - \lambda_2)$ is a normalized line shape function, which is either derived from theory or experimentally measured (determined from Fig. 2, for example). With λ_2 expressed in μm, $A = 1.0 \times 10^{-11}$ cm – $\mu m/W$.[20] The solutions of Eqs. (10) and (11) are:

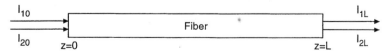

FIGURE 1 Beam geometry for stimulated Raman scattering in an optical fiber.

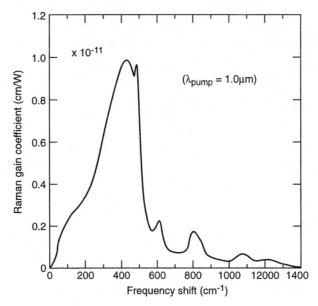

FIGURE 2   Raman gain spectrum in fused silica. (*Adapted from Ref. 22. © 1980 IEEE.*)

$$I_1(z) = \frac{\omega_1}{\omega_2} I_0 \exp(-\alpha z) \frac{\psi_r}{1 + \psi_r} \tag{13}$$

$$I_2(z) = I_0 \exp(-\alpha z) \frac{1}{1 + \psi_r} \tag{14}$$

In these equations, $I_0 = I_{20} + (\omega_1/\omega_2)I_{10}$, where I_{10} and I_{20} are the Stokes and pump intensities at the fiber input. The coupling parameter ψ_r assumes different forms, depending upon whether the input Stokes intensity I_{10} is present or not. If I_{10} is present, and if its magnitude is much greater than light from spontaneous Raman scattering, we have:

$$\psi_r = \frac{\omega_1}{\omega_2} \frac{I_{10}}{I_{20}} \exp(G_0) \tag{15}$$

When no Stokes input is present, the signal builds up from spontaneous Raman scattering, and the coupling parameter in this case becomes:

$$\psi_r = \frac{\hbar \omega_2 \Delta \omega_r}{4\sqrt{\pi}} \frac{1}{I_{20} A_{\text{eff}}} G_2^{-1/2} \exp(G_2) \tag{16}$$

with the gain parameters defined through $G_0 = g_r I_0 L_{\text{eff}}$ and $G_2 = g_r I_{20} L_{\text{eff}}$ The effective length of the fiber accounts for the reduction of Stokes and pump intensities as a result of loss, and is defined as

$$L_{\text{eff}} = \int_0^L \exp(-\alpha z) dz = \frac{1 - \exp(-\alpha L)}{\alpha} \tag{17}$$

The effective area of a single-mode fiber A_{eff} is calculated through πr_0^2, where r_0 is the mode field radius. For a multimode fiber, A_{eff} is usually taken as the core area, assuming that the power is uniformly distributed over the core. The power in the fiber is then $P_{1,2} = I_{1,2} A_{eff}$.

Two basic issues concerning SRS are of interest in fiber communication systems. First, pump-to-Stokes coupling provides a mechanism for crosstalk from short- to long-wavelength channels. This will occur most efficiently if the channel frequency spacing is in the vicinity of that associated with the maximum Raman gain. The Raman gain peak at approximately 500 cm^{-1} corresponds to a frequency spacing of 15 THz, meaning that operation at 1.55 μm produces a Stokes wave of about 1.67 μm wavelength. Two-channel operation at these wavelengths would lead to a maximum allowable signal level of about 50 mW.[21] In WDM systems, within the 1.53- to 1.56-μm erbium-doped fiber amplifier window, channel spacings on the order of 100 GHz are used. Raman gain is thus considerably reduced, but is still sufficient to cause appreciable crosstalk, which can lead to system penalties of between 1 and 3 dB depending on the number of channels.[22] Second, and of more importance to single-wavelength systems, is the conversion to Stokes power from the original signal—a mechanism by which signal power can be depleted. A related problem is walkoff[23] occurring between the signal and Stokes pulses, since these will have different group delays. Walkoff is a means for aliasing to occur in digital transmission, unless the signal is filtered at the output. If pulses are of subpicosecond widths, additional complications arise due to the increased importance of SPM and XPM.[24] In any event, an upper limit must be placed on the signal power if significant conversion to Stokes power is to be avoided. In single-wavelength systems, where crosstalk is not an issue, pulse peak powers must be kept below about 500 mW to avoid significant SRS conversion.[25]

A useful criterion is the so-called critical condition (or Raman threshold), defined as the condition under which the output Stokes and signal powers are equal. This occurs when $\psi_r = 1$, which, from Eq. (16), leads to $G_2 \approx 16$. SRS can also be weakened by taking advantage of the gain reduction that occurs as signal (pump) wavelengths increase, as shown in Eq. (12). For example, operation at 1.55 μm yields less SRS for a given signal power than operation at 1.3 μm.

Apart from the need to reduce SRS, the effect can be used to advantage in wavelength conversion and in amplification. Fiber Raman lasers have proven to be good sources of tunable radiation and operate at multiple Stokes wavelengths.[26] Specifically, a Stokes wave can serve as a pump to generate an additional (higher-order) Stokes wave at a longer wavelength.[27] Fiber Raman amplifiers have been demonstrated as repeaters in 1.3-μm wavelength systems.[28]

3.4 STIMULATED BRILLOUIN SCATTERING

The stimulated Brillouin scattering process (SBS) involves the input of a single intense optical wave at frequency ω_2, which initiates a copropagating acoustic wave at frequency ω_p. The acoustic wave is manifested as a traveling index grating in the fiber, which back-diffracts a portion of the original input. The backward (Stokes) wave is Doppler-shifted to a lower frequency ω_1 and is proportional to the phase conjugate of the input.[29] The backward wave is amplified as it propagates, with the gain increasing with increasing input (pump) power.

The beam interaction geometry is shown in Fig. 3. Usually, the Stokes wave builds up spontaneously, but can be inputted at the far end. The effect can be understood by considering a case in which counter-propagating Stokes and pump waves exist that together form a moving interference pattern whose velocity is proportional to the difference frequency $\omega_2 - \omega_1$. Coupling between the waves will occur via SBS when the interference pattern velocity is in the vicinity of the acoustic wave velocity v_p. It is the interference pattern that forms and reinforces the acoustic wave through electrostriction. With a single input, spontaneous scattering from numerous shock waves occurs, with preferential feedback from the acoustic wave that matches the condition just described. With the Stokes wave generated (although it is initially weak), the acoustic wave is reinforced, and so backscattering increases.

FIGURE 3   Beam geometry for stimulated Brillouin scattering in an optical fiber.

In terms of the wave vector magnitudes, the condition for phase matching is given by $k_p = k_1 + k_2$. Since the sound frequency is much less than those of the two optical waves, we can write $k_p \approx 2k_2$. Then, since $k_p = \omega_p/v_p$, it follows that $\omega_p \approx 2n\omega_2 v_p/c$, where n is the refractive index (assumed to be the same value at both optical frequencies). The Brillouin frequency shift under phase-matched conditions thus becomes

$$\omega_2 - \omega_1 \approx 2n\omega_2 \frac{v_p}{c} \tag{18}$$

This yields a value of about 11 GHz, with $v_p \approx 6$ km/s in fused silica and $\lambda_2 = 1.55$ μm.

The process can be described by the nonlinear polarization produced by the product of complex fields, E_1, E_2^*, and E_2; this yields a polarization at ω_1 that propagates with wave vector k_1 in the direction of the Stokes wave. Another polarization, describing back-coupling from Stokes to pump, involves the product $E_1 E_1^* E_2$. Substituting fields and polarizations into the wave equation yields the following coupled equations that describe the evolution of the optical intensities with distance (pp. 214–220 of Ref. 18):

$$\frac{dI_1}{dz} = -g_b I_1 I_2 + \alpha I_1 \tag{19}$$

$$\frac{dI_2}{dz} = -g_b I_1 I_2 - \alpha I_2 \tag{20}$$

where α is the linear loss coefficient. The Brillouin gain is given by

$$g_b = g_{b0}\left(1 + \frac{4(\omega_1 - \omega_{10})^2}{v_p^2 \alpha_p^2}\right)^{-1} \tag{21}$$

where ω_{10} is the Stokes frequency at precise phase matching, α_p is the loss coefficient for the acoustic wave, and the peak gain g_{b0} is a function of the material parameters. The Brillouin line width, defined as the full width at half-maximum of g_b, is $\Delta\omega_b = v_p \alpha_p$. In optical fibers, $\Delta f_b = \Delta\omega_b/2\pi$ is typically between 10 and 30 MHz (p. 374 of Ref. 16) and $g_{b0} = 4.5 \times 10^{-9}$ cm/W.[20] Signal bandwidths in high-data-rate communication systems greatly exceed the Brillouin line width, and so SBS is typically too weak to be considered a source of noise or signal depletion. This is to be compared to stimulated Raman scattering, which supports considerable gain over approximately 5 THz. Consequently, SRS is a much more serious problem in high-data-rate systems.

Using analysis methods similar to those employed in SRS, a critical condition (or threshold) can be defined for SBS, at which the backscattered power is equal to the input power:[30]

$$\frac{\omega_1 k_B T \Delta\omega_b}{4\sqrt{\pi}\omega_p I_{20} A_{\text{eff}}} G_b^{-3/2} \exp(G_b) = 1 \tag{22}$$

where k_B is Boltzmann's constant and T is the temperature in degrees Kelvin. The gain parameter is:

$$G_b = g_b I_{20} L_{\text{eff}} \tag{23}$$

with L_{eff} as defined in Eq. (17). Equation (22) is approximately satisfied when $G_b \approx 21$.[30] In practice, the backscattered power will always be less than the input power, since pump depletion will occur. Nevertheless, this condition is used as a benchmark to determine the point at which SBS becomes excessive in a given system.[31] In one study, it was found that $G_b \approx 21$ yields the pump power required to produce an SBS output that is at the level of Rayleigh backscattering.[32] Pump powers required to achieve threshold can be on the order of a few milliwatts for CW or narrowband signals, but these increase substantially for broadband signals.[33] Reduction of SBS is accomplished in practice by lowering the input signal power (I_{20}) or by taking advantage of the reduction in g_b that occurs when signal bandwidths ($\Delta\omega$) exceed the Brillouin line width. Specifically, if $\Delta\omega \gg \Delta\omega_b$,

$$g_b(\Delta\omega) \approx g_b \frac{\Delta\omega_b}{\Delta\omega} \tag{24}$$

3.5 FOUR-WAVE MIXING

The term *four-wave mixing* in fibers is generally applied to wave coupling through the electronic nonlinearity in which at least two frequencies are involved and in which frequency conversion is occurring. The fact that the electronic nonlinearity is involved distinguishes four-wave mixing interactions from stimulated scattering processes because in the latter the medium was found to play an active role through the generation or absorption of optical phonons (in SRS) or acoustic phonons (in SBS). If the nonlinearity is electronic, bound electron distributions are modified according to the instantaneous optical field configurations. For example, with light at two frequencies present, electron positions can be modulated at the difference frequency, thus modulating the refractive index. Additional light will encounter the modulated index and can be up- or downshifted in frequency. In such cases, the medium plays a passive role in the interaction, as it does not absorb applied energy or release energy previously stored. The self- and cross-phase modulation processes also involve the electronic nonlinearity, but in those cases, power conversion between waves is not occurring—only phase modulation.

As an illustration of the process, consider the interaction of two strong waves at frequencies ω_1 and ω_2, which mix to produce a downshifted (Stokes) wave at ω_3 and an upshifted (anti-Stokes) wave at ω_4. The frequencies have equal spacing, that is, $\omega_1 - \omega_3 = \omega_2 - \omega_1 = \omega_4 - \omega_2$ (Fig. 4). All fields assume the real form:

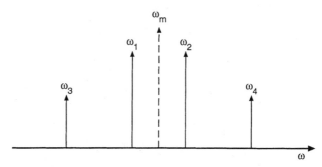

FIGURE 4 Frequency diagram for four-wave mixing, showing pump frequencies (ω_1 and ω_2) and sideband frequencies (ω_3 and ω_4).

$$\mathcal{E}_j = \tfrac{1}{2}E_{oj}\exp[i(\omega_j t - \beta_j z)] + c.c. \ j = 1 - 4 \tag{25}$$

The nonlinear polarization will be proportional to $\mathcal{E}^3$, where $\mathcal{E} = \mathcal{E}_1 + \mathcal{E}_2 + \mathcal{E}_3 + \mathcal{E}_4$. With all fields copolarized, complex nonlinear polarizations at ω_3 and ω_4 appear that have the form:

$$P_{NL}^{\omega_3} = \tfrac{3}{4}\varepsilon_0\chi^{(3)}E_{01}^2 E_{02}^*\exp[i(2\omega_1 - \omega_2)t]\exp[-i(2\beta^{\omega_1} - \beta^{\omega_2})z] \tag{26}$$

$$P_{NL}^{\omega_4} = \tfrac{3}{4}\varepsilon_0\chi^{(3)}E_{02}^2 E_{01}^*\exp[i(2\omega_2 - \omega_1)t]\exp[-i(2\beta^{\omega_2} - \beta^{\omega_1})z] \tag{27}$$

where $\omega_3 = 2\omega_1 - \omega_2$, $\omega_4 = 2\omega_2 - \omega_1$ and $\chi^{(3)}$ is proportional to the nonlinear refractive index n_2'. The significance of these polarizations lies not only in the fact that waves at the sideband frequencies ω_3 and ω_4 can be generated, but that preexisting waves at those frequencies can experience gain in the presence of the two pump fields at ω_1 and ω_2. The sideband waves will contain the amplitude and phase information on the pumps, thus making this process an important crosstalk mechanism in multiwavelength communication systems. Under phase-matched conditions, the gain associated with FWM is more than twice the peak gain in SRS.[34]

The wave equation, when solved in steady state, yields the output intensity at either one of the sideband frequencies.[35] For a medium of length L, having loss coefficient α, the sideband intensities are related to the pump intensities through

$$I^{\omega_3} \propto \left(\frac{n_2 L_{\text{eff}}}{\lambda_m}\right)^2 I^{\omega_2}(I^{\omega_1})^2 \eta \exp(-\alpha L) \tag{28}$$

$$I^{\omega_4} \propto \left(\frac{n_2 L_{\text{eff}}}{\lambda_m}\right)^2 I^{\omega_1}(I^{\omega_2})^2 \eta \exp(-\alpha L) \tag{29}$$

where L_{eff} is defined in Eq. (17), and where

$$\eta = \frac{\alpha^2}{\alpha^2 + \Delta\beta^2}\left(1 + \frac{4\exp(-\alpha L)\sin^2(\Delta\beta L/2)}{(1 - \exp(-\alpha L))^2}\right) \tag{30}$$

Other FWM interactions can occur, involving products of intensities at three different frequencies rather than two as demonstrated here. In such cases, the output wave intensities are increased by a factor of 4 over those indicated in Eqs. (28) and (29).

One method of suppressing four-wave mixing in WDM systems includes the use of unequal channel spacing.[36] This assures, for example, that $\omega_3 \neq 2\omega_1 + \omega_2$, where ω_1, ω_2, and ω_3 are assigned channel frequencies. Other methods involve phase-mismatching the process in some way. This is accomplished by increasing $\Delta\beta$, which has the effect of decreasing η in Eqs. (28) and (29). Note that in the low-loss limit, where $\alpha \to 0$, Eq. (30) reduces to $\eta = (\sin^2(\Delta\beta L/2))/(\Delta\beta L/2)^2$. The $\Delta\beta$ expressions associated with wave generation at ω_3 and ω_4 are given by

$$\Delta\beta(\omega_3) = 2\beta^{\omega_1} - \beta^{\omega_2} - \beta^{\omega_3} \tag{31}$$

and

$$\Delta\beta(\omega_4) = 2\beta^{\omega_2} - \beta^{\omega_1} - \beta^{\omega_4} \tag{32}$$

It is possible to express Eqs. (31) and (32) in terms of known fiber parameters by using a Taylor series for the propagation constant, where the expansion is about frequency ω_m as indicated in Fig. 4, where $\omega_m = (\omega_2 + \omega_1)/2$.

$$\beta \approx \beta_0 + (\omega - \omega_m)\,\beta_1 + \tfrac{1}{2}(\omega - \omega_m)^2\,\beta_2 + \tfrac{1}{6}(\omega - \omega_m)^3\beta_3 \tag{33}$$

In Eq. (33), β_1, β_2, and β_3 are, respectively, the first, second, and third derivatives of β with respect to ω, evaluated at ω_m. These in turn relate to the fiber dispersion parameter D (ps/nm·km) and its first derivative with respect to wavelength through $\beta_2 = -(\lambda_m^2/2\pi c)D(\lambda_m)$ and $\beta_3 = (\lambda_m^3/2\pi^2 c^2)[D(\lambda_m) + (\lambda_m/2)(dD/d\lambda)|_{\lambda_m}]$ where $\lambda_m = 2\pi c/\omega_m$. Using these relations along with Eq. (33) in Eqs. (31) and (32) results in:

$$\Delta\beta(\omega_3, \omega_4) \approx 2\pi c \frac{\Delta\lambda^2}{\lambda_m^2}\left[D(\lambda_m) \pm \frac{\Delta\lambda}{2} \frac{dD}{d\lambda}|_{\lambda_m} \right] \tag{34}$$

where the plus sign is used for $\Delta\beta(\omega_3)$, the minus sign is used for $\Delta\beta(\omega_4)$, and $\Delta\lambda = \lambda_1 - \lambda_2$. Phase matching is not completely described by Eq. (34), since cross-phase modulation plays a subtle role, as discussed on pp. 410–411 of Ref. 16. Nevertheless, Eq. (34) does show that the retention of moderate values of dispersion D is a way to reduce FWM interactions that would occur, for example, in WDM systems. As such, modern commercial fiber intended for use in WDM applications will have values of D that are typically in the vicinity of 2 ps/nm·km.[37] With WDM operation in conventional dispersion-shifted fiber (with the dispersion zero near 1.55 μm), having a single channel at the zero dispersion wavelength can result in significant four-wave mixing.[38] Methods that were found to reduce four-wave mixing in such cases include the use of cross-polarized signals in dispersion-managed links[39] and operation within a longer-wavelength band near 1.6 μm[40] at which dispersion is appreciable and where gain-shifted fiber amplifiers are used.[41]

Examples of other cases involving four-wave mixing include single-wavelength systems, in which the effect has been successfully used in a demultiplexing technique for TDM signals.[42] In another case, coupling through FWM can occur between a signal and broadband amplified spontaneous emission (ASE) in links containing erbium-doped fiber amplifiers.[43] As a result, the signal becomes spectrally broadened and exhibits phase noise from the ASE. The phase noise becomes manifested as amplitude noise under the action of dispersion, producing a form of modulation instability.

An interesting application of four-wave mixing is spectral inversion. Consider a case that involves the input of a strong single-frequency pump wave along with a relatively weak wave having a spectrum of finite width positioned on one side of the pump frequency. Four-wave mixing leads to the generation of a wave whose spectrum is the "mirror image" of that of the weak wave, in which the mirroring occurs about the pump frequency. Figure 5 depicts a representation of this, where four frequency components comprising a spectrum are shown along

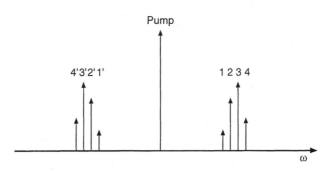

FIGURE 5 Frequency diagram for spectral inversion using four-wave mixing with a single pump frequency.

with their imaged counterparts. An important application of this is pulses that have experienced broadening with chirping after propagating through a length of fiber exhibiting linear group dispersion.[44] Inverting the spectrum of such a pulse using four-wave mixing has the effect of reversing the direction of the chirp (although the pulse center wavelength is displaced to a different value). When the spectrally inverted pulse is propagated through an additional length of fiber having the same dispersive characteristics, the pulse will compress to nearly its original input width. Compensation for nonlinear distortion has also been demonstrated using this method.[45]

3.6   CONCLUSION

An overview of fiber nonlinear effects has been presented here in which emphasis is placed on the basic concepts, principles, and perspectives on communication systems. Space is not available to cover the more subtle details of each effect or the interrelations between effects that often occur. The text by Agrawal[16] is recommended for further in-depth study, which should be supplemented by the current literature. Nonlinear optics in fibers and in fiber communication systems comprises an area whose principles and implications are still not fully understood. It thus remains an important area of current research.

3.7   REFERENCES

1. L. F. Mollenauer and P. V. Mamyshev, "Massive Wavelength-Division Multiplexing with Solitons," *IEEE J. Quantum Electron.* **34:**2089–2102 (1998).

2. C. M. de Sterke, B. J. Eggleton, and P. A. Krug, "High-Intensity Pulse Propagation in Uniform Gratings and Grating Superstructures," *IEEE J. Lightwave Technol.* **15:**1494–1502 (1997).

3. L. Clark, A. A. Klein, and D. W. Peckham, "Impact of Fiber Selection and Nonlinear Behavior on Network Upgrade Strategies for Optically Amplified Long Interoffice Routes," *Proceedings of the 10th Annual National Fiber Optic Engineers Conference,* vol. 4, 1994.

4. Y. R. Shen, *The Principles of Nonlinear Optics,* Wiley-Interscience, New York, 1984, p. 28.

5. R. H. Stolen and H. W. K. Tom, "Self-Organized Phase-Matched Harmonic Generation in Optical Fibers," *Opt. Lett.* **12:**585–587 (1987).

6. R. W. Boyd, *Nonlinear Optics,* Academic Press, San Diego, 1992, pp. 159 ff.

7. R. H. Stolen and C. Lin, "Self-Phase Modulation in Silica Optical Fibers," *Phys. Rev. A* **17:**1448–1453 (1978).

8. G. Bellotti, A. Bertaina, and S. Bigo, "Dependence of Self-Phase Modulation Impairments on Residual Dispersion in 10-Gb/s-Based Terrestrial Transmission Using Standard Fiber," *IEEE Photon. Technol. Lett.* **11:**824–826 (1999).

9. M. Stern, J. P. Heritage, R. N. Thurston, and S. Tu, "Self-Phase Modulation and Dispersion in High Data Rate Fiber Optic Transmission Systems," *IEEE J. Lightwave Technol.* **8:**1009–1015 (1990).

10. D. Marcuse and C. R. Menyuk, "Simulation of Single-Channel Optical Systems at 100 Gb/s," *IEEE J. Lightwave Technol.* **17:**564–569 (1999).

11. S. Reichel and R. Zengerle, "Effects of Nonlinear Dispersion in EDFA's on Optical Communication Systems," *IEEE J. Lightwave Technol.* **17:**1152–1157 (1999).

12. M. Karlsson, "Modulational Instability in Lossy Optical Fibers," *J. Opt. Soc. Am. B* **12:**2071–2077 (1995).

13. R. Hui, K. R. Demarest, and C. T. Allen, "Cross-Phase Modulation in Multispan WDM Optical Fiber Systems," *IEEE J. Lightwave Technol.* **17:**1018–1026 (1999).

14. S. Bigo, G. Billotti, and M. W. Chbat, "Investigation of Cross-Phase Modulation Limitation over Various Types of Fiber Infrastructures," *IEEE Photon. Technol. Lett.* **11:**605–607 (1999).

15. D. Marcuse, "Selected Topics in the Theory of Telecommunications Fibers," in *Optical Fiber Telecommunications II,* Academic Press, San Diego, 1988, p. 102.

16. G. P. Agrawal, *Nonlinear Fiber Optics,* 2d ed., Academic Press, San Diego, 1995, pp. 242–247.

17. G. Herzberg, *Infra-Red and Raman Spectroscopy of Polyatomic Molecules,* Van Nostrand, New York, 1945, pp. 99–101.

18. J. A. Buck, *Fundamentals of Optical Fibers,* Wiley-Interscience, New York, 1995, pp. 207–213.

19. F. L. Galeener, J. C. Mikkelsen Jr., R. H. Geils, and W. J. Mosby, "The Relative Raman Cross Sections of Vitreous SiO_2, GeO_2, B_2O_3, and P_2O_5," *Appl. Phys. Lett.* **32**:34–36 (1978).

20. R. H. Stolen, "Nonlinear Properties of Optical Fibers," in S. E. Miller and A. G. Chynoweth (eds.), *Optical Fiber Telecommunications,* Academic Press, New York, 1979.

21. A. R. Chraplyvy, "Optical Power Limits in Multi-Channel Wavelength Division Multiplexed Systems due to Stimulated Raman Scattering," *Electron. Lett.* **20**:58–59 (1984).

22. F. Forghieri, R. W. Tkach, and A. R. Chraplyvy, "Effect of Modulation Statistics on Raman Crosstalk in WDM Systems," *IEEE Photon. Technol. Lett.* **7**:101–103 (1995).

23. R. H. Stolen and A. M. Johnson, "The Effect of Pulse Walkoff on Stimulated Raman Scattering in Fibers," *IEEE J. Quantum Electron.* **22**:2154–2160 (1986).

24. C. H. Headley III and G. P. Agrawal, "Unified Description of Ultrafast Stimulated Raman Scattering in Optical Fibers," *J. Opt. Soc. Am. B* **13**:2170–2177 (1996).

25. R. H. Stolen, J. P. Gordon, W. J. Tomlinson, and H. A. Haus, "Raman Response Function of Silica Core Fibers," *J. Opt. Soc. Am. B* **6**:1159–1166 (1988).

26. L. G. Cohen and C. Lin, "A Universal Fiber-Optic (UFO) Measurement System Based on a Near-IR Fiber Raman Laser," *IEEE J. Quantum Electron.* **14**:855–859 (1978).

27. K. X. Liu and E. Garmire, "Understanding the Formation of the SRS Stokes Spectrum in Fused Silica Fibers," *IEEE J. Quantum Electron.* **27**:1022–1030 (1991).

28. T. N. Nielsen, P. B. Hansen, A. J. Stentz, V. M. Aguaro, J. R. Pedrazzani, A. A. Abramov, and R. P. Espindola, "8×10 Gb/s 1.3-μm Unrepeated Transmission over a Distance of 141 km with Raman Post- and Pre-Amplifiers," *IEEE Photonics Technology Letters,* vol. 10, pp. 1492–1494 (1998).

29. A. Yariv, *Quantum Electronics,* 3d ed., Wiley, New York, 1989, pp. 513–516.

30. R. G. Smith, "Optical Power Handling Capacity of Low Loss Optical Fibers as Determined by Stimulated Raman and Brillouin Scattering," *Appl. Opt.* **11**:2489–2494 (1972).

31. A. R. Chraplyvy, "Limitations on Lightwave Communications Imposed by Optical Fiber Nonlinearities," *IEEE J. Lightwave Technol.* **8**:1548–1557 (1990).

32. X. P. Mao, R. W. Tkach, A. R. Chraplyvy, R. M. Jopson, and R. M. Derosier, "Stimulated Brillouin Threshold Dependence on Fiber Type and Uniformity," *IEEE Photon. Technol. Lett.* **4**:66–68 (1992).

33. C. Edge, M. J. Goodwin, and I. Bennion, "Investigation of Nonlinear Power Transmission Limits in Optical Fiber Devices," *Proc. IEEE* **134**:180–182 (1987).

34. R. H. Stolen, "Phase-Matched Stimulated Four-Photon Mixing in Silica-Fiber Waveguides," *IEEE J. Quantum Electron.* **11**:100–103 (1975).

35. R. W. Tkach, A. R. Chraplyvy, F. Forghieri, A. H. Gnauck, and R. M. Derosier, "Four-Photon Mixing and High-Speed WDM Systems," *IEEE Journal of Lightwave Technology,* vol. 13, pp. 841–849 (1995).

36. F. Forghieri, R. W. Tkach, and A. R. Chraplyvy, and D. Marcuse, "Reduction of Four-Wave Mixing Crosstalk in WDM Systems Using Unequally-Spaced Channels," *IEEE Photon. Technol. Lett.* **6**:754–756 (1994).

37. AT&T Network Systems data sheet 4694FS-Issue 2 LLC, "TrueWave Single Mode Optical Fiber Improved Transmission Capacity," December 1995.

38. D. Marcuse, A. R. Chraplyvy, and R. W. Tkach, "Effect of Fiber Nonlinearity on Long-Distance Transmission," *IEEE J. Lightwave Technol.* **9**:121–128 (1991).

39. E. A. Golovchenko, N. S. Bergano, and C. R. Davidson, "Four-Wave Mixing in Multispan Dispersion Managed Transmission Links," *IEEE Photon. Technol. Lett.* **10**:1481–1483 (1998).

40. M. Jinno et al, "1580nm Band, Equally-Spaced 8X10 Gb/s WDM Channel Transmission Over 360km (3X120km) of Dispersion-Shifted Fiber Avoiding FWM Impairment," *IEEE Photon. Technol. Lett.* **10**:454–456 (1998).

41. H. Ono, M. Yamada, and Y. Ohishi, "Gain-Flattened Er^{3+}-Doped Fiber Amplifier for A WDM Signal in the 1.57–1.60μm Wavelength Region," *IEEE Photon. Technol. Lett.* **9**:596–598 (1997).

42. P. O. Hedekvist, M. Karlsson, and P. A. Andrekson, "Fiber Four-Wave Mixing Demultiplexing with Inherent Parametric Amplification," *IEEE J. Lightwave Technol.* **15**: 2051–2058 (1997).

43. R. Hui, M. O'Sullivan, A. Robinson, and M. Taylor, "Modulation Instability and Its Impact on Multispan Optical Amplified IMDD Systems: Theory and Experiments," *IEEE J. Lightwave Technol.* **15**: 1071–1082 (1997).

44. A. H. Gnauck, R. M. Jopson, and R. M. Derosier, "10 Gb/s 360 km Transmission over Dispersive Fiber Using Midsystem Spectral Inversion," *IEEE Photon. Technol. Lett.* **5**:663–666 (1993).

45. A. H. Gnauck, R. M. Jopson, and R. M. Derosier, "Compensating the Compensator: A Demonstration of Nonlinearity Cancellation in a WDM System," *IEEE Photon. Technol. Lett.* **7**:582–584 (1995).

CHAPTER 4
SOURCES, MODULATORS, AND DETECTORS FOR FIBER-OPTIC COMMUNICATION SYSTEMS

Elsa Garmire
Dartmouth College
Hanover, New Hampshire

4.1 INTRODUCTION

Optical communication systems utilize fiber optics to transmit the light that carries the signals. Such systems require optoelectronic devices as sources and detectors of such light, and they need modulators to impress the telecommunication signals onto the light. This chapter outlines the basics of these devices. Characteristics of devices designed for both high-performance, high-speed telecommunication systems (*telecom*) and for low-cost, more modest performance data communication systems (*datacom*) are presented. Sources for telecom are edge-emitting lasers, including double heterostructure (DH), quantum well (QW), strained layer (SL), distributed feedback (DFB), and distributed Bragg reflector (DBR) lasers. Operating characteristics of these edge-emitting lasers include threshold, light-out versus current-in, spatial, and spectral characteristics. The transient response includes relaxation oscillations, turn-on delay, and modulation response. The noise characteristics are described by relative intensity noise (RIN), signal-to-noise ratio (SNR), mode partition noise (in multimode lasers), and phase noise (which determines linewidth). Frequency chirping broadens the linewidth, described in the small and large signal regime; external optical feedback may profoundly disturb the stability of the lasers and may lead to coherence collapse.

Semiconductor lasers usually have a laser cavity in the plane of the semiconductor device, and emit light out through a cleaved edge in an elliptical output pattern. This output is not ideally suited to coupling into fibers, which have circular apertures. Low-cost systems, such as datacom, put a premium on simplicity in optical design. These systems typically use multimode fibers and surface-emitting light-emitting diodes (LEDs). The LEDs are less temperature dependent than lasers and are more robust, but they typically are slower and less efficient. Those LEDs applicable to fiber optics are described here, along with their operating and transient response characteristics. Edge-emitting LEDs have some niche fiber-optic applications and are briefly described.

Recently, vertical cavity surface-emitting lasers (VCSELs) have been developed, which have vertical laser cavities that emit light normal to the plane of the semiconductor device. Fibers couple more easily to these surface-emitting sources, but their laser performance is usually degraded compared to that of the edge-emitting sources. This chapter outlines typical VCSEL designs (material, optical, and electrical); their spatial, spectral, and polarization characteristics; and their light-out versus current-in characteristics. While most VCSELs are GaAs-based, rapid progress is being made toward long-wavelength InP-based VCSELs.

The most common modulators used in fiber-optic systems today are external lithium niobate modulators. These are usually used in Y-branch interferometric modulators, created by phase modulation from the electro-optic effect. These modulators are introduced here, along with a discussion of high-speed modulation, losses, and polarization dependence, and a brief description of optical damage and other modulator geometries. These devices provide chirp-free modulation that can be made very linear for applications such as cable TV.

An alternative modulator uses semiconductors, particularly quantum wells. This design has the advantage of allowing for more compact devices and monolithic integration. Typically, these are intensity modulators using electroabsorption. By careful design, the chirp in these modulators can be controlled and even used to counteract pulse spreading from chromatic dispersion in fibers. The quantum-confined Stark effect is described, along with the *pin* waveguides used as modulators and techniques for their integration with lasers. Their operating characteristics as intensity modulators, their chirp, and improvements available by using strained quantum wells are presented.

Some semiconductor modulators use phase change rather than absorption change. The electro-optic effect in III-V semiconductors is discussed, along with the enhanced refractive index change that comes from the quantum-confined Stark effect, termed *electrorefraction*. Particularly large refractive index changes result if available quantum well states are filled by electrons. The field-dependent transfer of electrons in and out of quantum wells in a barrier, reservoir, and quantum well electron transfer (BRAQWET) structure enables a particularly large refractive index change modulation. Phase-change modulators based on this principle can be used in interferometers to yield intensity modulators.

Detectors used in fiber systems are primarily *pin* diodes, although short descriptions of avalanche photodetectors (APDs) and metal-semiconductor-metal (MSM) detectors are provided. The geometry, sensitivity, speed, dark current, and noise characteristics of the most important detectors used in fiber systems are described.

Most of the devices discussed in this chapter are based on semiconductors, and their production relies on the ability to tailor the material to design specifications through *epitaxial growth*. This technology starts with a bulk crystal substrate (usually the binary compounds GaAs or InP) and employs the multilayered growth upon this substrate of a few micrometers of material with a different composition, called a *heterostructure*. *Ternary* layers substitute a certain fraction x for one of the two binary components. Thus, $Al_xGa_{1-x}As$ is a common ternary alloy used in laser diodes. Another common ternary is $In_xGa_{1-x}As$. Layers are *lattice matched* when the ternary layers have the same size lattice as the binary; otherwise, the epitaxial layer will have *strain*. Lattice-matched epitaxial layers require that the substituting atom be approximately the same size as the atom it replaces. This is true of Al and Ga, so that $Al_xGa_{1-x}As$ ternary layers are lattice matched to GaAs. The lowest-cost lasers are those based on GaAs substrates with $Al_xGa_{1-x}As$ ternary layers surrounding the active layer. These lasers operate at wavelengths near the bandgap of GaAs, about 850 nm, and are typically used in low-cost data communications (as well as in CD players).

The wavelengths required for laser sources in telecommunications applications are those at which the fiber has the lowest loss and/or dispersion, traditionally 1.55 and 1.3 μm. There is no binary semiconductor with a bandgap at these wavelengths, nor is there a lattice-matched ternary. The $In_xGa_{1-x}As$ ternary will be strained under compression when it is grown on either GaAs or InP, because indium is a much bigger atom than gallium, and arsenic is much bigger than phosphorus. The way to eliminate this strain is to use a fourth small atom to reduce the size of the lattice back to that of the binary. This forms a *quaternary*. The heterostructure most useful for fiber-optics applications is based on InP substrates. The quaternary $In_xGa_{1-x}As_yP_{1-y}$ is commonly used, with the compositions x and y chosen to simultaneously provide the desired wavelength and lattice match. These quaternary heterostructures are the basis for much of the long-wavelength technology: sources, modulators, and detectors.

Earlier volumes of this handbook discuss the basics of lasers (Vol. I, Chap. 13), LEDs (Vol. I, Chap. 12), modulators (Vol. II, Chap. 13), and detectors (Vol. I, Chaps. 15 to 17). The reader is referred there for general information. This chapter is specific to characteristics that are important for fiber communication systems.

4.2   *DOUBLE HETEROSTRUCTURE LASER DIODES*

Telecommunications sources are usually edge-emitting lasers, grown with an active laser layer that has a bandgap near either 1.55 or 1.3 µm. These are quaternary layers consisting of $In_xGa_{1-x}As_yP_{1-y}$, grown lattice-matched to InP. The materials growth and fabrication technology had to be developed specifically for telecommunication applications and is now mature. These lasers are more temperature sensitive than GaAs lasers, and this fact has to be incorporated into their use. For telecom applications they are often provided with a thermo-electric cooler and are typically provided with a monitoring photodiode in the laser package, in order to provide a signal for temperature and/or current control.

Today's telecom systems use single-mode fibers, which require lasers with a single spatial mode. In order to avoid dispersion over long distances, a single frequency mode is necessary. These requirements constrain the geometry of laser diodes (LDs) used for telecom applications, as discussed in the next section. Following sections discuss the operating characteristics of these LDs and their transient response and noise characteristics, both as isolated diodes and when subject to small reflections from fiber facets. The modulation characteristics of these diodes are discussed, along with frequency chirping. Advanced laser concepts, such as quantum well lasers, strained layer lasers, and lasers with distributed reflection (DFB and DBR lasers), are also introduced.

A typical geometry of an edge-emitting InGaAsP/InP laser is shown in Fig. 1. The active quaternary laser region is shown crosshatched. It is from this region that light will be emitted. Traditionally, these active regions have uniform composition and are lattice matched to the substrate. More advanced laser diodes, often used for telecom applications, have active regions containing one or more quantum wells and may be grown to incorporate internal strain in the active region. Both these characteristics are described in a separate section later in this chapter.

The design of a double heterostructure laser diode requires optimization of the issues discussed in the following subsections.

Injection of a Population Inversion into the Active Region

This is necessary so that stimulated emission can take place. This is done by placing the active region between *p* and *n* layers, and forward-biasing the resulting diode. Electrons are injected into the active region from the *n* side and holes are injected from the *p* side; they become *free carriers*. Efficient electrical injection requires high-quality ohmic contacts attached to the *n* and *p* layers; electrical current through the junction then drives the laser.

Confinement of Carriers Within the Plane of the Active Layer

This is done by growing the active region as a thin layer of thickness d and surrounding it with layers of wider bandgap material, as shown in Fig. 2a. In quaternary lasers, wider bandgap

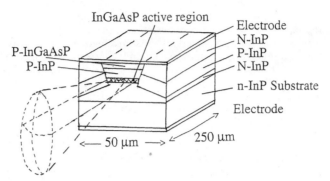

FIGURE 1   Typical geometry for an edge-emitting long-wavelength laser diode, as used in telecommunication systems. Light travels back and forth between cleaved mirror facets, confined to the active InGaAsP region by the buried heterostructure, and is emitted out of the crosshatched region, where it diffracts to the far-field. The current is confined to the stripe region by the current-blocking *npn* structure on either side.

material is provided by decreasing x and y relative to their values in the active region. Stimulated emission during electron-hole recombination in the narrow bandgap active layer provides the laser light. The thinner the active layer, the higher its gain. When the active layer thickness is as small as a few tens of nanometers, the free electron and hole energy levels become quantized in the growth direction, and the active layer becomes a *quantum well* (QW). Quantum wells have higher gain than bulk semiconductor active layers, and thus one or more quantum wells are often used as the active layers (see separate section later in this chapter).

Confinement of Light Near the Active Layer

Stimulated emission gain is proportional to the product of the carrier and photon densities, so that edge-emitting lasers require the highest possible light intensity. This is done by containing the light in an optical waveguide, with a typical light profile as shown in Fig. 2b. To achieve optical confinement, the layers surrounding the waveguide must have lower refractive index. It is fortunate that higher-bandgap materials that confine carriers also have smaller refractive index, and so the active layer automatically becomes a waveguide.

Proper optical confinement requires a single waveguide mode. This means that the waveguide layer must be thinner than the cutoff value for higher-order modes. The waveguide thickness d_g must be small enough that

$$d_g k_o \sqrt{n_g^2 - n_c^2} \equiv V < \pi \tag{1}$$

where n_g is the refractive index of the waveguide layer (usually the active layer), n_c is the refractive index of the surrounding cladding (usually the p and n layers), and $k_o = 2\pi/\lambda$, where λ is the free-space wavelength of the laser light. Typically, $n_g - n_c \sim 0.2$ and $d_g < 0.56\ \mu m$ for $\lambda = 1.3\ \mu m$. The parameter V is usually introduced to characterize a waveguide.

If the waveguide is too thin, however, the waveguided optical mode spreads out beyond the waveguide layer. The fraction of optical power Γ_g (called the *waveguide confinement factor*) that remains in the waveguide layer of thickness d_g is given approximately by[1]:

$$\Gamma_g = \frac{V^2}{V^2 + 2} \tag{2}$$

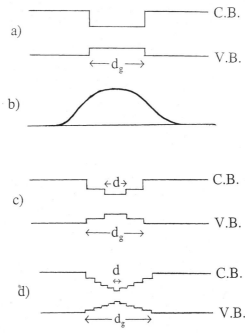

FIGURE 2 Conduction band (CB), valence band (VB), and guided optical mode as a function of position in the growth direction, near the active region in typical semiconductor laser geometries: (*a*) double heterostructure (DH) band structure, in which free carriers and light are both confined in the same region of small-bandgap material (of width d_g), surrounded by higher-bandgap cladding material; (*b*) near-field spatial profile for light guided in layer of width d_g; (*c*) separate confinement heterostructure (SCH) band structure, in which the free carriers are confined in a smaller active region (of width d) than the optical wave; and (*d*) graded index separate confinement heterostructure (GRINSCH), in which the composition of the cladding is graded in order to focus the light more tightly to the active region (deepest well) containing the free carriers.

As d_g becomes small, the confinement factor becomes small. When the carriers are confined in very thin layers, such as in quantum wells, the electrical carrier confinement layer cannot serve as an effective optical waveguide because the confinement factor is too small. Then a thicker waveguide region is used, and the photons and carriers are separately confined in a geometry called a *separate confinement heterostructure* (SCH), as shown in Fig. 2c. In this case the optical confinement factor, defined by the fraction of photons in the active layer of thickness d, is $\Gamma = \Gamma_g(d/d_g)$.

The light can be more effectively focused into a thin active layer by grading the refractive index in the separate confinement region, called a *graded index SCH* (GRINSCH) laser, shown in Fig. 2d. This graded refractive index is produced by growing material with varying bandgaps within the waveguide layer. Grading can be achieved by several discrete layers, as shown, or by grading many ultrathin layers with slight compositional differences. In either case, the focusing property of a GRINSCH structure can be approximated by fitting the graded refractive index to a parabolic refractive index profile $n(x)$ such that:

$$n(x)^2 = n_g^2 \left(1 - \frac{x^2}{x_o^2}\right)$$

where x_o is related to the curvature of the refractive index near $x = 0$: $x_o = (n_g/n'')^{1/2}$, where $n'' \equiv \partial^2 n/\partial x^2$ near $x = 0$. The mode guided by this profile has a gaussian beam-intensity profile:

$$I(x) = I_o \exp\left(\frac{-x^2}{w^2}\right)$$

where $w^2 = x_o/k_g$ and $k_g = 2\pi n_g/\lambda$

Limiting Carrier Injection to Stripe Geometry

Lasers are most efficient when the drive current is limited to the width of the optically active laser area. This requires defining a narrow stripe geometry electrode by means of a window etched in an isolating oxide layer or by ion implantation to render either side of the stripe resistive. More complex laser structures, such as those used in telecommunications applications, often define the conductive stripe electrode by using current-blocking *npn* layers grown on either side of the electrode, as shown in Fig. 1. The *npn* layers, consisting of back-to-back diodes, do not conduct current.

Injected carriers do not usually need lateral confinement, except to achieve the highest possible efficiency. Lateral free-carrier confinement will occur as a by-product of lateral optical confinement, which is discussed next.

Lateral Confinement of Light

The simplest laser diode structures do not specifically confine light laterally, except as the result of the stripe geometry carrier injection. These are called *gain-guided* lasers because high gain in the stripe region, due to the presence of free carriers, introduces a complex refractive index that guides the light laterally. Gain-guided lasers tend to be multimode (both lateral spatial modes and longitudinal frequency modes) unless the stripe is very narrow (<10 μm). In this case, the spatial far-field pattern has "rabbit ears," a double-lobed far-field pattern that is typically not very useful for coupling into single-mode fibers. Thus, gain-guided lasers are not usually used for telecommunications.

High-quality single-mode lasers for telecom applications typically require a means for creating a real refractive index difference laterally across the laser. The lowest threshold lasers use *buried heterostructure* (BH) lasers, the geometry shown in Fig. 1. After most of the layers are grown, the sample is taken out of the growth chamber and a stripe geometry mesa is etched. Then the sample is returned to the growth chamber, and one or more cladding layers with lower refractive index (higher bandgap) are grown, typically InP, as shown in Fig. 1. When the regrowth is planar, these are called *planar buried heterostructure* (PBH) lasers. The result is a real refractive index guide in the lateral dimension. The width of these *index-guided* laser stripes may be anything from 1 μm to more than 10 μm, depending on the refractive index difference between the active stripe and the lateral cladding material. Equation (1), which specifies the condition for single mode, applies here, with d_g as the width of the lateral index guide and n_c defined by the regrown material. A typical lateral width for low-threshold BH lasers is 3 μm.

A laser geometry that is much simpler to fabricate and has a higher reliability in production than that of BH lasers is the *ridge waveguide* (RWG) laser, shown in Fig. 3. The fabrication starts with the growth of a separate confinement heterostructure (sometimes with the addition of a thin etch-stop layer just after the top waveguide layer), followed by a stripe mesa etch down to the waveguide layer, finishing with planarization and contacting to the stripe. The etch leaves a ridge of *p*-cladding material above the waveguide layer, which causes *strip loading,* raising the effective refractive index locally in the stripe region, thereby creating lateral confinement of the light. Although the RWG laser is attractive because of its easy fabrication process, its threshold current is relatively high.

Retroreflection of Guided Light Along the Stripe

Light is usually reflected back and forth inside the laser cavity by Fresnel reflection from cleaved end facets. Since the waveguide refractive index is $n_g \sim 3.5$, the natural Fresnel reflectivity at an air interface, $R = [(n_g - 1)/(n_g + 1)]^2$, is ~0.3. This rather low reflectivity means that semiconductor lasers are high gain, requiring enough amplification that 70 percent of the light is regenerated on each pass through the active medium.

Relying on Fresnel reflection means that both facets emit light. The light emitted out the back facet may be recovered by including a high-reflectivity multilayer coating on the back facet, as is typically done in most telecom lasers. Sometimes a coating is also provided on the front facet in order to alter its reflectivity, typically to lower it, which increases the output power (as long as the gain is high enough to overcome the large loss upon reflection). The reflectivities must be such that the laser can obey the laser operating condition, which states that in a single round-trip through a laser of length L, the increase in optical power from gain must balance the reduction from finite reflectivity, so that their product is unity. That is,

$$R_1 R_2 \exp(2g_L L) = 1 \qquad (3)$$

where R_1 and R_2 are the reflectivities of the two facets and g_L is the *modal gain per unit length* (as experienced by the waveguided laser mode), with a subscript L to represent that the gain is measured with respect to length. If $R_1 = R_2 = 0.3$, then $g_L L = 1.2$. Typical laser diodes have lengths of 400 μm, so $g_L \sim 30$ cm^{-1}.

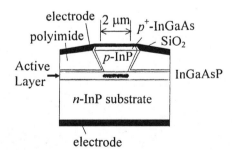

FIGURE 3 Geometry for a ridge waveguide (RWG) laser, fabricated by a single epitaxial growth followed by a mesa etch and planarization with polyimide. Light is confined to the region under the *p*-InP etched mesa by strip loading, which increases the effective refractive index in the waveguide region under the etched mesa.

In-plane retroreflection can also be achieved by using distributed feedback created from a grating impressed on top of the active layer. This method enables the construction of *distributed feedback* (DFB) lasers and *distributed Bragg reflector* (DBR) lasers, which are discussed later in a separate section.

Mounting so that Light is Edge-Emitted

Because the light is emitted out of the facet laterally, there must be a clear optical path for the light as it exits the laser. In many cases, the light is mounted with the active layer down, very close to the copper (or diamond) heat sink, in order to maximize cooling.[2] In this case, the laser chip must be placed at the very edge of the heat-sink block, as shown in Fig. 4a.

In some cases, the laser is mounted with its active region up with its substrate next to the heat sink. The edge alignment is not so critical in this case, but of course the laser light will still be emitted in a direction parallel to the plane of the heat sink. Because the thermal conductivity of the heat sink is much higher than that of the substrate, only the lowest threshold lasers, operating at moderate power levels, are operated with the active region up.

Suitable Packaging in a Hermetic Enclosure

Water vapor can degrade bare facets of a semiconductor laser when it is operating; therefore, LDs are usually passivated (i.e., their facets are coated with protective layers), and/or they are placed in sealed packages. The LD may be placed in a standard three-pin semiconductor device package, such as a TO-46 can with an optical window replacing the top of the can, as shown in Fig. 4a. The LD should be situated near the package window because the light diverges rapidly after it is emitted from the laser facet. The package window should be antireflection coated because any light reflected back into the laser can have serious consequences on the stability of the output (see Sec. 3.5).

Many high-end applications require an on-chip power monitor and/or a controllable thermoelectric cooler. In this case a more complex package will be used, typically a 14-pin "butterfly" package, often aligned to a fiber pigtail, such as is shown in Fig. 4b. In the less expensive datacom applications, nonhermetic packages may be acceptable with proper capping and passivation of the laser surfaces.

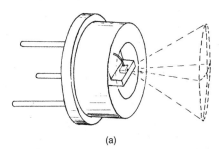
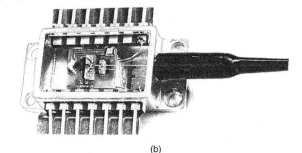

(a)   (b)

FIGURE 4   Packaging laser diodes: (*a*) typical hermetically sealed package showing heat sink and emission pattern for a laser diode with its active region placed down on a copper (or diamond) heat sink; and (b) typical butterfly package, showing laser in the middle, monitoring photodiode (behind), and fiber alignment chuck in front, all mounted on a thermoelectric cooler. Photo provided by Spectra-Diode Laboratories.

Fiber Pigtail Connection

Because light diverges at a rather large angle as it comes out of an edge-emitting laser (as discussed later), it is often desirable to use a laser provided with a *fiber pigtail,* which is a prealigned length of fiber that can be spliced or connected to the telecom fiber in the field. There will be an inevitable reduction in output power (compared to that of a laser with no pigtail) because of finite coupling efficiency into the pigtail, but the output will be immediately useful in a telecom system. The alternative to using a fiber pigtail is the use of a microlens—often a graded index (GRIN) lens, discussed elsewhere in this volume.

Long Life

Early lasers showed degradation with running time, but those problems have been solved, and it is expected that the semiconductor lasers used in telecom systems should last hundreds of thousands of hours. However, this requires that care be taken in their use. In particular, large reverse-bias static voltages can break down the *pn* diode. Thus, protection from electrostatic shock while handling and from reflected reverse-bias electrical currents during operation should be maintained. In addition, if LDs are driven with too much forward-bias current, the optical output can be so large that the light may erode the facet out of which it is emitted. Since the threshold is strongly temperature dependent, a laser driven at constant current that becomes too cold can emit too much light, with resulting optical damage. Thus, many telecom lasers have monitoring photodiodes to control the laser output and ensure that it stays within acceptable bounds.

4.3   OPERATING CHARACTERISTICS OF LASER DIODES

The principles of semiconductor laser operation are shown in Vol. I, Chap. 13 of this handbook. A forward-biased *pn* junction injects carriers into the active region. As the drive current increases, the carrier density in the active region increases. This reduces the absorption from an initially high value (at thermal equilibrium the absorption coefficient $\alpha \approx 500\ \mathrm{cm}^{-1}$) to zero, at which point the active layer becomes transparent at the prospective laser wavelengths. An active layer is characterized by its carrier density at transparency N_{tr}. Typically, $N_{\mathrm{tr}} \approx 10^{18}\ \mathrm{cm}^{-3}$. Above this carrier density, stimulated emission occurs, with a gain proportional to the diode carrier density above transparency. The gain depends on the detailed device design, taking into account the issues enumerated in the previous section and the materials involved. The gain is sizeable only in direct-band semiconductors (semiconductors based on the III-V or II-VI columns of the periodic table).

Laser Threshold

Threshold is given by the requirement that the round-trip optical gain due to stimulated emission must equal the round-trip optical loss due to the sum of the transmission out the end facets and any residual distributed loss. Gain occurs only for light that is actually in the active region, and not for the fraction of waveguided light that extends outside the active region. Typically, the *local* gain per unit length G_L is defined as that experienced locally by light inside the active region. (The *modal* gain per unit length is $g_L = \Gamma\, G_L$.) Near transparency, the gain depends linearly on carrier density N:

$$G_L = a_L\, \frac{N - N_{\mathrm{tr}}}{N_{\mathrm{tr}}} \qquad (4)$$

where a_L is the proportionality constant in units of length ($a_L \equiv N \, \partial G_L / \partial N$ near N_{tr}). When $N = 0$, $G_L = -a_L$, which is the loss per unit length in the unpumped active region (assuming the gain is linear in N). Typically, $a_L \sim 250$ cm^{-1}.

The current density (J) is related to the carrier density through

$$J = \frac{eNd}{\tau} \tag{5}$$

where τ is the lifetime of the electron-hole pairs. The transparency current density for $d = 0.15$ μm, $N_{tr} = 10^{18}$ cm^{-3}, and $\tau = 2$ ns is 1200 A/cm^2. The threshold condition can be found by taking the natural logarithm of Eq. (3):

$$g_{L,th} = G_{L,th}\Gamma = \alpha_i + \alpha_m \tag{6}$$

where α_m is the mirror reflectivity amortized over length, $2\alpha_m L = \ln(1/R_1 R_2)$; and α_i represents any internal losses for the laser mode, also amortized over length.

Combining Eqs. (2) through (6), along with the fact that a laser diode with stripe width w and length L will have a current $I = JwL$, gives

$$I_{th} = I_{tr} + \frac{ewN_{tr}}{\tau a_L} \left[\frac{1}{2} \ln\left(\frac{1}{R_1 R_2}\right) + \alpha_i L \right] d \left(1 + \frac{2}{V^2}\right) \tag{7}$$

where the waveguide V parameter is from Eq. (1) with $d = d_g$. Note that when the internal losses are small, the threshold current is independent of device length L, but depends on the reflectivity of the facets. Note also that the longer the spontaneous lifetime, the lower the threshold current density (although this may make a long turn-on delay, as discussed later). Finally, as expected by the relation between current and current density, a thinner stripe width w will lower the threshold current (consistent with appropriate spatial output, as discussed later). The current density at transparency N_{tr} is a basic property of the gain curve of the active region. It is smaller for quantum well lasers (discussed later) than for thicker active regions.

Note that because V is linearly proportional to d there is an optimal active layer thickness, a trade-off between increasing the carrier density as much as possible, but not so much as to lose optical confinement. The optimum thickness for 1.3-μm lasers is 0.15 μm; for 1.55-μm lasers it is comparable (0.15 – 0.18 μm). Threshold currents for broad-area DH lasers can be under ~500 A/cm^2 at 1.3 μm and ~1000 A/cm^2 at 1.55 μm. Confining carriers and light separately can beat this requirement, a trick used in designing quantum well lasers.

Light Out Versus Current In (the *L-I* Curve)

Below laser threshold only spontaneous emission is observed, which is the regime of the LED, as discussed in Sec. 4.8. In the spontaneous regime, the output varies linearly with input current and is emitted in all directions within the active region. As a result, a negligible amount of light is captured by the single-mode fiber of telecom below threshold.

Above threshold, the electrical power is converted to optical power. In general, the light will come out of both facets, and the amount of light reflected out the front facet depends on the rear facet reflectivity. When 100 percent mirror is placed on the back facet, the optical power at photon energy $h\nu$ (wavelength $\lambda = c/\nu$) emitted out the front facet is

$$\mathcal{P}_{out} = \frac{h\nu}{e} \frac{\alpha_m}{\alpha_m + \alpha_i} (I - I_{th} - I_L) \eta_i \tag{8}$$

where η_i is the *internal quantum efficiency*, which is the fraction of injected carriers that recombine by radiative recombination (usually close to unity in a well-designed semiconductor laser), and I_L is any leakage current. This equation indicates a linear dependence between light out and current above threshold (for constant quantum efficiency). The power out will drop by a factor of 2 if the back facet has a reflectivity equal to that of the front facet, since half the light will leave out the back.

From Eq. (8) can be calculated the *external slope efficiency* of the LD, given by $\partial\mathcal{P}_{out}/\partial I$. This allows the *differential quantum efficiency* η_D to be calculated:

$$\eta_D \equiv \frac{e}{h\nu}\frac{\partial\mathcal{P}_{out}}{\partial I} = \eta_i \frac{\alpha_m}{\alpha_m + \alpha_i} \tag{9}$$

This expression assumes that $\mathcal{P}_{out}$ includes the power out *both* facets.

The internal quantum efficiency depends on the modes of recombination for carriers. The rate of carrier loss is the sum of spontaneous processes, expressed in terms of carrier density divided by a lifetime τ_e, and stimulated emission, expressed in terms of gain per unit time G_T and photon density P:

$$R(N) = \frac{N}{\tau_e} + G_T(N)P \tag{10}$$

The spontaneous carrier lifetime is given by:

$$\frac{1}{\tau_e} = A_{nr} + BN + CN^2 \tag{11}$$

which includes spontaneous radiative recombination, given by BN. (The dependence on N results from needing the simultaneous presence of an electron and a hole, which have the same charge densities because of charge neutrality in undoped active regions.) The nonradiative recombination terms that decrease the quantum efficiency below unity are a constant term A_{nr} (that accounts for all background nonradiative recombination) and an Auger recombination term (with coefficient C) that depends on the square of the carrier density and comes from processes involving several carriers simultaneously. This term is particularly important in long-wavelength lasers where the Auger coefficient C is large. Stimulated emission is accounted for by gain in the time domain G_T, which depends on N (approximately linearly near threshold). The group velocity v_g converts gain per unit length G_L into a rate G_T (gain per unit time), $G_T \equiv v_g G_L$. We can define a gain coefficient in the time domain $a_T = v_g a_L$ so that

$$G_T = a_T \frac{N - N_{tr}}{N_{tr}} \tag{12}$$

The internal quantum efficiency in a laser is the fraction of the recombination processes that emit light:

$$\eta_i = \frac{BN^2 + G_T(N)P}{A_{nr}N + BN^2 + CN^3 + G_T(N)P} \tag{13}$$

Referring to Eq. (9), the external quantum efficiency depends on the sources of intrinsic loss. In long-wavelength lasers, this is primarily absorption loss due to intervalence band absorption. Another source of loss is scattering from roughness in the edges of the waveguide.

Figure 5 shows a typical experimental result for the light out of a laser diode as a function of applied current (the so-called *L-I* curve) for various temperatures. It can be seen that the linear relation between light out and current saturates as the current becomes large enough, particularly at high temperatures. Three main mechanisms have been proposed for the decrease in external slope efficiency with increasing current, each of which can be seen in the form of Eq. (9):

1. The leakage current increases with injection current.
2. Junction heating reduces recombination lifetime and increases threshold current.
3. The internal absorption increases with injection current.

When there is more than one laser mode (longitudinal or transverse) in the LD, the *L-I* curve has *kinks* at certain current levels. These are slight abrupt reductions in light out as the current increases. After a kink the external slope efficiency may be different, along with different spatial and spectral features of the laser. These multimode lasers may be acceptable for low-cost communication systems, but high-quality communication systems require single-mode lasers that do not exhibit such kinks in their *L-I* curves.

Temperature Dependence of Laser Properties

The long-wavelength lasers are more typically sensitive to temperature than are GaAs lasers. This sensitivity is usually expressed as an experimentally measured exponential dependence of threshold on temperature *T*:

$$I_{th}(T) = I_o \exp\left(\frac{T}{T_o}\right) \tag{14}$$

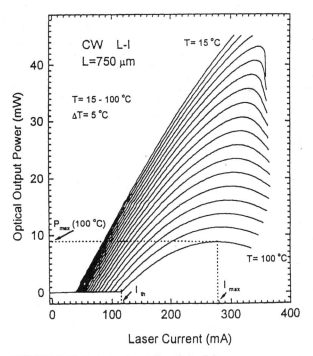

FIGURE 5   Typical experimental result for light out versus current in (the *L-I* curve). These results are for diodes operating at 1.3 μm, consisting of strained layer multiple quantum well InGaAsP lasers measured at a series of elevated temperatures.[3]

where T_o is a characteristic temperature (in degrees Kelvin) that expresses the measured thermal sensitivity. This formula is valid over only a limited temperature range, because it has no real physical derivation, but it has proved convenient and is often quoted. The data in Fig. 5 correspond to $T_o \approx 80$ K. The mechanisms for this sensitivity to temperature depend on the material system begin used. In long-wavelength double heterostructure lasers, T_o appears to be dominated by Auger recombination. However, in short-wavelength GaAs lasers and in strained layer quantum wells, where Auger recombination is suppressed, T_o is higher and is attributed to intervalence band absorption and/or carrier leakage over the heterostructure barrier, depending on the geometry. Typical long-wavelength DH lasers have T_o in the range of 50 to 70 K. Typical strained layer quantum well lasers have T_o in the range of 70 to 90 K, although higher T_o can be achieved by incorporating aluminum in barriers, with as high as 143 K reported.[4] This temperature dependence limits the maximum optical power that can be obtained because of the phenomenon of *thermal runaway,* as shown at the highest temperatures in Fig. 5. While the power is usually increased by increasing the current, the junction temperature also increases (due to ohmic losses), so the threshold may increase and the output power may tend to decrease.

Various means for increasing T_o have been explored. The most effective way to increase T_o has proven to be the use of tensile strained quantum wells (discussed in Sec. 3.6). The result has been to increase T_o from ~50 K to as high as 140 K, comparable to that measured in GaAs. In double heterostructures, losses by carrier leakage can be reduced by using a dual active region for double carrier confinement, which has been demonstrated to achieve T_o values as high as 180 K in 1.3-μm InP lasers.[5]

In practice, many long-wavelength lasers require thermoelectric coolers to moderate the temperature. The temperature dependence of long-wavelength lasers may limit their performance at high temperatures, which in turn limits where they can be used in the field.

Spatial Characteristics of Emitted Light

Light is emitted out of the facet of the laser diode after it has been guided in both directions. It will diverge by diffraction, more strongly in the out-of-plane dimension, where it has been more strongly waveguided. The diffracting output is sketched in Fig. 1. The spatial characteristics of the output can be estimated by fitting the guided light to a gaussian beam and then calculating the far-field pattern. The *out-of-plane* near-field profile for the lowest order mode in an optical confinement layer of width d_g can be fit to a gaussian distribution $\exp(-x^2/w^2)$ by[6]:

$$w = d_g \left(0.321 + \frac{2.1}{V^{3/2}} + \frac{4}{V^6} \right) \qquad \text{for } 1.8 < V < 6 \qquad (15)$$

where V is from Eq. (1). The far-field diffraction angle can be found from the Fourier transform multiplied by the obliquity factor, resulting in a slightly different gaussian fit. The gaussian half-angle in the far field is given by[6]:

$$\theta_{ff} = \tan^{-1} \left(\frac{\lambda}{\pi w_o} \right) \qquad (16)$$

where

$$w_o = d_g \left(0.31 + \frac{3.15}{V^{3/2}} + \frac{2}{V^6} \right) \qquad \text{for } 1.5 < V < 6.$$

Experimental data can be compared to the gaussian beam formula by remembering that the full-width half-maximum power FWHM $= w(2 \ln 2)^{1/2}$. For a typical strongly index-guided buried heterostructure laser, the far-field FWHM angle out of plane is ~1 rad and in-plane is ~1/2 rad. These angles are independent of current for index-guided lasers. Separate confinement heterostructure lasers can have smaller out-of-plane beam divergences, more typically ~30°.

Single-mode lasers that are index guided in the lateral direction (buried heterostructure and ridge waveguide) will obey the preceding equations, with lateral divergence angles varying from 30° to 10°, depending on design. This beam width will also be independent of current. When lasers are gain guided laterally, the spatial variation of the gain leads to a complex refractive index and a curved wavefront. The result is that the equivalent gaussian lateral beam seems to have been emitted from somewhere inside the laser facet. The out-of-plane beam, however, is still index guided and will appear to be emitted from the end facet. This means that the output of a gain-guided laser has *astigmatism,* which must be compensated for by a suitably designed external lens if the laser is to be focused effectively into a fiber (as discussed elsewhere in this handbook).

If the laser emits a diverging gaussian beam with waist w, a lens can be used to focus it into a fiber. An effective thin lens of focal length f placed a distance d_1 after the laser facet will focus to a new waist w' given by:

$$w'^2 = w^2 \frac{f^2}{b^2 + X_1^2} \tag{17}$$

where

$$X_1 \equiv d_1 - f$$
$$b \equiv \pi w^2 / \lambda$$

The distance d_2 from the lens to the new beam waist is given by:

$$X_2 = X_1 \frac{f^2}{X_1^2 + b^2} \tag{18}$$

where

$$X_2 \equiv d_2 - f$$

This new waist must be matched to the fiber mode. Because of the large numerical aperture of laser light, simple lenses exhibit severe spherical aberration. Fiber systems usually utilize pigtailed fiber, butt coupled as close as possible to the laser, without any intervening lens. Typical coupling efficiencies are only a few percent. Alternatively, a ball lens may be melted directly onto a fiber tip and placed near the laser facet. Sometimes *graded index (GRIN)* lenses are used to improve coupling into fibers.

Gain-guided lasers with electrode stripe widths of >5 μm usually emit multiple spatial modes in the in-plane direction. These modes interfere laterally, producing a spatial output with multiple maxima and nulls. Such spatial profiles are suitable for multimode fiber applications, but cannot be coupled into single-mode fibers with high efficiency. They will diffract at an angle given by setting w equal to the minimum near-field feature size. If the stripe is narrow enough, gain-guided lasers are always single mode, but the double-lobed far-field spatial profile (from the complex refractive index in the gain medium) cannot be conveniently coupled into single-mode fibers.

Spectral Characteristics of Laser Light

In principle, a Fabry-Perot laser has many frequency modes with frequencies v_m, given by requiring standing waves within the laser cavity. Since the mth mode obeys $m\lambda/2n = L$, where n is the refractive index experienced by the guided laser mode, then

$$\nu_m = \frac{mc}{2nL} \qquad (19)$$

Taking the differential, the frequency difference between modes is

$$\Delta\nu = \frac{c}{2n_{\text{eff}}L} \qquad (20)$$

where the *effective group refractive index* $n_{\text{eff}} = n + \nu(\partial n/\partial\nu)$. For typical semiconductor lasers, $n = 3.5$ and $n_{\text{eff}} = 4$, so that when $L = 250$ μm, the frequency difference between modes is $\Delta\nu = 150$ GHz, and since $\Delta\lambda = (\lambda^2/c)\Delta\nu$, when $\lambda = 1.5$ μm, the wavelength spacing is $\Delta\lambda \approx 1$ nm.

At any given instant in time, a single spatial mode emits in only one spectral mode. However, in multimode lasers, considerable *mode hopping* occurs, in which the LD jumps from one spectral mode to another very rapidly. Most spectral measurements are time averages and do not resolve this mode hopping, which can occur in nanoseconds or less. Explanations for the mode-hopping typically involve *spatial hole burning* or *spectral hole burning*. Hole burning occurs when the available carrier density is momentarily depleted, either spatially or spectrally. At that time an adjacent mode with a different (longitudinal or lateral) spatial profile or a different resonance wavelength may be more advantageous for laser action. Thus, the laser jumps to this new mode. The competition between different modes for available gain is a strong mechanism for creating lasers with multiple wavelength modes.

One way to provide a single spectral mode is to ensure a single (lateral) spatial mode. It has been found that single spatial mode lasers usually have single spectral modes, at least at moderate power levels. The only way to *ensure* a single-frequency LD is to ensure a single longitudinal mode by using distributed feedback, as discussed in Sec. 5.7.

Polarization

The emitted light from a typical semiconductor laser is usually linearly polarized in the plane of the heterostructure. While the gain in a semiconductor has no favored polarization dependence, the *transverse electric (TE) waveguide mode* (polarized in-plane) is favored for two reasons. First, the TE mode is slightly more confined than the *transverse magnetic* (TM) mode (polarized out-of-plane). Second, the Fresnel reflectivity off the cleaved end facets is strongly polarization sensitive. As waveguided light travels along the active stripe region, it can be considered to follow a zig-zag path, being totally internally reflected by the cladding layers. The total internal reflection angle for these waves is about 10° off the normal to the cleaved facets of the laser. This is enough to cause the *TM waveguide mode* to experience less reflectivity, while the TE-polarized mode experiences more reflectivity. Thus, laser light from LDs is traditionally polarized in the plane of the junction.

However, the introduction of strain (Sec. 4.6) in the active layer changes the polarization properties, and the particular polarization will depend on the details of the device's geometry. In addition, DFB and DBR lasers (Sec. 4.7) do not have strong polarization preferences, and they must be carefully designed and fabricated if well-defined single polarization is required.

4.4 *TRANSIENT RESPONSE OF LASER DIODES*

When laser diodes are operated by direct current, the output is constant and follows the *L-I* curve discussed previously. When the LD is rapidly switched, however, there are transient phenomena that must be taken into account. Such considerations are important for any high-speed communication system, especially digital systems. The study of these phenomena comes from solving the semiconductor rate equations.[7]

Turn-on Delay

When a semiconductor laser is turned on abruptly by applying forward-biased current to the diode, it takes time for the carrier density to reach its threshold value and for the photon density to build up, as shown in the experimental data of Fig. 6. This means that a laser has an unavoidable turn-on time. The delay time depends on applied current and on carrier lifetime, which depends on carrier density N, as shown in Eq. (11). Using a differential analysis, the turn-on time for a laser that is switched from an initial current I_i just below threshold to I just above threshold is

$$\tau_d = \tau'(N_{th}) \frac{I_{th} - I_i}{I - I_{th}} \tag{21}$$

where $\tau'(N)$ is a differential lifetime given by

$$\frac{1}{\tau'} = A_{nr} + 2BN + 3CN^2 \tag{22}$$

When $I_i = 0$ and $I \gg I_{th}$, the turn-on delay has an inverse current dependence:

$$\tau_d = \tau_e(N_{th}) \frac{I_{th}}{I} \tag{23}$$

When radiative recombination dominates, then $1/\tau_e \approx BN$ and $1/\tau' \approx 2BN \approx 2/\tau_e$, as seen by comparing the middle terms of Eqs. (11) and (22). For a 1.3-μm laser, $A_{nr} = 10^8$/s, $B = 10^{-10}$ cm^3/s, $C = 3 \times 10^{-29}$ cm^6/s, and $N_{th} \approx N_{tr} = 10^{18}$ cm^{-3}. Thus, $\tau_e = 5$ ns and a typical turn-on time at 1.5 times threshold current is 3 ns. The increase in delay time as the current approaches threshold is clearly seen in the data of Fig. 6. As a result, to switch a laser rapidly, it is necessary to switch it from just below threshold to far above threshold. However, Fig. 6 shows that under these conditions there are large transient oscillations, discussed next.

Relaxation Oscillations

An important characteristic of the output of any rapidly switched laser (not just semiconductor lasers) is the *relaxation oscillations* that can be observed in Fig. 6. These overshoots occur as the photon dynamics and carrier dynamics are coming into equilibrium. Such oscillations are characteristic of the nonlinear coupled laser rate equations and can be found by simple perturbation theory. These relaxation oscillations have a radian frequency Ω_R given, to first order, by[9]:

$$\Omega_R^2 = \frac{1 + \chi}{\tau_e \tau_p} \frac{I - I_{th}}{I_{th}} \tag{24}$$

where I is the current, I_{th} is the current at threshold, τ_p is the photon lifetime in the cavity, given by

$$v_g \tau_p = \frac{1}{\alpha_i + \alpha_m} \tag{25}$$

and

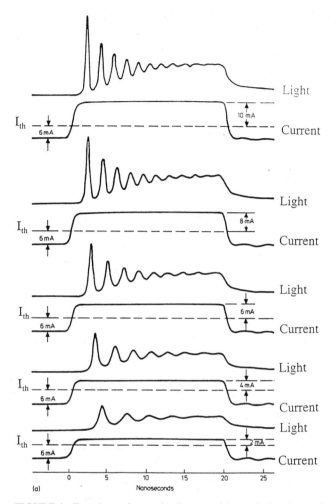

FIGURE 6 Experimental example of turn-on delay and relaxation oscillations in a laser diode when the operating current is suddenly switched
from 6 mA below the threshold current of 177 mA to varying levels above
threshold (from 2 to 10 mA). The GaAs laser diode was 50 μm long, with a
SiO_2 defined stripe 20 μm wide. Light output and current pulse are shown
for each case.[8]

$$\chi = \Gamma a_L v_g \tau_p = \Gamma a_T \tau_p = \Gamma \frac{a_L}{\alpha_i + \alpha_m} \qquad (26)$$

where a_L is from Eq. (4). The factor χ is the ratio of the unpumped absorption loss to the cavity loss. For semiconductor lasers, χ is on the order of 1 to 3. It can also be shown that $\chi = I_{tr}/$
$(I_{th} - I_{tr})$, where I_{tr} is the current at transparency.

When $\chi \approx 1$, at 1.5 times threshold current, where $(1 + \chi)(I - I_{th})/I_{th} \approx 1$, the time between successive relaxation oscillation maxima is approximately the geometric mean of the carrier and
photon lifetimes: $\Omega_R^2 \approx 1/\tau_e \tau_p$. Typical numbers for semiconductor lasers are $\tau_e = 10$ ns, $\tau_p = 3$ ps,

so at 1.5 times threshold current, the relaxation oscillation frequency is $f_R = \Omega_R/2\pi = 1$ GHz, and the time between the relaxation oscillation peaks is 1 ns.

The decay rate of these relaxation oscillations γ_R is given by

$$2\tau_e\gamma_R = 1 + \chi\,\frac{I - I_{\text{th}}}{I_{\text{tr}}} = 1 + (1 + \chi)\cdot\frac{I - I_{\text{th}}}{I_{\text{th}}} = 1 + \Omega_R^2\tau_e\tau_p \tag{27}$$

and is roughly 2 ns at twice threshold for typical heterostructure lasers. At 1.5 times threshold, when $\chi \approx 1$, $\gamma_R \approx 1/\tau_e$. The relaxation oscillations will last approximately as long as the spontaneous emission lifetime of the carriers.

This analysis employs several assumptions which do not seriously affect the relaxation oscillation frequency, but which will overestimate the time that relaxation oscillations will last. The analysis ignores *gain saturation,* which reduces gain with increased photon density P and is important at high optical powers. It also ignores the rate of spontaneous emission in the cavity R_{sp}, which is important at small optical powers. Finally, it ignores the impact of changing carrier density on spontaneous emission lifetime. A more exact formulation[10] includes these effects:

$$2\gamma_R = \frac{1}{\tau'} + P\left(\frac{\partial g_T}{\partial N} - \frac{\partial g_T}{\partial P}\right) + \frac{R_{sp}}{P} \tag{28}$$

where g_T is the modal gain per unit time. This more exact analysis increases the rate of decay, since the sign of $\partial g_T/\partial P$ is negative and also $1/\tau' \approx 2/\tau_e$. A more typical experimental decay rate for lasers at 1.3-μm wavelength is $\gamma \approx 3/\tau_e$.

The number of relaxation oscillations (before they die out) in an LD at 1.5 times threshold is proportional to $\Omega_R/\gamma_R \propto (\tau_e/\tau_p)$. The longer the carrier lifetime, the more relaxation oscillations will occur (because the carriers do not decay rapidly to steady state). Shorter carrier lifetimes also mean shorter turn-on times. Thus, achieving short carrier lifetimes by high carrier densities is important for high-speed semiconductor lasers. This can be achieved by using as small an active region as possible (such as quantum wells) and by reducing the reflectivity of the laser facets to raise the threshold carrier density.

The relaxation oscillations disappear if the current is just at threshold. However, we've also seen that under this situation the turn-on time becomes very long. It is more advantageous to turn the laser on fast, suffering the relaxation oscillations and using a laser designed to achieve a high decay rate, which means using the laser with the highest possible relaxation oscillation frequency.

Other useful forms for the relaxation oscillations are:

$$\Omega_R^2 = g_T g_T' P = \frac{g_T' P}{\tau_m} = g_T'\left(\frac{\mathscr{P}_{\text{out}}}{h\nu V_a}\,\frac{\alpha_m + \alpha_i}{\alpha_m}\right) \tag{29}$$

where $g_T' \equiv \partial g_T/\partial N = \Gamma a_T/N_{\text{tr}}$. These expressions can be found by inserting the following equations into Eq. (24):

$$I - I_{\text{th}} = \frac{\mathscr{P}_{\text{out}}}{h\nu}\,\frac{e}{\eta_D} \tag{30a}$$

$$\mathscr{P}_{\text{out}} = h\nu\left(\frac{P}{\tau_m}\right)V_a \tag{30b}$$

where $\tau_m = (v_g\alpha_m)^{-1}$ is the time it takes light to bleed out the mirror and

$$I_{\text{th}} = \frac{eN_{\text{th}}V_a}{\tau_e} \tag{30c}$$

$$g_T = \Gamma\, a_T \frac{(N - N_{\text{tr}})}{N_{\text{tr}}} \tag{30d}$$

so that $g_T' \equiv \partial g_T / \partial N = \Gamma\, a_T / N_{\text{tr}}$ and $N_{\text{tr}} = \Gamma a_T / g_T'$. Also

$$\chi = \frac{I_{\text{tr}}}{I_{\text{th}} - I_{\text{tr}}}, \quad \text{so} \quad 1 + \chi = \frac{I_{\text{th}}}{I_{\text{th}} - I_{\text{tr}}} \tag{30e}$$

$$\eta_D = \eta_i \left(\frac{\alpha_i}{\alpha_m + \alpha_i} \right) = \eta_i \left(\frac{\tau_p}{\tau_i} \right), \quad \text{assuming} \quad \eta_i \approx 1 \tag{30f}$$

Note that the relaxation oscillation frequency increases as the photon density increases, showing that smaller laser dimensions are better.

Relaxation oscillations can be avoided by biasing the laser just below threshold, communication systems often operate with a prebiased laser. In digital and high-speed analog systems, relaxation oscillations may limit speed and performance.

Modulation Response and Gain Saturation

The *modulation response* describes the amplitude of the modulated optical output as a function of frequency under small-signal current modulation. There is a resonance in the modulation response at the relaxation oscillation frequency, as indicated by the experimental data in Fig. 7. It is more difficult to modulate the laser, above the relaxation oscillation frequency. Carrying out a small-signal expansion of the rate equations around photon density P, the modulation response (in terms of current density J) is[12]:

$$\frac{\partial P}{\partial J} = \frac{(1/ed)(g_T'P + \beta_{\text{sp}}/\tau_e)}{(g_T'P/\tau_p + \beta_{\text{sp}}/\tau_e\tau_p - \omega^2) + j\omega\,(g_T'P + 1/\tau_e)} \tag{31}$$

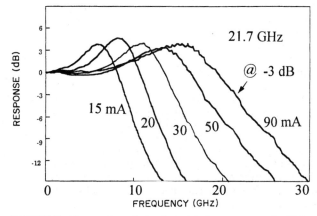

FIGURE 7 Measured small-signal modulation response of a high-speed DFB laser at several bias levels. Zero-dB modulation response is defined in terms of the low-frequency modulation response, Eq. (32).[11]

where β_{sp} is the fraction of spontaneous emission that radiates into the mode ($\beta_{sp} = R_{sp}\tau_e/N$); and, as before, τ_e is the spontaneous carrier lifetime, and $g_T' \equiv \partial g_T/\partial N = \Gamma a_T/N_{tr}$. This modulation response has the form of a second-order low-pass filter. Resonance occurs when $\omega^2 \approx g_T' P/\tau_p = \Omega_R^2$ (from Eq. (29), with negligible internal loss); that is, at the relaxation oscillation frequency.

The modulation response at a frequency well below the relaxation oscillation frequency can be expressed as the change in optical power $\mathscr{P}_{out}$ as a function of current I using the limit of Eq. (31) when $\omega \to 0$. From $\partial P/\partial J = \tau_p/ed$, and relating output power to photon density through $\mathscr{P}_{out} = (h\nu)(P/\tau_m)V_a$, the low frequency modulation response becomes

$$\frac{\partial \mathscr{P}_{out}}{\partial I} = \frac{h\nu}{\tau_m} V_a \frac{\partial P}{\partial J} \cdot \frac{1}{wL} = \frac{h\nu}{e} \cdot \frac{\tau_p}{\tau_m} = \frac{h\nu}{e} \cdot \frac{\alpha_m}{(\alpha_m + \alpha_i)} \tag{32}$$

which is expected from Eq. (8) when $\eta_i \to 1$.

The 3-dB modulation radian frequency bandwidth ω_B can be expressed in terms of the relaxation oscillation parameters by[13]:

$$\omega_B^2 = \Omega_R^2 - \gamma_R^2 + 2\sqrt{\Omega_R^2 (\Omega_R^2 + \gamma_R^2) + \gamma_R^4} \tag{33}$$

where the oscillation frequency Ω_R and damping rate γ_R are as previously described. The parameters are strongly power dependent and the bandwidth increases with optical power. When $\gamma_R \ll \Omega_R$, the 3-dB bandwidth $\omega_B \approx \sqrt{3}\,\Omega_R \propto \sqrt{P}$. At high optical powers the presence of gain saturation (reduced gain at high optical power densities) must be included; the modulation bandwidth saturates, and the limiting value depends on the way that the gain saturates with photon density. Using the following approximate expression for gain saturation:

$$g_T(N, P) = g_T' \frac{N - N_o}{\sqrt{1 + P/P_s}} \tag{34}$$

where N_o is the equilibrium carrier density and P_s is the saturation photon density, a simple expression can be found for the limiting value of the modulation bandwidth at high optical powers:

$$(\omega_{B, max})^2 = \frac{3g_T' P_s}{2\tau_p}$$

Typical numbers for a 1.55-μm InGaAsP laser are 20 to 40 GHz.

Frequency Chirping

When the carrier density in the active region is rapidly changed, the refractive index also changes rapidly, causing a frequency shift proportional to $\partial n/\partial t$. This broadens the laser linewidth from its original width of ~100 MHz into a double-peaked profile with a gigahertz linewidth, as shown in the experimental results of Fig. 8. The frequency spread is directly proportional to the dependence of the refractive index n on carrier density N. This is a complex function that depends on wavelength and degree of excitation, but for simplicity a Taylor expansion around the steady-state carrier density N_o can be assumed: $n = n_o + n_1(N - N_o)$, where $n_1 \equiv \partial n/\partial N$. The (normalized) ratio of this slope to that of the gain per unit length g_L is called the *linewidth enhancement factor* β_c.

$$\beta_c \equiv -2k_o \frac{\partial n/\partial N}{\partial g_L/\partial N} = -2k_o \frac{n_1}{g_L'} \tag{35}$$

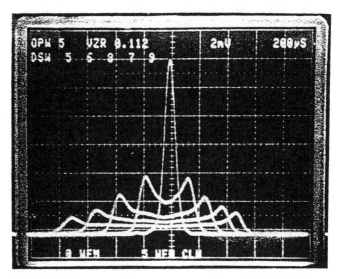

FIGURE 8 Time-averaged power spectra of 1.3 μm InGaAsP laser under sinusoidal modulation at 100 MHz. Horizontal scale is 0.05 nm per division. Spectrum broadens with increase in modulation current due to frequency chirping.[14]

The magnitude of the frequency spread between the double lobes of a chirped pulse, $2\delta\omega_{CH}$, can be estimated in the small-signal and large-signal regimes from analyzing the time dependence of a modulated pulse in terms of the sum of all frequency components.[15]

Small-Signal Modulation. For a modulation frequency ω_m that is less than the relaxation oscillation frequency, and assuming that $\gamma_R \ll \Omega_R$, a small modulation current I_m will cause a frequency chirp of magnitude

$$\delta\omega_{CH} = \frac{\beta_c I_m h_v}{2e\mathcal{P}_{out}}\left(\frac{\alpha_m}{\alpha_m + \alpha_i}\right)\sqrt{\omega_m^2 + \gamma_p^2} \qquad (36)$$

where $\gamma_p = R_{sp}/P - (\partial g_T/\partial P)P$ (remembering that $\partial g_T/\partial P$ is negative). The origin of chirp is the linewidth enhancement factor β_c. It will be largest for gain-guided devices where β_c is a maximum. The chirp will be smaller in lasers with $\alpha_m \ll \alpha_i$, such as will occur for long lasers, where mirror loss is amortized over a longer length, but such lasers will have a smaller differential quantum efficiency and smaller relaxation oscillation frequency. Typical numbers at 25-mA modulation current can vary from 0.2 nm for gain-guided lasers to 0.03 nm for ridge waveguide lasers.

Large-Signal Modulation. There is a transient frequency shift during large-signal modulation given by:

$$\delta\omega_{CH} = \frac{\beta_c}{2}\left(\frac{1}{P}\frac{\partial P}{\partial t}\right) \qquad (37a)$$

When a gaussian shape pulse is assumed, $\exp(-t^2/T^2)$, the frequency shift becomes

$$\delta\omega_{CH} \approx \beta_c/T \qquad (37b)$$

The importance of the linewidth enhancement factor β_c is evident from this equation; its existence will inevitably broaden modulated laser linewidths.

4.5 NOISE CHARACTERISTICS OF LASER DIODES

Noise in LDs results from fluctuations in spontaneous emission and from the carrier generation-recombination process (shot noise). To analyze the response of LDs to noise, one starts with rate equations, introduces Langevin noise sources as small perturbations, and linearizes (performs a small-signal analysis). Finally, one solves in the frequency domain using Fourier analysis.[16, 17] Only the results are given here.

Relative Intensity Noise (RIN)

Noise at a given frequency is described in terms of *relative intensity noise,* defined by:

$$\text{RIN} = \frac{S_P(\omega)}{P_T^2} \tag{38}$$

where $S_P(\omega)$ is the photon noise spectral density (noise per unit frequency interval), and P_T is the total photon number, $P_T = PV_a$. The solution to the analysis for RIN is:

$$\text{RIN} = \frac{2\beta_{sp}I_{th}}{ePV_a} \cdot \frac{1/\tau'^2 + \omega^2 + (\partial g_T/\partial N)^2\, P/(\beta_{sp}V_a)}{[(\Omega_R - \omega)^2 + \gamma_R^2][(\Omega_R + \omega)^2 + \gamma_R^2]} \tag{39}$$

where β_{sp} is the fraction of spontaneous emission emitted into the laser cavity, and is related to the spontaneous emission rate R_{sp} by $\beta_{sp}\,(I_{th}/eV_a) = R_{sp}$. As before, the photon density P can be related to the optical power out both facets by $\mathscr{P}_{out} = (h\nu)\, PV_a/\tau_m$. Note the significant enhancement of noise near the relaxation oscillation frequency $\omega = \Omega_R$ where the noise has its maximum value. An example of RIN as a function of frequency is shown in Fig. 9, for both low power and high power, showing that the RIN goes up as the total optical power decreases.

At low frequencies, and for $\gamma_R \ll \Omega_R$, the noise is proportional to the inverse fourth power of the relaxation oscillation frequency. Clearly, it is advantageous to use as high a relaxation oscillation frequency as possible to reduce RIN. Since the relaxation oscillation frequency is proportional to the square root of the power P, the RIN increases as $1/P^3$ as the power decreases. Inserting the expression for Ω_R into Eq. (39) gives:

$$\text{RIN}_{lf} = \frac{2\,\beta_{sp}I_{th}\tau_m^2 V_a^2}{eP_T^3\,(\partial g_T/\partial N)^2} \left[\frac{1}{\tau'^2} + \left(\frac{\partial g_T}{\partial N}\right)^2 \frac{P}{\beta_{sp}V_a} \right] \tag{40}$$

Usually, the first term dominates. It can be seen that the volume of the active laser region V_a should be as small as possible, consistent with maintaining a significant power out.

Signal-to-Noise Ratio (SNR)

The *signal-to-noise ratio* (SNR) can be found in terms of the relaxation oscillation parameters using the expression for RIN (which assumes $\tau_e\Omega_R \gg \gamma_R\Omega_R \gg 1$) and the total photon number:

$$(\text{SNR})^2 = \frac{2\gamma_R e}{\beta_{sp}I_{th}}\, P_T = \frac{2\gamma_R e\tau_m}{\beta_{sp}I_{tr}h\nu}\, \mathscr{P}_{out} \tag{41}$$

As expected, the SNR increases with smaller spontaneous emission and larger laser power.

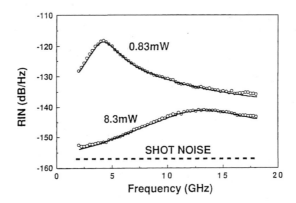

FIGURE 9 Measured relative intensity noise as a function of frequency in a multiple quantum well 1.5-μm laser diode, for optical power near threshold and high above threshold. The shot noise level for the higher power measurement is also shown.[18]

Far above threshold, inserting the value for the decay rate $\gamma_R \approx \tau_e/3$ gives

$$(\text{SNR})^2 = \frac{2e\tau_e\tau_m}{3\beta_{sp}I_{th}h\nu}\,\mathcal{P}_{\text{out}} \tag{42}$$

Gain saturation at high powers eventually limits the SNR to about 30 dB; while at powers of a few milliwatts it is 20 dB, with intensity fluctuation typically close to 1 percent.

Mode Partition Noise in Multimode Lasers

The preceding discussion of noise holds qualitatively for multimode lasers as long as all the laser modes are included. However, measurements made on any one mode show much more noise, particularly at low frequencies. This is due to the mode-hopping discussed previously, and is referred to as *mode partition noise*. That is, the power partitions itself between different laser modes in a way that keeps the overall intensity relatively constant, as shown by the solid line in Fig. 10. The power in each mode is not a steady function of time, because the power distribution among the modes changes with time. Whenever the distribution changes, the power output undergoes fluctuation, leading to a noise term on a nominally stable output. This leads to the enhanced RIN on the dominant mode in Fig. 10. Even an output whose spectrum looks nominally single mode, as shown in the inset of Fig. 10, can have a large RIN on the dominant mode. This is because the spectrum is time averaged. A side mode does not contain 5 percent of the power, for example; it contains 100 percent of the power for 5 percent of the time. This causes the very large RIN observed. The solution to avoiding this noise is to insist on a single longitudinal mode by using distributed feedback. Since lasers for telecommunication applications are typically single mode, we will not consider mode partition noise further. It becomes important for data communications based on multimode lasers, however, and it is crucial to gather all the light into the fiber.

Phase Noise—Linewidth

The fundamental linewidth of a laser is given by the stochastic process of spontaneous emission, as first derived by Schawlow and Townes in the very early days of lasers. In a semicon-

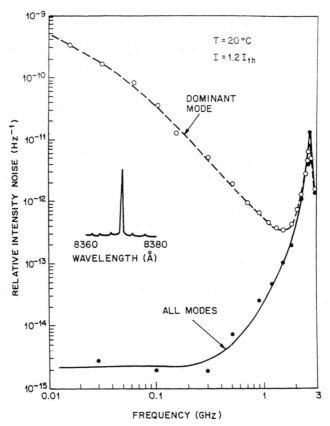

FIGURE 10 Effect of mode partition noise on relative intensity noise in multimode lasers. Experimentally observed intensity-noise spectra in all modes (solid curve) or in dominant mode (dashed curve). Inset shows spectrum of average mode power.[19]

ductor laser, additional noise enters from the stochastic process of carrier injection. Because the refractive index is a function of the carrier density, changes in carrier density cause changes in refractive index, which in turn create phase noise.

The formula for the radian frequency linewidth of a semiconductor laser includes the *linewidth enhancement factor* β_c (defined in Eq. (35):

$$\delta\omega = (1 + \beta_c^2)\,\delta\omega_o \tag{43}$$

where the original Schawlow-Townes linewidth is given by

$$\delta\omega_o = \frac{R_{sp}}{2P} = \frac{\beta_{sp}I_{th}h\nu\tau_m}{2e\mathcal{P}_{out}} \tag{44}$$

Typical values of the linewidth enhancement factor are $\beta_c = 5$. It can be seen that the linewidth decreases inversely as the laser power increases. However, as shown in the experimental data in Fig. 11, at high enough power (above 10 mW) the linewidth narrowing saturates at ~1 to 10 MHz and then begins to broaden again at even higher power levels. It is also possible to

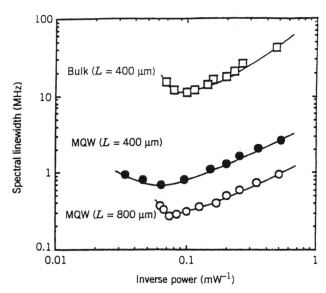

FIGURE 11   Linewidth of DFB lasers as a function of inverse power, comparing bulk active regions and multiple quantum well active regions.[20]

reduce the linewidth by using QWs and increasing the cavity length (to decrease N_{th} and increase P).

External Optical Feedback and Coherence Collapse

Semiconductor lasers are extremely sensitive to weak time-delayed feedback, such as from reflections off the front ends of fiber pigtails. These fed-back signals can result in mode hopping, strong excess noise, and chaotic behavior in the *coherence collapse* regime. Some of the features of feedback-induced noise are outlined here.

Regimes of Feedback.   The following provides a useful classification scheme[21]:

Regime I.   At the lowest levels of feedback, narrowing or broadening of the emission line is observed, depending on the phase of the feedback.

Regime II.   At higher levels of feedback, mode hopping between different external cavity modes may appear.

Regime III.   Further increasing the levels of feedback, the laser is observed to operate in the lowest linewidth mode.

Regime IV.   At yet higher feedback levels, satellite modes appear, separated from the main mode by the relaxation oscillation frequency. These grow as the feedback increases and the laser line eventually broadens. This regime does not depend on the distance from the laser to the reflector. This is the regime of *coherence collapse*.

Regime V.   A regime of stable operation that can be reached only with antireflection coating on the laser facet to ensure the largest possible coupling back into the laser.

These regimes of feedback are characterized by the value of a *feedback parameter C,* given by:

$$C = \sqrt{f_{ext}} \, C_e \, \frac{\tau_{ext}}{\tau_L} \, \sqrt{1 + \beta_c^2} \qquad (45)$$

where f_{ext} is the ratio of the externally reflected power that enters back into the laser divided by the emitted power. Also, the external coupling factor $C_e = (1 - R)/\sqrt{R}$, where R is the reflectivity of the laser facet. As before, β_c is the linewidth enhancement factor. The external round-trip time delay is τ_{ext}, and the laser round-trip time is τ_L. The regimes have the following values of the feedback parameter:

Regime I. $C < 1$
Regime II. $C > 1$
Regime III. $C \gg 1$
Regime IV. This is the so-called *coherence collapse* regime, where C is even larger.

Fig. 12 gives an example of the linewidth of a semiconductor laser versus the parameter C. A quantitative discussion of these regimes follows.[22]

Assume that the coupling efficiency from the laser into the fiber is η. Then, because feedback requires a double pass, the fraction of emitted light fed back into the laser is $f_{ext} = \eta^2 R_e$, where R_e is the reflectivity from the end of the fiber. The external reflection changes the overall reflectivity and therefore the required gain for threshold, depending on its phase ϕ_{ext}. Possible modes are defined by the threshold gain and the phase condition that requires an effective external round-trip phase for fed-back light $\delta\phi_L = m\pi$. But a change in the threshold gain also changes the refractive index and the phase through the linewidth enhancement factor β_c. The phase of the returning light is:

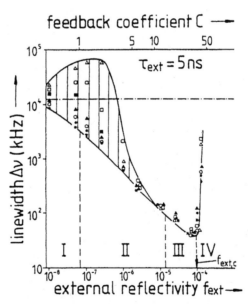

FIGURE 12 Linewidth versus feedback for a semiconductor laser and the corresponding feedback regimes.[22]

$$\delta\phi_L = \frac{\tau_L}{\tau_{ext}} [(\omega - \omega_{th})\tau_{ext} + C \sin(\omega\tau_{ext} + \tan^{-1}\beta_c)] \tag{46}$$

where ω_{th} is the frequency of the solitary laser at threshold.

Regime I.   For very weak feedback, $C < 1$ and there is only one solution when $\delta\phi_L$ is set equal to $m\pi$, so that the frequency of the mode of the solitary laser ω is at most slightly changed. The line will be narrowed or broadened as the external reflection adds to or subtracts from the output of the laser.

The linewidth is:

$$\Delta\omega = \frac{\Delta\omega_o}{[1 + C \cos(\omega\tau_{ext} + \tan^{-1}\beta_c)]^2} \tag{47}$$

with maximum and minimum values given by:

$$\Delta\omega_{min} = \frac{\Delta\omega_o}{(1 + C)^2} \tag{47a}$$

$$\Delta\omega_{max} = \frac{\Delta\omega_o}{(1 - C)^2} \tag{47b}$$

This is regime I.

The system performance moves toward regime II as $C \rightarrow 1$. Note that at $C = 1$ the maximum value predicts an infinite linewidth. This indicates that even very small feedback can cause wide spectral response, as long as $C \sim 1$.

Regime II.   For higher feedback with $C > 1$, several solutions with $\delta\phi_L = m\pi$ may exist. Linewidth broadening occurs because the single external cavity mode now has split into a dual mode, accompanied by considerable phase noise. Mode hopping gives linewidth broadening and intensity noise. This is a low-frequency noise with a cutoff frequency of about 10 MHz.

Regime III.   As the system performance moves toward regime III with increasing feedback, the mode splitting increases up to a frequency $\Delta\omega = 1/\tau_{ext}$ and the cutoff frequency for mode hopping noise decreases until only one of the split modes survives. To understand which mode survives, it is important to realize that in regime III, stable and unstable modes alternate with increasing phase. Because $\beta_c \neq 0$, the mode with the best phase stability (corresponding to the minimum linewidth mode) does not coincide with the mode with minimum threshold gain. However, in feedback regime III, the mode with the minimum linewidth rather than the mode with the minimum gain is the predominant lasing mode. This has been understood by analyzing the importance of phase stability to laser operation. This minimum linewidth mode remains relatively stable in regime III and is at the emission frequency of the laser without feedback. The laser operates in the lowest linewidth mode as long as the inverse of the linewidth of the solitary laser is larger than the external cavity round-trip time. In this mode of operation, the laser is stably phase locked to the feedback.

Regime IV.   The stable linewidth solution of regime III collapses as the fraction of power fed back f_{ext} increases to a critical value. There is considerable discussion of the physical mechanism that leads to this coherence collapse. The existence of this regime has been demonstrated by simulation, through numerical solution of the rate equations. Fitting to experimental results and theoretical analyses indicates that the onset of coherence collapse occurs when the feedback is larger than a critical value given by[23]:

$$C \geq C_{\text{crit}} = 2\gamma_R \tau_{\text{ext}} \frac{1 + \beta_c^2}{\beta_c^2} \tag{48}$$

where γ_R is the damping rate of the relaxation oscillations, as previously defined. As the feedback level approaches the critical value C, undamped relaxation oscillations appear, and oscillations of carrier density induce the phase of the field to oscillate through the linewidth enhancement factor β_c. To obtain an analytical result, it must be assumed that the external cavity round-trip time is larger than the time for relaxation oscillations to damp out.

As the feedback increases, the relaxation oscillation ceases to be damped, as a result of the interaction between the field amplitude in the semiconductor cavity and the carrier density, which shows up as a phase difference between the field in the semiconductor cavity and in the feedback field. The onset of coherence collapse is determined by the feedback parameter at which the relaxation oscillation ceases to be damped.

Regime V. This is a regime of stable operation that can only be achieved with an anti-reflection-coated laser output facet (such as a bare diode in an external cavity), and is not of concern here.

Cavity Length Dependence and RIN. In some regimes the regions of stability depend on the length of the external cavity, that is, the distance from the extra reflection to the laser diode. These regions have been mapped out for two different laser diodes, as shown in Fig. 13. The qualitative dependence on distance to reflection should be the same for all lasers.

The RIN is low for weak to moderate levels of feedback but increases tremendously in regime IV. The RIN and the linewidth are strongly related (see Fig. 12); the RIN is suppressed in regimes III and V.

Low-Frequency Fluctuations. When a laser operating near threshold is subject to a moderate amount of feedback, chaotic behavior evolves into *low-frequency fluctuations* (*LFF*). During LFF the average laser intensity shows sudden dropouts, from which it gradually recovers, only to drop out again after some variable time, typically on the order of tens of external cavity round-trips. This occurs in regimes of parameter space where at least one stable external cavity mode exists, typically at the transition between regimes IV and V. Explanations differ as to the cause of LFF, but it appears to originate in strong intensity pulses that occur during the buildup of average intensity, as a form of mode locking, being frustrated by the drive toward maximum gain. Typical frequencies for LFF are 20 to 100 MHz, although feedback from reflectors very close to the laser has caused LFF at frequencies as high as 1.6 GHz.

Conclusions. Semiconductor laser subject to optical feedback exhibits a rich and complex dynamic behavior that can enhance or degrade the laser's performance significantly. Feedback can occur through unwanted back reflections—for instance, from a fiber facet—and can lead to a severe degradation of the spectral and temporal characteristics, such as in the coherence collapse regime or in the LFF regime. In both regimes, the laser intensity fluctuates erratically and the optical spectrum is broadened, showing large sidebands. Because these unstable regimes can occur for even minute levels of feedback, optical isolators or some other means of prevention are usually used.

4.6 QUANTUM WELL AND STRAINED LASERS

Quantum Well Lasers

We have seen that the optimum design for low-threshold LDs uses the thinnest possible active region to confine free carriers, as long as the laser light is waveguided. When the active

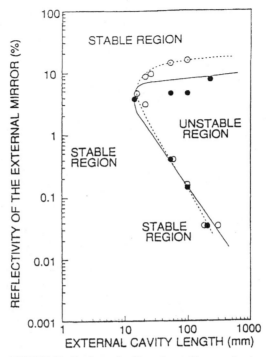

FIGURE 13 Regimes of stable and unstable operation for two laser diodes ($\bigcirc$ and $\bullet$) when subject to external feedback at varying distances and of varying amounts.[24]

layer has a thickness less than a few tens of nanometers (hundreds of angstroms), it becomes a *quantum well* (QW). That is, the layer is so thin that the confined carriers have energies that are quantized in the growth direction *z*, as described in Vol. 1, Chap. 13 of this handbook. This changes the density of states and the gain (and absorption) spectrum. While bulk semiconductors have an absorption spectrum near the band edge that increases with photon energy above the bandgap energy E_g as $(h\nu - E_g)^{1/2}$, quantum wells have an absorption spectrum that is steplike in photon energy at each of the allowed quantum states. Riding on this steplike absorption is a series of exciton resonances at the absorption steps that occur because of the Coulomb interaction between free electrons and holes, which can be seen in the spectra of Fig. 14. These abrupt absorption features result in much higher gain for quantum well lasers than for bulk semiconductor lasers. The multiple spectra in Fig. 14 record the reduction in absorption as the QW states are filled with carriers. When the absorption goes to zero, transparency is reached. Figure 14 also shows that narrower wells push the bandgap to higher energies, a result of quantum confinement. The QW thickness is another design parameter in optimizing lasers for telecommunications.

Because a single quantum well (SQW) is so thin, its optical confinement factor is small. It is necessary either to use multiple QWs (separated by heterostructure barriers that contain the electronic wave functions within individual wells) or to use a guided wave structure that focuses the light into a SQW. The latter is usually a GRIN structure, as shown in Fig. 2d. Band diagrams as a function of distance in the growth direction for typical quantum well separate confinement heterostructures are shown in Fig. 15. The challenge is to properly confine carriers and light using materials that can be reliably grown and processed by common crystal growth methods.

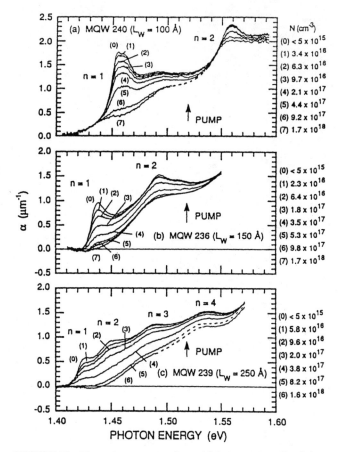

FIGURE 14 Absorption spectrum for multiple quantum wells of three different well sizes, for varying levels of optically induced carrier density, showing the decrease in absorption toward transparency. Note the stronger excitonic resonances and increased bandgap with smaller well size.[25]

Quantum wells have provided significant improvement over bulk active regions, as originally observed in GaAs lasers. In InP lasers, Auger recombination and other losses come into play at the high carrier densities that occur in quantum confined structures, which tends to degrade laser performance. However, it has been found that providing strain in the active region can improve the performance of QW InGaAsP lasers to a level comparable with GaAs lasers. Strained QW lasers are described in the next section.

The LD characteristics described in Secs. 5.2 to 5.5 hold for QW lasers as well as for bulk lasers. The primary difference is that the gain is larger and the optical confinement factor will be much smaller, because the light is not well confined in a single thin QW active region. The optical confinement factor in a typical QW of thickness d is dominated by the second term in the denominator of Eq. (2). When multiple quantum wells (MQWs) are used, d_g can be the thickness of the entire region containing the MQWs and their barriers, but Γ must now be multiplied by the filling factor Γ_f of the quantum wells within the MQW region—that is, if there are N_w wells, each of thickness d_w, then $\Gamma_f = N_w d_w / d_g$. When a GRINSCH structure is used, the optical confinement factor depends on the curvature of its refractive gradient near the center of the guide.

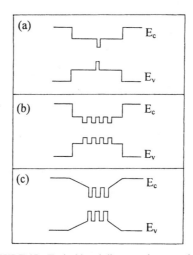

FIGURE 15 Typical band diagrams (energy of conduction band E_c and valence band E_v versus growth direction) for quantum wells in separate confinement laser heterostructures: (*a*) single quantum well; (*b*) multiple quantum wells; and (*c*) graded index separate confinement heterostructure (GRINSCH) and multiple quantum wells.

There are subtle differences in performance between different geometries, depending on how many QWs are used and the extent to which a GRINSCH structure is dominant. The lowest threshold current densities have been reported for the highest Q cavities (longest lengths or highest reflectivities) using single QWs. However, for lower Q cavities the lowest threshold current densities are achieved with MQWs, even though they require higher carrier densities to achieve threshold. This is presumably because Auger recombination depends on the cube of the carrier density, so that SQW lasers will have excess losses with their higher carrier densities. In general, MQWs are a better choice in long-wavelength lasers, while SQWs have the advantage in GaAs lasers. However, with MQW lasers it is important to realize that the transport of carriers moving from one well to the next during high-speed modulation must be taken into account. In addition, improvements through the use of strained layer QWs make single QW devices more attractive.

Strained Layer Quantum Well Lasers

Active layers containing *strained quantum wells* have proven to be an extremely valuable advance in high-performance long-wavelength InP lasers. They have lower thresholds, enhanced differential quantum efficiency η_D, larger characteristic temperature T_o, reduced linewidth enhancement factor β_c (less chirp), and enhanced high-speed characteristics (larger relaxation oscillation frequency Ω_R) compared to unstrained QW and bulk devices. This results from the effect of strain on the energy-versus-momentum band diagram. Bulk semiconductors have two valence bands that are degenerate at the potential well minimum, as shown in Fig. 16. They are called *heavy-hole* and *light-hole* bands, since the smaller curvature means a heavier effective mass. Quantum wells lift this degeneracy, and interaction between the two bands near momentum $k = 0$ causes a local distortion in the formerly parabolic bands, also shown in Fig. 16. As a result, the heavy hole effective mass becomes smaller, more nearly approaching that

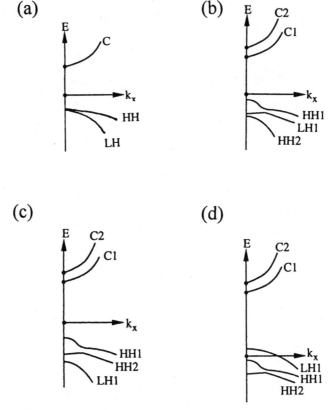

FIGURE 16   The effect of strain on the band diagram (energy E versus in-plane momentum k_x) of III-V semiconductors: (*a*) no strain, showing the degeneracy of the heavy holes HH and light holes LH at $k_x = 0$; (*b*) quantum wells, showing the separately quantized conduction bands (C_1 and C_2) and removal of the valence band degeneracy, with the lowest energy heavy holes HH_1 no longer having the same energy as the lowest energy light holes LH_1 at $k = 0$; (*c*) compressive strain, with enhanced separation between the light-hole and the lowest heavy-hole band; and (*d*) tensile strain, with light holes having the lowest energy.[26]

of the conduction band. This allows population inversion to become more efficient, increasing the differential gain; this is one factor in the reduced threshold of QW lasers.[26]

Strain additionally alters this structure in a way that can improve performance even more. Compressive strain in the QW moves the heavy-hole and light-hole valence bands further apart and further reduces the hole effective mass. Strain also decreases the heavy-hole effective mass by a factor of two or more, further increasing the differential gain and reducing the threshold carrier density. Higher differential gain also results in a smaller linewidth enhancement factor. Tensile strain moves the heavy-hole and light-hole valence bands closer together. In fact, at one particular tensile strain value these bands become degenerate at $k = 0$. Further tensile strain results in the light hole having the lowest energy at $k = 0$. These lasers will be polarized TM, because of the angular momentum properties of the light-hole band. This polarization has a larger optical matrix element, which can enhance the gain over some wavelength regions.

In addition to the heavy- and light-hole bands, there is an additional, higher-energy valence band (called the *split-off band*) which participates in Auger recombination and intervalence band absorption, both of which reduce quantum efficiency. In unstrained material there is a near-resonance between the bandgap energy and the difference in energy between the heavy-hole and split-off valence bands, which enhances these mechanisms for nonradiative recombination. Strain removes this near-degeneracy and reduces those losses that are caused by Auger recombination and intervalence band absorption. This means that incorporating strain is essential in long-wavelength laser diodes intended to be operated at high carrier densities. The reliability of strained layer QW lasers is excellent, when properly designed. However, strain does increase the intraband relaxation time, making the gain compression factor worse, so strained lasers tend to be more difficult to modulate at high speed.

Specific performance parameters are strongly dependent on the specific material, amount of strain, size and number of QWs, and device geometry, as well as the quality of crystal growth. Calculations show that compressive strain provides the lowest transparency current density, but tensile strain provides the largest gain (at sufficiently high carrier densities), as shown in Fig. 17. The lowest threshold lasers, then, will typically be compressively strained. Nonetheless, calculations show that, far enough above the band edge, the differential gain is 4 times higher in tensile compared to compressive strain. This results in a smaller linewidth enhancement factor, even if the refractive index changes per carrier density are larger. It has also been found that tensile strain in the active region reduces the Auger recombination, decreasing the losses introduced at higher temperatures. This means that T_o can increase with strain, particularly tensile strain. Performance at 1.55 μm comparable with that of GaAs lasers has been demonstrated using strained layer QWs. Deciding between compressively and tensilely strained QWs will be a matter of desired performance for specific applications.

Threshold current densities under 200 A/cm^2 have been reported at 1.55 μm; T_o values on the order of 140 K have been reported, 3 times better than bulk lasers. Strained QW lasers have improved modulation properties compared with bulk DH lasers. Because the gain coefficient can be almost double, the relaxation oscillation frequency is expected to be almost 50 percent higher, enhancing the modulation bandwidth and decreasing the relative intensity noise for the same output power. Even the frequency chirp under modulation will be less, because the linewidth enhancement factor is less. The typical laser geometry, operating characteristics, transient response, noise, frequency chirping, and the effects of external optical feedback are all similar in the strained QW lasers to what has been described previously for bulk lasers. Only the experimentally derived numerical parameters will be somewhat different; strained long-wavelength semiconductor lasers have performance parameters comparable to those of GaAs lasers. One difference is that the polarization of the light emitted from

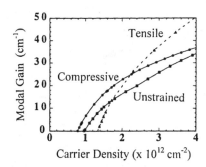

FIGURE 17   Modal gain at 1.55 μm in InGaAs QW lasers calculated as a function of the carrier density per unit area contained in the quantum well. Well widths were determined by specifying wavelength.[27]

strained lasers may differ from that emitted from bulk lasers. As explained in Sec. 3.3, the gain in bulk semiconductors is independent of polarization, but lasers tend to be polarized in-plane because of higher facet reflectivity for that polarization. The use of quantum wells causes the gain for the TE polarization to be slightly (~10 percent) higher than for the TM polarization, so lattice-matched QW lasers operate with in-plane polarization. Compressive strain causes the TE polarization to have significantly more gain than the TM polarization (typically 50 to 100 percent more), so these lasers are also polarized in-plane. However, tensile strain severely depresses the TE gain, and these lasers have the potential to operate in TM polarization.

Typical 1.3- and 1.5-μm InP lasers today use from 5 to 15 wells that are grown with internal strain. By providing strain-compensating compressive barriers, there is no net buildup of strain. Typical threshold current densities today are ~1000 A/cm^2, threshold currents ~10 mA, T_o ~ 50 to 70 K, maximum powers ~40 mW, differential efficiencies ~0.3 W/A, and maximum operating temperatures ~70°C before the maximum power drops by 50 percent. There are trade-offs on all these parameters; some can be made better at the expense of some of the others.

4.7 DISTRIBUTED FEEDBACK (DFB) AND DISTRIBUTED BRAGG REFLECTOR (DBR) LASERS

Rather than cleaved facets for feedback, some lasers use distributed reflection from corrugated waveguide surfaces. Each groove provides some slight reflectivity, which adds up coherently along the waveguide at the wavelength given by the corrugation. This has two advantages. First, it defines the wavelength (by choice of grating spacing) and can be used to fabricate single-mode lasers. Second, it is an in-plane technology (no cleaves) and is therefore compatible with monolithic integration with modulators and/or other devices.

Distributed Bragg Reflector (DBR) Lasers

The *distributed Bragg reflector* (DBR) laser replaces one or both laser facet reflectors with a waveguide diffraction grating located outside the active region, as shown in Fig. 18. The

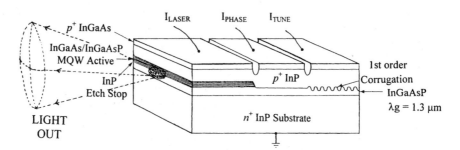

FIGURE 18 Schematic for DBR laser configuration in a geometry that includes a phase portion for phase tuning and a tunable DBR grating. Fixed-wavelength DBR lasers do not require this tuning region. Designed for 1.55-μm output, light is waveguided in the transparent layer below the MQW that has a bandgap at a wavelength of 1.3 μm. The guided wave reflects from the rear grating, sees gain in the MQW active region, and is partially emitted and partially reflected from the cleaved front facet. Fully planar integration is possible if the front cleave is replaced by another DBR grating.[28]

reflectivity of a Bragg mirror is the square of the reflection coefficient (given here for the assumption of lossless mirrors)[29]:

$$r = \frac{\kappa}{\delta - iS \coth (SL)} \tag{49}$$

where κ is the *coupling coefficient* due to the corrugation (which is real for corrugations that modify the effective refractive index in the waveguide, but would be imaginary for periodic modulations in the gain and could, indeed, be complex). Also, δ is a *detuning parameter* that measures the offset of the optical wavelength λ from that defined by the grating periodicity Λ. When the grating is used in the mth order,

$$\delta = \frac{2\pi n_g}{\lambda} - \frac{m\pi}{\Lambda} \tag{50}$$

where n_g is the effective group refractive index of the waveguide mode, and m is any integer. Also, S is given by:

$$S^2 = \kappa^2 - \delta^2 \tag{51}$$

The Bragg mirror has its maximum reflectivity on resonance when $\delta \to 0$ and the wavelength λ_m is determined by the mth order of the grating spacing Λ:

$$\Lambda = \frac{m\lambda_m}{2n_g} \tag{52}$$

The reflection coefficient on resonance is $r_{max} = -i \tanh (KL)$ and the Bragg reflectivity is:

$$R_{max} = \tanh^2 (KL) \tag{53}$$

where K is the coupling per unit length, $K = |\kappa|$, and is larger for deeper corrugations or when the refractive index difference between the waveguide and the cladding is larger. The reflectivity falls off as the wavelength moves away from resonance and the detuning increases.

When off resonance far enough that $|\delta| > |\kappa|$, it is more practical to define:

$$\sigma^2 = \delta^2 - \kappa^2 \tag{54}$$

and the reflectivity has the form:

$$R = \frac{(KL)^2}{(\delta L)^2 + (\sigma L)^2 \cot^2 (\sigma L)} \tag{55}$$

Note that when $\sigma \to 0$, $\delta = K$ and $R \to (KL)^2/[1 + (KL)^2]$. For moderate values of the grating coupling KL, this value of the reflectivity is not very different from that given by Eq. (53). Thus, $\sigma^2 > 0$ over most of the detuning range.

The half-width of the resonance can be found by noting that the reflectivity goes to zero when $\sigma L = \pi$, where the cotangent goes to infinity. This occurs at a cutoff detuning δ_c given by $\delta_c^2 L^2 = \pi^2 + K^2 L^2$. This fact allows us to define a reflection resonance half-width as $\delta_c/2$ and the full width as δ_o. The width of the resonance is constant ($\delta_c = \pi/L$) when $KL \ll \pi$, but broadens for large KL. Typical numbers are $2 < KL < 5$, so it is reasonable to take $\delta_c \approx \pi/L$. The detuning is related to the wavelength bandwidth of the mirror by differentiating Eq. (50): $\Delta \delta = 2\pi n_g (\Delta \lambda/\lambda_o^2)$. Then the wavelength bandwidth for $\delta_c = \pi/L$ is $\Delta \lambda = \lambda^2/(2Ln_g)$ and the width of the resonance is 0.5 nm (when $L = 500$ μm, $\lambda = 1.3$ μm, and $KL \ll \pi$). This narrow resonance, fixable by choosing the grating spacing and variable by varying the refractive index (with, for example, carrier injection) makes the DBR laser very favorable for use in optical communication systems.

The characteristics of Fabry-Perot lasers previously described still hold for DBR lasers, except that the narrow resonance can ensure that these lasers are single mode, even at high excitation levels.

Distributed Feedback (DFB) Lasers

When the corrugation is put directly on the active region or its cladding, this is called *distributed feedback* (*DFB*). One typical example is shown in Fig. 19. As before, the grating spacing is chosen such that, for a desired wavelength near λ_o, $\Lambda = m\lambda_o/2n_g$, where now n_g is the effective group refractive index of the laser mode inside its waveguiding active region, and m is any integer. A laser operating under the action of this grating has feedback that is distributed throughout the laser gain medium. In this case, Eq. (49) is generalized to allow for the gain: $\delta = \delta_o + ig_L$, where g_L is the laser gain and $\delta_o = 2\pi n_g/\lambda - 2\pi n_g/\lambda_o$. Equations (49) to (54) remain valid, understanding that now δ is complex.

The laser oscillation condition requires that after a round-trip inside the laser cavity, a wave must have the same phase that it started out with, so that successive reflections add in phase. Thus, the phase of the product of the complex reflection coefficients (which now include gain) must be an integral number of 2π. This forces r^2 to be a positive real number. So, laser oscillation requires that:

$$r^2 > 0 \tag{56}$$

On resonance $\delta_o = 0$ and $S_o^2 = \kappa^2 + g_L^2$, so that S_o is pure real for simple corrugations (κ real). Since the denominator in Eq. (49) is now pure imaginary, r^2 is negative and the round-trip condition of Eq. (56) cannot be met. Thus, there is no on-resonance solution to a simple DFB laser with a corrugated waveguide and/or a periodic refractive index.

DFB Threshold. We look for an off-resonance solution to the DFB laser with a corrugated waveguide in the active region (κ real). A laser requires sufficient gain that the reflection coefficient becomes infinite. That is,

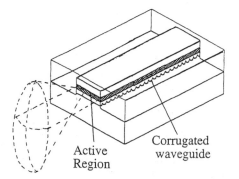

FIGURE 19 Geometry for a DFB laser, showing a buried grating waveguide that forms the separate confinement heterostructure laser, which was grown on top of a grating-etched substrate. The cross-hatched region contains the MQW active layer. A stripe mesa is etched and regrown to provide a buried heterostructure laser. Reflection from the cleaved facets must be suppressed by means of an antireflection coating.

$$\delta_{th} = iS_{th} \coth (S_{th}L) \tag{57}$$

where

$$S_{th}^2 = \kappa^2 - \delta_{th}^2 \tag{58}$$

By simple algebraic manipulation, Eq. (57) can be written as:

$$\exp (2S_{th}) \frac{S_{th} + i\delta_{th}}{S_{th} - i\delta_{th}} = -1 \tag{59}$$

Multiplying and dividing by $S_{th} + i\delta_{th}$ gives:

$$\exp (2S_{th}) \frac{(S_{th} + i\delta_{th})^2}{S_{th}^2 + \delta_{th}^2} = -1 \tag{60}$$

The denominator is κ^2, which, for pure corrugations, is K^2. For large gain, $\delta_{th}^2 \gg K^2$, so that Eq. (58) gives $S_{th} = i\delta_{th} = i\delta_o - g_L$. Inserting this in the numerator, Eq. (60) becomes[30]:

$$\exp (2S_{th}) \frac{4(g_L - i\delta_o)^2}{K^2} = -1. \tag{61}$$

This is a complex eigenvalue equation that has both a real and an imaginary part, which give both the detuning δ_o and the required gain g_L. Equating the phases gives:

$$2 \tan^{-1} \left(\frac{\delta_o}{g_L} \right) - 2\delta_o L + \delta_o L \frac{K^2}{g_L^2 + \delta_o^2} = (2m + 1)\pi \tag{62}$$

There is a series of solutions, depending on the value of m.
    For the largest possible gains,

$$\delta_o L = -(m + \tfrac{1}{2}) \pi \tag{63}$$

There are two solutions, $m = -1$ and $m = 0$, giving $\delta_o L = -\pi/2$ and $\delta_o L = +\pi/2$. These are two modes equally spaced around the Bragg resonance. Converting to wavelength units, the mode detuning becomes $\delta_o L = -2\pi n_g L(\delta\lambda/\lambda^2)$, where $\delta\lambda$ is the deviation from the Bragg wavelength. Considering $\delta_o L = \pi/2$, for $L = 500$ μm, $n_g = 3.5$, and $\lambda = 1.55$ μm, this corresponds to $\delta\lambda = 0.34$ nm. The mode spacing is twice this, or 0.7 nm.
    The required laser gain is found from the magnitude of Eq. (61) through

$$\frac{K^2}{4} = (g_L^2 L^2 + \delta_o^2 L^2) \exp (-2g_L L) \tag{64}$$

For detuning $\delta_o L = -\pi/2$, the gain can be found by plotting Eq. (64) as a function of gain g_L, which gives $K(g_L)$, which can be inverted to give $g_L(K)$.
    These results show that there is a symmetry around $\delta_o = 0$, so that there will tend to be *two* modes, equally spaced around λ_o. Such a multimode laser is not useful for communication systems, so something must be done about this. The first reality is that there are usually cleaved facets, at least at the output end of the DFB laser. This changes the analysis from that given here, requiring additional Fresnel reflection to be added to the analysis. The additional reflection will usually favor one mode over the other, and the DFB will end up as a single mode. However, there is very little control over the exact positioning of these additional cleaved facets with respect to the grating, and this has not proven to be a reliable way to achieve single-mode operation. The most common solution to this multimode problem is to use a *quarter-wavelength-shifted grating*, as shown in Fig. 20. Midway along the grating, the phase changes by $\pi/2$ and the two-mode degeneracy is lifted. This is the way that DFB lasers are made today.

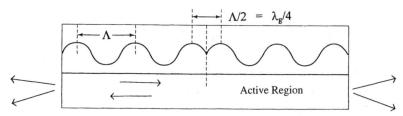

FIGURE 20 Side view of a quarter-wavelength-shifted grating, etched into a separate confinement waveguide above the active laser region. Light with wavelength in the medium λ_g sees a $\pi/4$ phase shift, resulting in a single-mode DFB laser operating on line-center.

Quarter-Wavelength-Shifted Grating. Introducing an additional phase shift of π to the round-trip optical wave enables an on-resonance DFB laser. Thus, light traveling in each direction must pass through an additional phase shift of $\pi/2$. This is done by interjecting an additional phase region of length $\Lambda/2$, or $\lambda/4n_g$, as shown in Fig. 20. This provides an additional $\pi/2$ phase in Eq. (63), so that the high-gain oscillation condition becomes:

$$\delta_o L = -m\pi \qquad (65)$$

Now there is a unique solution at $m = 0$, given by Eq. (64) with $\delta_o = 0$:

$$KL = g_L L \exp(-g_L L) \qquad (66)$$

Given a value for KL, the gain can be calculated. Alternatively, the gain can be varied, and the coupling coefficient used with that gain can be calculated. It can be seen that if there are internal losses α_i, the laser must have sufficient gain to overcome them as well: $g_L + \alpha_i$.

Quarter-wavelength-shifted DFB lasers are commonly used in telecommunications applications. There are a variety of ways in which the DFB corrugations are placed with respect to the active layer. Most common is to place the corrugations laterally on either side of the active region, where the evanescent wave of the guided mode experiences sufficient distributed feedback for threshold to be achieved. Alternative methods place the corrugations on a thin cladding above the active layer. Because the process of corrugation may introduce defects, it is traditional to avoid corrugating the active layer directly. Once a DFB laser has been properly designed, it will be single mode at essentially all power levels and under all modulation conditions. Then the single-mode laser characteristics described in the early part of this chapter will be well satisfied. However, it is crucial to avoid reflections from fibers back into the laser, because instabilities may arise, and the output may cease to be single mode.

A different technique that is sometimes used is to spatially modulate the gain. This renders κ complex and enables an on-resonance solution for the DFB laser, since S will then be complex on resonance. Corrugation directly on the active region makes this possible, but care must be taken to avoid introducing centers for nonradiative recombination.

There have been more than 35 years of research and development in semiconductor lasers for telecommunications. Today it appears that the optimal sources for telecommunications applications are strained quantum well distributed feedback lasers at 1.3 or 1.55 μm.

4.8 LIGHT-EMITTING DIODES (LEDS)

Sources for low-cost fiber communication systems, such as are used for communicating data, are typically light-emitting diodes (LEDs). These may be *edge-emitting LEDs* (*E-LEDs*), which resemble laser diodes, or, more commonly, *surface-emitting LEDs* (*S-LEDs*), which emit light from the surface of the diode and can be butt-coupled to multimode fibers.

When a PN junction is forward biased, electrons are injected from the N region and holes are injected from the P region into the active region. When free electrons and free holes co-exist with comparable momentum, they will combine and may emit photons of energy near that of the bandgap, resulting in an LED. The process is called *injection* (or *electro-*) *luminescence,* since injected carriers recombine and emit light by spontaneous emission. A semiconductor laser diode below threshold acts as an LED. Indeed, a semiconductor laser without mirrors is an LED. Because LEDs have no threshold, they usually are not as critical to operate and are usually much less expensive. Also, they do not need the optical feedback of lasers (in the form of cleaved facets or distributed feedback). Because the LED operates by spontaneous emission, it is an incoherent light source, typically emitted from a larger aperture (out the top surface) with a wider far-field angle and a much wider wavelength range (30 to 50 nm). In addition, LEDs are slower to modulate than laser diodes. Nonetheless, they can be excellent sources for inexpensive multimode fiber communication systems. Also, LEDs have the advantages of simpler fabrication procedures, lower cost, and simpler drive circuitry. They are longer lived, exhibit more linear input-output characteristics, are less temperature sensitive, and are essentially noise-free electrical-to-optical converters. The disadvantages are lower power output, smaller modulation bandwidths, and distortion in fiber systems because of the wide wavelength band emitted. Some general characteristics of LEDs are discussed in Vol. 1, Chap. 12 of this handbook (pp. 12.36–12.37).

In fiber communication systems, LEDs are used for low-cost, high-reliability sources typically operating with graded index multimode fibers (core diameters approximately 62 μm) at data rates up to 622 Mb/s. The emission wavelength will be at the bandgap of the active region in the LED; different alloys and materials have different bandgaps. For medium-range distances up to ~10 km (limited by modal dispersion), LEDs of InGaAsP grown on InP and operating at $\lambda = 1.3$ μm offer low-cost, high-reliability transmitters. For short-distance systems, up to 2 km, GaAs-based LEDs operating near 850 nm wavelength are used, because they have the lowest cost, both to fabricate and to operate, and the least temperature dependence. The link length is limited to ~2 km because of chromatic dispersion in the fiber and the finite linewidth of the LED. For lower data rates (a few megabits per second) and short distances (a few tens of meters), very inexpensive systems consisting of red-emitting LEDs with GaAlAs or GaInP active regions emitting at 650 nm can be used with plastic fibers and standard silicon detectors. The 650-nm wavelength is a window in the absorption in acrylic plastic fiber, where the loss is ~0.3 dB/m.

A typical GaAs LED heterostructure is shown in Fig. 21. The forward-biased *pn* junction injects electrons and holes into the GaAs active region. The AlGaAs cladding layers confine the carriers in the active region. High-speed operation requires high levels of injection (and/or doping) so that the recombination rate of electrons and holes is very high. This means that the active region should be very thin. However, nonradiative recombination increases at high carrier concentrations, so there is a trade-off between internal quantum efficiency and speed. Under some conditions, LED performance is improved by using quantum wells or strained layers. The improvement is not as marked as with lasers, however.

Spontaneous emission causes light to be emitted in all directions inside the active layer, with an internal quantum efficiency that may approach 100 percent in these direct band semiconductors. However, only the light that gets out of the LED and into the fiber is useful in a communication system, as illustrated in Fig. 21a. The challenge, then, is to collect as much light as possible into the fiber end. The simplest approach is to butt-couple a multimode fiber to the LED surface as shown in Fig. 21a (although more light is collected by lensing the fiber tip or attaching a high-index lens directly on the LED surface). The alternative is to cleave the LED, as in a laser (Fig. 1), and collect the waveguided light that is emitted out the edge. Thus, there are two generic geometries for LEDs: *surface-emitting* and *edge-emitting.* The edge-emitting geometry is similar to that of a laser, while the surface-emitting geometry allows light to come out the top (or bottom). Its inexpensive fabrication and integration process makes

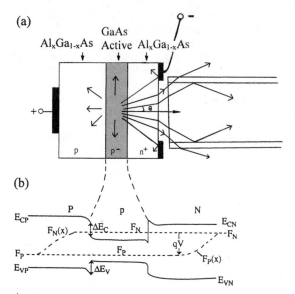

FIGURE 21 Cross-section of a typical GaAs light-emitting diode (LED) structure: (*a*) surface-emitting LED aligned to a multimode fiber, indicating the small fraction of spontaneous emission that can be captured by the fiber; (*b*) energy of the conduction band E_c and valence band E_v as a function of depth through the LED under forward bias V, as well as the Fermi energies that indicate the potential drop that the average electron sees.

the surface-emitting LED the most common type for inexpensive data communication; it will be discussed first. The edge-emitting LEDs have a niche in their ability to couple with reasonable efficiency into single-mode fibers. Both LED types can be modulated at bit rates up to 622 Mb/s, an ATM standard, but many commercial LEDs have considerably smaller bandwidths.

Surface-Emitting LEDs

The geometry of a surface-emitting LED butt-coupled to a multimode graded index fiber is shown Fig. 21a. The coupling efficiency is typically small, unless methods are employed to optimize it. Because light is spontaneously emitted in all internal directions, only half of it is emitted toward the top surface, so that often a mirror is provided to reflect back the downward-traveling light. In addition, light emitted at too great an angle to the surface normal is totally internally reflected back down and is lost. The critical angle for total internal reflection between the semiconductor of refractive index n_s and the output medium (air or plastic encapsulant) of refractive index n_o is given by $\sin \theta_c = n_o/n_s$. Because the refractive index of GaAs is $n_s \sim 3.3$, the internal critical angle with air is $\theta_c \sim 18°$. Even with encapsulation, the angle is only $27°$. A butt-coupled fiber can accept only spontaneous emission at those external angles that are smaller than its numerical aperture. For a typical fiber $NA \approx 0.25$, this corresponds to an external angle (in air) of $14°$, which corresponds to $4.4°$ inside the GaAs. This means that the cone of spontaneous emission that can be accepted by the fiber is only ~0.2 percent of the entire spontaneous emission. Fresnel reflection losses makes this number even smaller. Even including all angles, less than 2 percent of the total internal spontaneous emission will come out the top surface of a planar LED.

The LED source is incoherent, a Lambertian emitter, and follows the law of imaging optics: a lens can be used to reduce the angle of divergence of LED light, but will enlarge the apparent source. The use of a collimating lens means that the LED source diameter must be proportionally smaller than the fiber into which it is to be coupled. Unlike a laser, the LED has no modal interference, and the output of a well-designed LED has a smooth Lambertian intensity distribution that lends itself to imaging.

The coupling efficiency can be increased in a variety of ways, as shown in Fig. 22. The LED can be encapsulated in materials such as plastic or epoxy, with direct attachment to a focusing lens (Fig. 22a). Then the output cone angle will depend on the design of this encapsulating lens; the finite size of the emitting aperture and resulting aberrations will be the limiting consideration. In general, the user must know both the area of the emitting aperture and the angular divergence in order to optimize coupling efficiency into a fiber. Typical commercially available LEDs at 850 nm for fiber-optic applications have external half-angles of ~25° without a lens and ~10° with a lens, suitable for butt-coupling to multimode fiber.

Additional improvement can be achieved by lensing the pigtailed fiber to increase its acceptance angle (Fig. 22b). An alternative is to place a microlens between the LED and the fiber (Fig. 22c). Perhaps the most effective geometry for capturing light is the integrated domed surface fabricated directly on the back side of an InP LED, as shown in Fig. 22d. Because the refractive index of encapsulating plastic is <1.5, compared to 3.3 of the semi-conductor, only a semiconductor dome can entirely eliminate total internal reflection. Integrated semiconductor domes require advanced semiconductor fabrication technology, but have been proven effective. In GaAs diodes the substrate is absorptive, but etching a well and inserting a fiber can serve to collect backside emission. For any of these geometries, improvement in efficiency of as much as a factor of two can be obtained if a mirror is provided to reflect backward-emitted light forward. This mirror can be either metal or a dielectric stack at the air-semiconductor interface, or it can be a DBR mirror grown within the semiconductor structure.

Current must be confined to the surface area of emission, which is typically 25 to 75 μm in diameter. This is done by constricting the flow of injection current by mesa etching or by using an oxide-defined (reflective) electrode. Regrowth using *npn* blocking layers or semi-insulating material in the surrounding areas (as in lasers) has the advantage of reducing thermal heating. Surface-emitting LEDs require that light be emitted out of the surface in a gaussian-like pattern; it must not be obscured by the contacting electrode. Typically, a highly conductive cap layer brings the current in from a ring electrode; alternatively, when light is collected out of the substrate side rather than the top side, electrical contact may be made to the substrate.

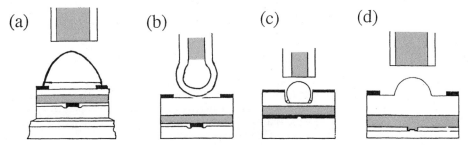

FIGURE 22   Typical geometries for coupling from LEDs into fibers: (*a*) hemispherical lens attached with encapsulating plastic; (*b*) lensed fiber tip; (*c*) microlens aligned through use of an etched well; and (*d*) spherical semiconductor surface formed on the substrate side of the LED.

Typical operating specifications for a surface-emitting LED at 1.3 μm pigtailed to a 62-μm core graded index fiber might be 15 μW at 100 mA input current, for ~0.02 percent efficiency,[31] with a modulation capability of 622 Mb/s. A factor of 2.5 times improvement in power can be achieved with a comparable reduction in speed. The LEDs are typically placed in lensed TO-18 cans, and a lens micromachined on the back of the InP die is used to achieve this output coupling efficiency. At 1.55 μm, the specifications are for 7 times less power and 3 times less speed.

Recently, improved S-LED performance has been obtained by using resonant cavities to reduce the linewidth and increase the bandwidth that can be transmitted through fibers. These devices have integral mirrors grown above and below the active region that serve to resonate the spontaneous emission. As such, they look very much like VCSELs below threshold (Sec. 4.9).

Edge-Emitting LEDs

Edge-emitting LEDs (E-LEDs or EELEDs) have a geometry that is similar to that of a conventional laser diode (Fig. 1), but without a feedback cavity. That is, light travels back and forth in the plane of the active region of an E-LED and it is emitted out one anti-reflection coated cleaved end. As in a laser, the active layer is 0.1 to 0.2 μm thick. Because the light in an E-LED is waveguided in the out-of-plane dimension and is lambertian in-plane, the output radiation pattern will be elliptical, with the largest divergence in-plane with a full width at half-maximum (FWHM) angle of 120°. The out-of-plane guided direction typically radiates with a 30° half-angle. An elliptical collimating lens will be needed to optimally couple light into a fiber. The efficiency can be doubled by providing a reflector on the back facet of the E-LED, just as in the case of a laser.

Edge-emitting LEDs can be coupled into fibers with greater efficiency because their source area is smaller than that of S-LEDs. However, the alignment and packaging is more cumbersome than with S-LEDs. Typically, E-LEDs can be obtained already pigtailed to fibers. Edge-emitting diodes can be coupled into *single-mode* fiber with modest efficiency. A single-mode fiber pigtailed to an E-LED can typically transmit 30 μW at 150 mA drive at 1 V, for an overall efficiency of 0.04 percent. This efficiency is comparable to the emission of surface-emitting lasers into *multimode* fiber with 50 times the area. Because of their wide emission wavelength bandwidth, E-LEDs are typically used as low-coherence sources for fiber sensor applications, rather than in communications applications.

Operating Characteristics of LEDs

In an LED, the output optical power P_{opt} is linearly proportional to the drive current; the relation defines the output efficiency η:

$$P_{out} = \frac{\eta h \nu I}{e} \tag{67}$$

This efficiency is strongly affected by the geometry of the LED. The power coupled into a fiber is further reduced by the coupling efficiency between the LED emitter and the fiber, which depends on the location, size, and numerical aperture of the fiber as well as on the spatial distribution of the LED output light and the optics of any intervening lens. The internal quantum efficiency (ratio of emitted photons to incident electrons) is usually close to 100 percent.

Figure 23 shows a typical result for power coupled into a graded index multimode fiber as a function of current for various temperatures. The nonlinearity in the light out versus current, which is much less than in a laser diode, nevertheless causes some nonlinearity in the modulation of LEDs. This LED nonlinearity arises both from material properties and device configuration; it may be made worse by ohmic heating at high drive currents. The residual nonlinearity is an important characteristic of any LED used in communication systems. Edge emitters are typically less linear because they operate nearer the amplified spontaneous limit.

There is ~10 percent reduction in output power for a 25°C increase in temperature (compared to ~50 percent reduction for a typical laser). Unlike a laser, there is no temperature-dependent threshold. Also, the geometric factors that determine the fraction of light emitted from the LED are not temperature dependent. Nonetheless, the InP-based LEDs have a stronger temperature dependence than GaAs-based LEDs, because of the larger presence of nonradiative recombination, particularly at the high injection levels required by high-speed LEDs.

The spectrum of the incoherent light emitted from an LED is roughly gaussian with a FWHM around 40 nm in the case of a typical GaAs/AlGaAs LED operating around 0.8 μm. This bandwidth, along with chromatic dispersion in graded index fibers, limits the distance over which these LEDs can be used in fiber systems. InGaAsP/InP LEDs have wider linewidths (due to alloy scattering, heavy doping, and temperature fluctuations), which depend on the details of their design. As temperature increases, the peak of the spectrum shifts to longer wavelength and the spectrum widens. The variation of the central wavelength with temperature is ~5 meV/°C. However, at 1.3 μm, graded index fibers have negligible chromatic dispersion, so this usually is not a problem; if it is, heat sinking and/or cooling can be provided. Resonant cavity LEDs can provide narrower linewidths, but are more difficult to fabricate.

LEDs do not suffer from the catastrophic optical damage that lasers do, because of their lower optical power densities. However, they do degrade with time. Lifetimes of 10^6 to 10^9 hours can be expected. Because degradation processes have an exponential dependence on temperature, LED life can be shortened by operating at excessive temperatures. Using concepts of thermally accelerated life testing, the power out P varies with time t as:

$$P(t) = P(0) \exp(-qt) \tag{68}$$

where $q = q_o \exp(-W_a/k_B T)$, with W_a as the activation energy, k_B as Boltzman's constant, and T as temperature. In GaAs LEDs, W_a is 0.6 to 1 eV. Of course, this assumes that the LEDs are placed in a proper electrical circuit.

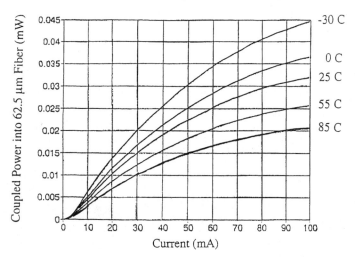

FIGURE 23 Optical power coupled from an *InGaAsP S–* LED into graded index fiber at 1.3 μm wavelength as a function of drive current, for several temperatures.[32]

LED light is typically unpolarized, since there is no preferred polarization for spontaneous emission.

Transient Response

Most LEDs respond in times faster than 1 μs; with optimization, they can reach the nanosecond response times needed for optical communication systems. To achieve the 125 Mb/s rate of the fiber distributed data interface (FDDI) standard requires a maximum rise time and fall time of 3.5 ns; to achieve the 622 Mb/s rate of the asynchronous transfer mode (ATM) standard, the necessary times drop to 0.7 ns.

The speed of an LED is limited by the lifetime of injected carriers; it does not have the turn-on delay of lasers, nor any relaxation oscillations, but it also does not have the fast decay of stimulated emission. The LED intrinsic frequency response (defined as the ratio of the AC components of the emitted light to the current) is[33]:

$$r(\omega) = (1 + \omega^2 \tau^2)^{-1/2} \tag{69}$$

where τ is the minority carrier lifetime in the injected region. It can be seen that high-speed LEDs require small minority carrier lifetimes. The square-root dependence comes out of solving the rate equations.

When the active region is doped more highly than the density of injected carriers (the low-injection regime), the lifetime τ_L is determined by the background doping density N_o:

$$\frac{1}{\tau_L} = BN_o \tag{70}$$

The lifetime decreases as the doping increases. The challenge is to provide high levels of doping without increasing the fraction of nonradiative recombination. The fastest speeds that are usually obtained are ~1 ns, although doping with beryllium (or carbon) at levels as high as 7×10^{19} cm^{-3} has allowed speeds to increase to as much as 0.1 ns, resulting in a cutoff frequency of 1.7 GHz (at the sacrifice of some efficiency).[34]

When operating in the high-injection regime, the injected carrier density N can be much larger than the doping density, and $1/\tau_H = BN$. But N is created by a current density J such that $N = J\tau/ed$. Combining these two equations:

$$\frac{1}{\tau_H} = \left(\frac{BJ}{ed}\right)^{1/2} \tag{71}$$

The recombination time may be reduced by thinning the active region and by increasing the drive current. However, too much injection may lead to thermal problems, which in turn may cause modulation nonlinearity. LEDs with thin active layers operated in the high-injection regime will have the fastest response. Bandwidths in excess of 1 GHz have been achieved in practical LEDs.

Because LEDs have such wide wavelength spectra, frequency chirping is negligible. That is, LEDs cannot be modulated fast enough for their wavelengths to be affected by the modulation. Because LEDs do not have optical cavities, as do lasers, they will not have modal interference and noise. Also, there will not be strong feedback effects coming from external fiber facets, such as the coherence collapse. Because of their inherent light-current linearity, the modulation response of LEDs should be a direct measure of their frequency response. They add no noise to the circuit, and they add distortion only at the highest drive levels.

Drive Circuitry and Packaging

The LED is operated under sufficient forward bias to flatten the bands of the *pn* junction. This voltage depends on the bandgap and doping and is typically between 1 and 2 V. The current will be converted directly to light; typically, ~100 mA is required to produce a few milliwatts of output, with a series resistor used to limit the current.

The LED is modulated by varying the drive current. A typical circuit might apply the signal to the base circuit of a transistor connected in series with the LED and a current-limiting resistor. The variation in current flowing through the LED (and therefore in the light out) is proportional to the input voltage in the base circuit. LEDs are typically mounted on standard headers such as TO-18 or TO-46 cans; SMA and ST connectors are also used. The header is covered by a metal cap with a clear glass top through which light can pass.

4.9 VERTICAL CAVITY SURFACE-EMITTING LASERS (VCSELS)

The *vertical cavity surface-emitting laser* (VCSEL) has advantages for low-cost data transmission. The use of a laser means that multigigahertz modulation is possible, and the stimulated emission is directional, rather than the isotropic spontaneous emission of LEDs. Because the light is emitted directly from the surface, single or multimode fiber can be directly butt-coupled with an inexpensive mounting technology, and the coupling efficiency can be very high. The VCSELs can also be fabricated in linear arrays that can be coupled inexpensively to linear arrays of fibers for parallel fiber interconnects with aggregate bit rates of several gigabits per second, amortizing the alignment cost over the number of elements in the array. VCSELs lend themselves to two-dimensional arrays as well, which makes them attractive to use with smart pixels. The planar fabrication of VCSELs allows for wafer-scale testing, another cost savings.

The VCSEL requires mirrors on the top and bottom of the active layer, forming a vertical cavity, as shown in Fig. 24. These lasers utilize the fact that a DBR (multilayer quarter-wavelength dielectric stack) can make a very high reflectance mirror. Thus, the very short path length through a few quantum wells (at normal incidence to the plane) is sufficient to reach threshold.

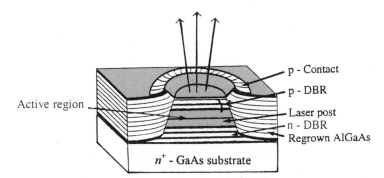

FIGURE 24 One example of a vertical cavity surface emitting laser (VCSEL) geometry. This is a passive antiguide region (PAR) VCSEL.[35] Light is reflected up and down through the active region by the two DBR mirrors. After the laser post is etched, regrowth in the region outside the mesa provides a high-refractive-index AlGaAs*nipi* region to stop current flow and to provide excess loss to higher-order modes.

In the 1990s, the only commercial VCSELs were based on GaAs: either GaAs active regions that emit at 850 nm, or strained InGaAs active regions that emit at 980 nm. The former are of greater interest in low-cost communication systems because they are compatible with inexpensive silicon detectors. This section describes the design of VCSELs and some of their key characteristics.

Number of Quantum Wells

A single quantum well of GaAs requires ~100 A/cm^2 to achieve transparency; N wells require N times this current. To keep the threshold current less than 1 kA/cm^2, then, means less than 10 QWs. The VCSEL provides an optical standing wave which, in GaAs, has a period of ~120 nm. The gain region should be confined to the quarter-wavelength region at the peak of the optical standing wave, a region of about 60 nm. Thus, a typical active region might consist of 3 QWs of 10 nm thickness, each separated by ~10 nm. The lowest threshold VCSELs are single quantum wells of InGaAs grown on GaAs, sacrificing power for threshold.

Mirror Reflectivity

When the mirror reflectivity R in a laser is very high, such that $R = 1 - \varepsilon$, a simple expression for the threshold gain-length product $G_L L$ is

$$G_L L = \varepsilon_1 \varepsilon_2 \qquad (72)$$

Typical GaAs lasers have gains $G_L \sim 1000$ cm^{-1}. For a quantum well thickness of 10 nm, the gain per quantum well is 10^{-3} and reflectivities of ~98 percent for each mirror should be sufficient to achieve threshold for 3 QW. Very often, however, in order to lower the threshold much higher reflectivities are used, particularly on the back mirror.

The on-resonance Bragg mirror reflectivity is the square of the reflection coefficient r, given by:

$$r = \frac{1 - (n_f/n_i)(n_l/n_h)^{2N}}{1 + (n_f/n_i)(n_l/n_h)^{2N}} \qquad (73)$$

where there are N pairs of quarter-wavelength layers that alternate high-index and low-index (n_h and n_l, respectively), and n_f and n_i are the refractive index of the final and initial media, respectively.[36]

For high-reflectance Bragg mirrors, the second term in the numerator and denominator is small, and the reflectivity can be simplified to:

$$\varepsilon = 1 - R = 1 - r^2 = 4 \left(\frac{n_f}{n_i}\right)\left(\frac{n_l}{n_h}\right)^{2N} \qquad (74)$$

Higher reflectivity (smaller ε) is provided by either more layer pairs or a larger refractive index difference between the two compositions in the layer pairs. Also, Eq. (74) shows that internal mirrors ($n_f = n_i$) will have a smaller reflectivity than external mirrors ($n_f = 1$) for the same number of layer pairs. If the layer pair is GaAs ($n \sim 3.6$) and AlAs ($n \sim 3.0$), a mirror consisting of 15 layer pairs will have an internal reflectivity $R = 98$ percent and external reflectivity $R = 99.5$ percent. Thirty layer pairs will increase the internal mirror reflectivity to 99.96 percent. Bragg mirrors with a smaller fraction of AlAs in the low-index layers will require more layer pairs to achieve the same reflectivity.

Some advanced technologies reduce the number of required layer pairs by selectively oxidizing the AlAs layers to lower their refractive index to $n \sim 1.5$. Using such techniques, reflectivities as high as 99.95 and 99.97 percent can be achieved from mirrors grown with only 7 interior pairs and 5 outside pairs, respectively; these mirrors can be used in VSCELs, but do not easily conduct current.

Electrical Injection

There is difficulty in injecting carriers from the top electrode down through the Bragg reflector, even if it is n-doped, because the GaAs layers provide potential wells that trap carriers. Furthermore, n-doping increases the optical loss in the mirrors. Possible solutions include reducing the AlAs concentration to < 60 percent; using graded compositions rather than abrupt layer pairs; using lateral carrier injection (which increases the operating voltage); using a separately deposited dielectric mirror on top of a transparent electrode; or accepting the high resistivity of the Bragg mirror and operating the laser at relatively high voltage.

The major issue for VCSELs, then, is to inject carriers efficiently, without resistive loss and without carrier leakage. Because resistance in n-doped mirrors is less than in p-doped mirrors, typically the top mirror is doped n-type and carrier injection comes from a top electrode. Light is emitted through a window hole in this top electrode. Carrier injection into the active region often requires rather high voltages because it may be difficult to drive carriers across the Bragg mirrors. Transverse current injection typically requires even higher voltages, although this method has been proven useful when highly conductive layers are grown just above and below the active region.

Some VCSELs use GRINSCH structures (similar to the composition used in edge emitters) to reduce the resistivity in the active region. Typical thresholds for VCSELs are about 3.5 V. Because the drive is limited by resistance, thresholds are typically given as voltages, rather than currents.

Planar VCSELs of fairly large diameter (>10 µm) are straightforward to make, and are useful when a low threshold is not required and multispatial mode is acceptable. Ion implantation outside the VCSEL controls the current in this region; the light experiences gain guiding and perhaps thermal lensing. Smaller diameters (3 to 10 µm) require etching mesas vertically through the Bragg mirror in order to contain the laser light that tends to diffract away.

Higher injection efficiency is obtained by defining the active region through an oxide window just above the active layer. This uses a selective lateral oxidation process that can double the maximum conversion efficiency to almost 60 percent. A high-aluminum fraction AlGaAs layer (~98 percent) is grown. A mesa is etched to below that layer. Then a long, soaking, wet-oxidization process selectively creates a ring of native oxide that stops vertical carrier transport. The chemical reaction moves in from the side of an etched pillar and is stopped when the desired diameter is achieved. Such a current aperture confines current only where needed. Threshold voltages of <6 V are common in diameters ~12 µm. This geometry is shown in Fig. 25. This oxide-defined current channel increases the efficiency, but tends to cause multiple transverse modes due to relatively strong oxide-induced index guiding. Single-mode requirements force the diameter to be very small (below 4 to 5 µm).

Spatial Characteristics of Emitted Light

Single transverse mode remains a challenge for VCSELs, particularly at the larger diameters. When VCSELs are modulated, lateral spatial instabilities tend to set in, and spatial hole burning causes transverse modes to jump. This can introduce considerable modal noise in coupling VCSEL light into fibers. Techniques for mode selection include incorporating a spatial filter, using an antiguide structure where the losses are much higher for higher order modes, or using sidewall scattering losses that are higher for higher-order modes. The requirement is that the mode selective losses must be large enough to overcome the effects of spatial hole burning.

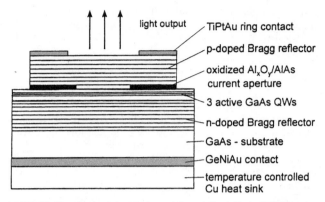

light output
TiPtAu ring contact
p-doped Bragg reflector
oxidized Al$_x$O$_y$/AlAs current aperture
3 active GaAs QWs
n-doped Bragg reflector
GaAs - substrate
GeNiAu contact
temperature controlled Cu heat sink

FIGURE 25 Cross-sectional view of an oxidized GaAs VCSEL. An AlGaAs layer with high aluminum content grown just above the active region is chemically oxidized into Al$_x$O$_y$ by a process that moves in from the edge of the etched mesa with time. Controlling the oxidization rate and time results in a suitable current aperture to obtain high conversion efficiency.[37]

One approach to achieving single transverse mode output is to include a passive antiguide region (PAR), the geometry shown in Fig. 24.[35] The surrounding region has been etched and the sides backfilled with material of *higher* refractive index. This provides an *antiguide* for the laser, which has low loss only for the lowest order transverse mode. A single mode with a FWHM mode size of 7.4 μm (which matches single-mode fibers) can be achieved at 2.4 times threshold with VCSEL diameters of 15 μm. Current blocking outside the active area can be achieved by regrowing an *nipi*-doped antiguide. Typical thresholds for such lasers are 2 V (at 3 mA). A single-mode output of 1.7 mW with an input of 6.6 mA was reported, with more than 20 dB higher-order spatial mode suppression. Fixed polarization along one of the crystal orientations was observed during single-mode operation and attributed to asymmetry introduced in the etching and regrowth process. These structures have slightly higher thresholds than other geometries, but offer single-mode operation.

Other low-cost means of confining current are either proton implantation or etching mesas and then planarizing with polyimide. In both these cases, the regions surrounding the mesa will have a lower refractive index, which will cause the VCSEL to be a real index guide, which will tend toward multimode operation. This may introduce *modal noise* into fiber communication systems.

When the QWs are composed of InGaAs, the VCSELs will emit at 980 nm, and they can be designed to be bottom emitting, since the substrate is transparent. However, inexpensive silicon detectors can no longer be used at this wavelength, so these VCSELs offer fewer advantages in optical communication systems.

Light Out versus Current In

The VCSEL will, in general, have similar *L-I* performance to edge-emitting laser diodes, with some small differences. Because the acceptance angle for the mode is higher than in edge-emitting diodes, there will be more spontaneous emission, which will show up as a more graceful turn-on of light out versus voltage in. As previously mentioned, the operating voltage is 2 to 3 times that of edge-emitting lasers. Thus, Eq. (8) must be modified to take into account the operating voltage drop across the resistance R of the device. The operating power efficiency is:

$$P_{\text{eff}} = \eta_D \, \frac{I_{\text{op}} - I_{\text{th}}}{I_{\text{th}}} \, \frac{V_g}{V_g + I_{\text{op}}R} \tag{75}$$

Single-mode VCSELs of small diameter would typically have a 5 μm radius, a carrier injection efficiency of 80 to 90 percent, an internal optical absorption loss $\alpha_i L$ of 0.003, an optical scattering loss of 0.001, and a net transmission through the front mirror of 0.005 to 0.0095. Carrier losses reducing the quantum efficiency are typically due to spontaneous emission in the wells, spontaneous emission in the barriers, Auger recombination, and carrier leakage.

Typical VCSELs designed for a compatibility with single-mode fiber incorporate an 8-μm proton implantation window and 10-μm-diameter window in the top contact. Such diodes may have threshold voltages of ~3 V and threshold currents of a few milliamps. These lasers may emit up to ~2 mW maximum output power. Devices will operate in zero-order transverse spatial mode with gaussian near-field profile when operated with DC drive current less than about twice the threshold. Output optical powers in single mode as high as 4.4 mW have been reported.

When there is emission in more than one spatial mode, or with both polarizations, there will usually be kinks in the *L-I* curve, as with multimode edge-emitting lasers.

Spectral Characteristics

Since the laser cavity is short, the longitudinal modes are much farther apart in wavelength, typically $\delta\lambda \sim 50$ nm, so only one longitudinal mode will appear, and there is longitudinal mode purity. The problem is with spatial modes, since at higher power levels the laser does not operate in a single spatial mode. Each spatial mode will have slightly different wavelengths, perhaps 0.01 to 0.02 nm apart. There is nothing in a typical VCSEL that selects a given polarization state. Thus, the VCSEL tends to oscillate in both polarization states, also with slightly different wavelengths.

When modulated, lateral spatial instabilities may set in, and spatial hole burning may cause transverse modes to jump. This can cause spectral broadening. In addition, external reflections can cause instabilities and increased relative intensity noise, just as in edge-emitting lasers.[38] For very short cavities, such as between the VCSEL and a butt-coupled fiber (with ~4 percent reflectivity), instabilities do not set in, but the output power can be affected by the additional mirror, which forms a Fabry-Perot cavity with the output mirror and can reduce or increase its effective reflectivity, depending on the round-trip phase difference. When the external reflection comes from ~1 cm away, bifurcations and chaos can be introduced with a feedback parameter $F > 10^{-4}$, where $F = C_e \sqrt{f_{\text{ext}}}$, with C_e and f_{ext} as defined in the discussion surrounding Eq. (45). For $R_o = 0.995$, $R_{\text{ext}} = 0.04$, the feedback parameter $F \sim 10^{-3}$, and instabilities can be observed if one is not careful about back-reflections.

Polarization

Most VCSELs exhibit linear but random polarization states, which may wander with time (and temperature) and may have slightly different emission wavelengths. These unstable polarization characteristics are due to the in-plane crystalline symmetry of the quantum wells grown on (100) oriented substrates. Polarization-preserving VCSELs require breaking the symmetry by introducing anisotropy in the optical gain or loss. Some polarization selection may arise from an elliptical current aperture. The strongest polarization selectivity has come from growth on (311) GaAs substrates, which causes anisotropic gain.

VCSELs at Other Wavelengths

Long-wavelength VCSELs at 1.3 and 1.55 μm have been limited by their poor high-temperature characteristics and by the low reflectivity of InP/InGaAsP Bragg mirrors due to low index contrast between lattice-matched layers grown on InP. These problems have been overcome by using the same InGaAsP/InP active layers as in edge-emitting lasers, but providing mirrors another way: dielectric mirrors, wafer fusion, or metamorphic Bragg reflectors. Dielectric mirrors have limited thermal dissipation and require lateral injection, although carrier injection through a tunnel junction has shown promise. More success has been achieved by wafer-fusing GaAs/AlGaAs Bragg mirrors (grown lattice-matched onto GaAs) to the InP lasers. Wafer fusion occurs when pressing the two wafers together (after removing oxide off their surfaces) at 15 atm and heating to 630°C under hydrogen for 20 min. Typically one side will have an integrally grown InP/InGaAsP lattice-matched DBR (GaAlAsSb/AlAsSb mirrors also work). Mirrors can be wafer-fused on both sides of the VCSEL by etching away the InP substrate and one of the GaAs substrates. An integrated fabrication technology involves growing metamorphic GaAs/AlGaAs Bragg reflectors directly onto the InP structure. These high-reflectivity mirrors, grown by molecular beam epitaxy, have a large lattice mismatch and a high dislocation density. Nonetheless, because current injection is based on majority carriers, these mirrors can still be conductive, with high enough reflectivity to enable promising long-wavelength VCSELs.[39]

4.10   *LITHIUM NIOBATE MODULATORS*

The most direct way to create a modulated optical signal for communications applications is to directly modulate the current driving the laser diode. However, as discussed in the sections on lasers, this may cause turn-on delay, relaxation oscillation, mode-hopping, and/or chirping of the optical wavelength. Therefore, an alternative often used is to operate the laser in a continuous manner and to place a modulator after the laser. This modulator turns the laser light on and off without impacting the laser itself. The modulator can be butt-coupled directly to the laser, located in the laser chip package and optically coupled by a microlens, or remotely attached by means of a fiber pigtail between the laser and modulator.

Lithium niobate modulators have become one of the main technologies used for high-speed modulation of continuous-wave (CW) diode lasers, particularly in applications (such as cable television) where extremely linear modulation is required, or where chirp is to be avoided at all costs. These modulators operate by the electro-optic effect, in which the applied electric field changes the refractive index. Integrated optic waveguide modulators are fabricated by diffusion into a lithium niobate substrate. The end faces are polished and butt-coupled (or lens-coupled) to a single-mode fiber pigtail (or to the laser driver itself). This section describes the electro-optic effect in lithium niobate, its use as a phase modulator and an intensity modulator, considerations for high-speed operation, and the difficulties in achieving polarization independence.[40]

The most commonly used modulator is the *Y*-branch interferometric modulator shown in Fig. 26, discussed in a following subsection. The waveguides that are used for these modulators are fabricated in lithium niobate either by diffusing titanium into the substrate from a metallic titanium strip or by using ion exchange. The waveguide pattern is obtained by photolithography. The standard thermal indiffusion process takes place in air at 1050°C over 10 h. An 8-µm-wide strip of titanium 50 nm thick creates a fiber-compatible single mode at 1.3 µm. The process introduces ~1.5 percent titanium at the surface, with a diffusion profile depth of ~4 µm. The result is a waveguide with increased extraordinary refractive index of 0.009 at the surface. The ordinary refractive index change is ~0.006. A typical modulator will use aluminum electrodes 2 cm long, etched on either side of the waveguides, with a gap of 10 µm.

In the case of ion exchange, the lithium niobate sample is immersed in a melt containing a large proton concentration (typically benzoic acid or pyrophosphoric acid at >170°C), with some areas protected from diffusion by masking; the lithium near the surface of the substrate is replaced by protons, which increases the refractive index. The ion-exchange process changes only the extraordinary polarization; that is, only light polarized parallel to the *Z* axis

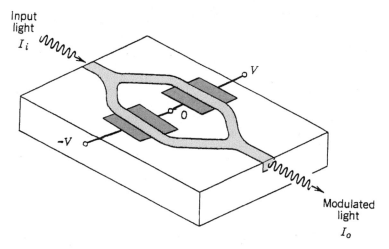

FIGURE 26 Y-branch interferometric modulator in the "push-pull" configuration. Center electrodes are grounded. Light is modulated by applying positive or negative voltage to the outer electrodes.

is waveguided. Thus, it is possible in lithium niobate to construct a polarization-independent modulator with titanium indiffusion, but not with proton-exchange. Nonetheless, ion exchange makes possible a much larger refractive index change (~0.12), which provides more flexibility in modulator design. Annealing after diffusion can reduce insertion loss and restore the electro-optic effect. Interferometric modulators with moderate index changes ($\Delta n < 0.02$) are insensitive to aging at temperatures of 95°C or below. Using higher index change devices, or higher temperatures, may lead to some degradation with time. Tapered waveguides can be fabricated easily by ion exchange for high coupling efficiency.[41]

Electro-Optic Effect

The *electro-optic effect* is the change in refractive index that occurs in a noncentrosymmetric crystal in the presence of an applied electric field. The linear electro-optic effect is represented by a third-rank tensor. However, using symmetry rules it is sufficient to define a reduced tensor r_{ij}, where $i = 1 \ldots 6$ and $j = x, y, z$, denoted as 1, 2, 3. Then, the linear electro-optic effect is traditionally expressed as a linear change in the inverse refractive index squared (see Vol. II, Chap. 13 of this handbook):

$$\Delta\left(\frac{1}{n^2}\right)_i = \sum_j r_{ij} E_j \qquad j = x, y, z \tag{76}$$

where E_j is the component of the applied electric field in the jth direction. The applied electric field changes the index ellipsoid of the anisotropic crystal into a new form based on Eq. (76):

$$a_1 x^2 + a_2 y^2 + a_3 z^2 + 2a_4 yz + 2a_5 xz + 2a_6 xy = 1 \tag{77}$$

where the diagonal elements are given by:

$$a_1 = \frac{1}{n_x^2} + \Delta\left(\frac{1}{n^2}\right)_1 \qquad a_2 = \frac{1}{n_y^2} + \Delta\left(\frac{1}{n^2}\right)_2 \qquad a_3 = \frac{1}{n_z^2} + \Delta\left(\frac{1}{n^2}\right)_3$$

and the cross terms are given by

$$a_4 = \Delta\left(\frac{1}{n^2}\right)_4, \quad a_5 = \Delta\left(\frac{1}{n^2}\right)_5, \quad a_6 = \Delta\left(\frac{1}{n^2}\right)_6,$$

The presence of cross terms indicates that the ellipsoid is rotated and the lengths of the principal dielectric axes have changed.

Diagonalizing the ellipsoid of Eq. (77) will give the new axes and values. The general case is treated in Vol. II, Chap. 13. In lithium niobate, the material of choice for electro-optic modulators, the equations are simplified because the only nonzero components and their magnitudes are[42]:

$$r_{33} = 31 \times 10^{-12} \text{ m/V} \qquad r_{13} = r_{23} = 8.6 \times 10^{-12} \text{ m/V}$$

$$r_{51} = r_{42} = 28 \times 10^{-12} \text{ m/V} \qquad r_{22} = -r_{12} = -r_{61} = 3.4 \times 10^{-12} \text{ m/V}$$

The crystal orientation is usually chosen so as to obtain the largest electro-optic effect. This means that if the applied electric field is along Z, then light polarized along Z sees the largest field-induced change in refractive index. Since $\Delta(1/n^2)_3 = \Delta(1/n_z)^2 = r_{33}E_z$, performing the difference gives

$$\Delta n_z = -\frac{n_z^3}{2} r_{33}E_z\Gamma \tag{78}$$

We have included a filling factor Γ (also called an *optical-electrical field overlap parameter*) to include the fact that the applied field may not be uniform as it overlaps the waveguide, resulting in an effective field that is somewhat less than 100 percent of the maximum field.

In the general case for the applied electric field along Z, the only terms in the index ellipsoid will be $\Delta(1/n^2)_1 = r_{13}E_z = \Delta(1/n^2)_2 = r_{23}E_z$, and $\Delta(1/n^2)_3 = r_{33}E_z$. This means that the index ellipsoid has not rotated, its axes have merely changed in length. Light polarized along any of these axes will see a pure phase modulation. Because r_{33} is largest, polarizing the light along Z and providing the applied field along Z will provide the largest phase modulation. Light polarized along either X or Y will have the same (although smaller) index change, which might be a better direction if polarization-independent modulation is desired. However, this would require that light enter along Z, which is the direction in which the field is applied, so it is not practical.

As another example, consider the applied electric field along Y. In this case the nonzero terms are

$$\Delta\left(\frac{1}{n^2}\right)_1 = r_{12}E_y \quad \Delta\left(\frac{1}{n^2}\right)_2 = r_{22}E_y = -r_{12}E_y \quad \Delta\left(\frac{1}{n^2}\right)_4 = r_{42}E_y \tag{79}$$

It can be seen that now there is a YZ cross-term, coming from r_{42}. Diagonalization of the perturbed index ellipsoid finds new principal axes, only slightly rotated about the Z axis. Therefore, the principal refractive index changes are essentially along the X and Y axes, with the same values as $\Delta(1/n^2)_1$ and $\Delta(1/n^2)_2$ in Eq. (79). If light enters along the Z axis without a field applied, both polarizations (X and Y) see an ordinary refractive index. With a field applied, both polarizations experience the same phase change (but opposite sign). We later describe an interferometric modulator that does not depend on the sign of the phase change. This modulator is polarization independent, using this crystal and applied-field orientation, at the expense of operating at somewhat higher voltages, because $r_{22} < r_{33}$.

Since lithium niobate is an insulator, the direction of the applied field in the material depends on how the electrodes are applied. Fig. 27 shows a simple phase modulator. Electrodes that straddle the modulator provide an in-plane field as the field lines intersect the

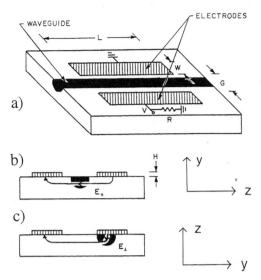

FIGURE 27 (*a*) Geometry for phase modulation in lithium niobate with electrodes straddling the channel waveguide. (*b*) End view of (*a*), showing how the field in the channel is parallel to the surface. (*c*) End view of a geometry placing one electrode over the channel, showing how the field in the channel is essentially normal to the surface.

waveguide, as shown in Fig. 27*b*. This requires the modulator to be *Y-cut* LiNbO$_3$ (the *Y* axis is normal to the wafer plane), with the field lines along the *Z* direction; *X-cut* LiNbO$_3$ will perform similarly. Figure 27*c* shows a modulator in *Z-cut* LiNbO$_3$. In this case, the electrode is placed over the waveguide, with the electric field extending downward through the waveguide (along the *Z* direction). The field lines will come up at a second, more distant electrode. In either case, the field may be fringing and nonuniform, which is why the filling factor Γ has been introduced.

Phase Modulation

Phase modulation is achieved by applying a field to one of the geometries shown in Figure 27. The field is roughly *V/G*, where *G* is the gap between the two electrodes. For an electrode length *L*, the phase shift is:

$$\Delta\phi = \Delta n_z kL = -\frac{n_o^3}{2} r_{33} \left(\frac{V}{G}\right) \Gamma kL \qquad (80)$$

The refractive index for bulk LiNbO$_3$ is given by[43]:

$$n_o = 2.195 + \frac{0.037}{[\lambda \, (\mu m)]^2}$$

and

$$n_e = 2.122 + \frac{0.031}{[\lambda \, (\mu m)]^2}$$

Inserting numbers for a wavelength of 1.55 µm, $n_o = 2.21$. When $G = 10$ µm and $V = 5$ V, a π phase shift is expected in a length $L \sim 1$ cm.

It can be seen from Eq. (80) that the electro-optic phase shift depends on the product of the length and voltage. Longer modulators can use smaller voltages to achieve π phase shift. Shorter modulators require higher voltages. Thus, phase modulators typically use the product of the voltage required to reach π times the length as the figure of merit. The modulator just discussed has a 5 V · cm figure of merit.

The electro-optic phase shift has a few direct uses, such as providing a *frequency shifter* (since $\partial\phi/\partial t \propto v$). However, in communication systems this phase shift is generally used in an interferometric configuration to provide intensity modulation, discussed next.

Y-Branch Interferometric (Mach-Zehnder) Modulator

The *interferometric modulator* is shown schematically in Fig. 26. This geometry allows wave-guided light from the two branches to interfere, forming the basis of an intensity modulator. The amount of interference is tunable by providing a relative phase shift on one arm with respect to the other. Light entering a single-mode waveguide is equally divided into the two branches at the *Y* junction, initially with zero relative phase difference. The guided light then enters the two arms of the waveguide interferometer, which are sufficiently separated that there is no coupling between them. If no voltage is applied to the electrodes, and the arms are exactly the same length, the two guided beams arrive at the second *Y* junction in phase and enter the output single-mode waveguide in phase. Except for small radiation losses, the output is equal in intensity to the input. However, if a π phase difference is introduced between the two beams via the electro-optic effect, the combined beam has a lateral amplitude profile of odd spatial symmetry. This is a second-order mode and is not supported in a single-mode waveguide. The light is thus forced to radiate into the substrate and is lost. In this way, the device operates as an electrically driven optical intensity on-off modulator. Assuming perfectly equal splitting and combining, the fraction of light transmitted is:

$$\eta = \left[\cos\left(\frac{\Delta\phi}{2}\right) \right]^2 \tag{81}$$

where $\Delta\phi$ is the difference in phase experienced by the light in the different arms of the interferometer: $\Delta\phi = \Delta n\, kL$, where $k = 2\pi/\lambda$, Δn is the difference in refractive index between the two arms, and L is the path length of the refractive index difference. The voltage at which the transmission goes to zero ($\Delta\phi = \pi$) is usually called V_π. By operating in a push-pull manner, with the index change increasing in one arm and decreasing in the other, the index difference Δn is twice the index change in either arm. This halves the required voltage.

Note that the transmitted light is periodic in phase difference (and therefore voltage). The response depends only on the integrated phase shift and not on the details of its spatial evolution. Therefore, nonuniformities in the electro-optically induced index change that may occur along the interferometer arms do not affect the extinction ratio. This property has made the interferometric modulator the device of choice in communications applications.

For analog applications, where linear modulation is required, the modulator is prebiased to the quarter-wave point (at voltage $V_b = \pi/2$), and the transmission efficiency becomes linear in $V - V_b$ (for moderate excursions):

$$\eta = \frac{1}{2}\left[1 + \sin\frac{\pi(V - V_b)}{2V_\pi} \right] \approx \frac{1}{2} + \frac{\pi}{4}\frac{(V - V_b)}{V_\pi} \tag{82}$$

The electro-optic effect depends on the polarization. For the electrode configuration shown here, the applied field is in the plane of the lithium niobate wafer, and the polarization of the light to be modulated must also be in that plane. This will be the case if a TE-polarized semiconductor laser is butt-coupled (or lens-coupled) with the plane of its active region parallel to the lithium niobate wafer, and if the wafer is Y-cut. Polarization-independent modulation requires a different orientation, to be described later. First, however, we discuss the electrode requirements for high-speed modulation.

High-Speed Operation

The optimal modulator electrode design depends on how the modulator is to be driven. Because the electrode is on the order of 1 cm long, the fastest devices require traveling wave electrodes rather than lumped electrodes. Lower-speed modulators use lumped electrodes, in which the modulator is driven as a capacitor terminated in a parallel resistor matched to the impedance of the source line. The modulation speed depends primarily on the RC time constant determined by the electrode capacitance and the terminating resistance. To a smaller extent, the speed also depends on the resistivity of the electrode itself. The capacitance per unit length is a critical design parameter. This depends on the material dielectric constant and the electrode gap-to-width ratio G/W. The capacitance-to-length ratio decreases and the bandwidth-length product increases essentially logarithmically with increasing G/W. At $G/W = 1$, $C/L = 2.3$ pF/cm and $\Delta f_{RC} L = 2.5$ GHz $\cdot$ cm. The tradeoff is between large G/W to reduce capacitance and a small G/W to reduce drive voltage and electrode resistance. The ultimate speed of lumped electrode devices is limited by the electric transit time, with a bandwidth-length product of 2.2 GHz $\cdot$ cm. The way to achieve higher speed modulation is to use traveling wave electrodes.

The traveling wave electrode is a miniature transmission line. Ideally, the impedance of this coplanar line is matched to the electrical drive line and is terminated in its characteristic impedance. In this case, the modulator bandwidth is determined by the difference in velocity between the optical and electrical signals (velocity mismatch or walk-off), and any electrical propagation loss. Because of competing requirements between a small gap to reduce drive voltage and a wide electrode width to reduce RF losses, as well as reflections at any impedance transition, there are subtle trade-offs that must be considered in designing traveling-wave devices.

Lithium niobate modulators that operate at frequencies out to 8 GHz at 1.55 μm wavelength are commercially available, with operating voltages of <4 V.[44] Typical modulators have <5 dB insertion loss and >20 dB extinction ratio. To operate near quadrature, which is the linear modulation point, a bias voltage of ~10 V is required. Direct coupling from a laser or polarization-maintaining fiber is required, since these modulators are not independent of polarization. Traveling wave modulators operating well beyond 20 GHz have been reported in the research literature.

Insertion Loss

Modulator insertion loss can be due to Fresnel reflection at the lithium niobate–air interfaces, which can be reduced by using antireflection coatings or index matching (which only helps, but does not eliminate this loss, because of the very high refractive index of lithium niobate). The other cause of insertion loss is mode mismatch. The diffusion process must make a deep waveguide. Typically, the waveguide will be 9 μm wide and 5 μm deep. While the in-plane mode can be gaussian and can match well to the fiber mode, the out-of-plane mode tends to be asymmetric, and its depth must be carefully optimized. In an optimized modulator, the coupling loss per face is about 0.35 dB and the propagation loss is about 0.3 dB/cm. This result includes a residual index-matched Fresnel loss of 0.12 dB.

Misalignment can also cause insertion loss. An offset of 2 μm typically increases the coupling loss by 0.25 dB. The angular misalignment must be maintained below 0.5° in order to keep the excess loss below 0.25 dB.[40]

Propagation loss comes about from absorption, metallic overlay, scattering from the volume or surface, bend loss, and excess loss in the Y-branches. Absorption loss at 1.3- and 1.55-μm wavelengths appears to be <0.1 dB/cm. Bend loss can be large, unless any curvature of guides is small. The attenuation coefficient in a bend has the form:

$$\alpha = C_1 \exp(-C_2 R) \tag{83}$$

where $C_1 = 15$ mm^{-1} and $C_2 = 0.4$ mm^{-1} in titanium indiffused lithium niobate, at wavelengths around 1.3 to 1.5 μm. This means that a 5-mm-long section of constant radius 20 mm will introduce only 0.1 dB of excess loss.[45]

A final source of loss in Y-branches is excess radiation introduced by sharp transitions. These branches must be fabricated carefully to avoid such losses, since the tolerances on waveguide roughness are critically small.

Polarization Independence

As previously shown, if the light is incident along the Z axis and the field is along the Y axis, then light polarized along X and Y experience the same phase shift, but opposite signs. An X-cut crystal, with an in-plane field along Y, therefore, provides polarization-independent interferometric modulation at the sacrifice of somewhat higher half-wave voltage (e.g., 17 V).[46] Because of the difficulty of achieving exactly reproducible lengths in the two arms of the Y-branch interferometer, it has been found useful to do a postfabrication phase correction using laser ablation.

Photorefractivity and Optical Damage

Lithium niobate exhibits *photorefractivity,* also called *optical damage* when it is a nuisance. This phenomenon is a change in refractive index as a result of photoconduction originating in weak absorption by deep traps and a subsequent redistribution of charges within the lithium niobate. Because the photoconductive crystal is electro-optic, the change in electric field resulting from charge motion shows up as a change in refractive index, altering the phase shift as well as the waveguiding properties. While photorefractivity seriously limits the performance of lithium niobate modulators at shorter wavelengths (even at 850 nm),[47] it is not a serious concern at 1.3 and 1.55 μm.

However, partial screening by photocarriers may cause a drift in the required bias voltage of modulators, and systems designers may need to be sensitive to this.

Delta-Beta Reversal Modulators

Early designs for modulators used a configuration entitled the *delta-beta reversal modulator.* This is based on the concept of the *directional coupler.* When two parallel waveguides are situated close enough that their evanescent fields overlap, light couples between them. If they are identical, light can oscillate completely between them, similar to the coupling of energy between two coupled pendula. When they are not identical, the coupling occurs more rapidly, and there is not complete transfer of energy between the two guides. A modulator can be built, then, by using a field applied to one guide to destroy their synchronicity and therefore their coupling.[48] This has not proven to be practical, however, both because of fabrication difficulties and because of residual effects due to photorefractivity. These modulators are not discussed further here.

4.11 ELECTROABSORPTION MODULATORS FOR FIBER-OPTIC SYSTEMS

When modulators are composed of III-V semiconductors, they can be integrated directly on the same chip as the laser, or placed external to the laser chip. External modulators may be butt-coupled to the laser, coupled by means of a microlens, or coupled by means of a fiber pigtail.

Electroabsorption. Semiconductor modulators typically use *electroabsorption,* the electric field dependence of the absorption near the band edge of a semiconductor. Electroabsorption is particularly strong in quantum wells (QWs), where it is called the *quantum-confined Stark effect (QCSE).* An example of the frequency dependence of the QCSE is shown in Fig. 28. The absorption spectrum of QWs exhibits a peak at the *exciton resonance.* When a field is applied, the exciton resonance moves to longer wavelengths, becomes weaker, and broadens. This means that the absorption increases with field on the long-wavelength side, as the exciton resonance moves to longer wavelengths. At wavelengths closer to the exciton resonance, the absorption will first increase with field, then plateau, and finally decrease, as the field continues to grow. At wavelengths shorter than the zero-field exciton resonance, the absorption will decrease with increasing field, as the resonance moves to longer wavelengths.

While electroabsorption in QWs is much larger than in bulk, due to the sharpness of the excitonic-enhanced absorption edge, the useful absorption change must be multiplied by the filling factor of the QW in the waveguide, which reduces its effective magnitude. Under some conditions, electroabsorption near the band edge in bulk semiconductors (typically called the *Franz-Keldysh effect*) may also be useful in electroabsorption modulators.

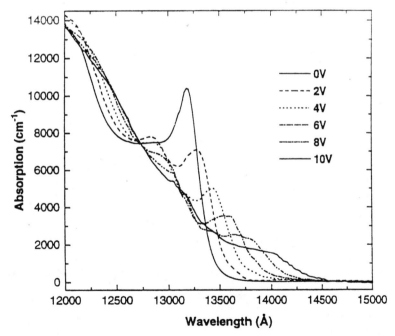

FIGURE 28 Spectrum of quantum-confined Stark effect (QCSE) in InAsP/InP strained MQWs. The absorption changes with applied field.[49]

Waveguide Modulators. When light traverses a length of QW material, the transmission will be a function of applied voltage. An electroabsorption modulator consists of a length of waveguide containing QWs. The waveguide is necessary to confine the light to the QW region so that it does not diffract away. Thus, low-refractive-index layers must surround the layer containing the QWs. Discrete electroabsorption modulators are typically made by using geometries very similar to those of edge-emitting lasers (Fig. 1). They are cleaved, antireflection coated and then butt-coupled to the laser chip. They are operated by a reverse bias, rather than the forward bias of a laser. Alternatively, the modulator is integrated on the laser chip, with the electroabsorption modulator region following a DFB or DBR laser in the optical train, as shown in Fig. 29. This figure shows the simplest electroabsorption modulator, with the same MQW composition as the DFB laser. This ridge waveguide device has been demonstrated with a 3-dB bandwidth of 30 GHz. The on-off contrast ratio is 12.5 dB for a 3-V drive voltage in a 90-μm-long modulator.[50] The use of the same QW is possible by setting the grating that determines the laser wavelength to well below the exciton resonance. Because of the inherently wide gain spectrum exhibited by strained layer MQWs, this detuning is possible for the laser and allows it to operate in the optimal wavelength region for the electroabsorption modulator.

Other integrated electroabsorption modulators use a QW composition in the electroabsorption region that is different from that of the laser medium. Techniques for integration are discussed later.

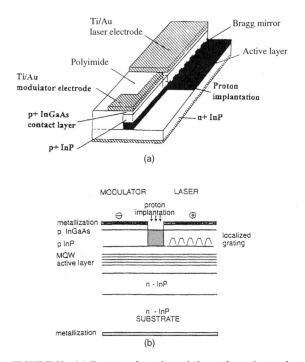

FIGURE 29 (*a*) Geometry for a channel electroabsorption modulator (foreground) integrated on the same chip with a DFB laser (background, under the Bragg mirror). (*b*) Side view, showing how the same MQW active layer can be used under forward bias with a grating to provide a DFB laser, and in a separate region under reverse bias for modulation, with the two regions electrically separated by proton implantation.[50]

Intensity Modulation by Electroabsorption

In an electroabsorption waveguide modulator of length L, where the absorption is a function of applied field E, the transmission is a function of field: $T(E) = \exp[-\alpha(E)L]$, where α is the absorption per unit length, averaging the QW absorption over the entire waveguide. (That is, α is the QW absorption multiplied by the filling factor of the QW in the waveguide.) Performance is usually characterized by two quantities: *insertion loss* (throughput at high transmission) and *contrast ratio* (ratio of high transmission to low transmission). Assume that the loss in the QW, initially at low value α_-, increases by $\delta\alpha$. The *contrast ratio* is given by:

$$\text{CR} \equiv \frac{T_{\text{high}}}{T_{\text{low}}} = \exp(\delta\alpha L). \tag{84}$$

The *insertion loss* is given by

$$A \equiv 1 - T_{\text{high}} = 1 - \exp(-\alpha_- L) \approx \alpha_- L \tag{85}$$

A long path length L means a high contrast ratio but also a large insertion loss and large capacitance, which results in a slower speed. Choosing the most practical length for any given application requires trading off the contrast ratio against insertion loss and speed.

To keep a moderate insertion loss, waveguide lengths should be chosen so that $L \approx 1/\alpha_-$. This sets the contrast ratio as

$$\text{CR} = \exp\left(\frac{\delta\alpha}{\alpha_-}\right)$$

The contrast ratio depends on the ratio of the change in absorption to the absorption in the low-loss state; this fact is used to design the QW composition and dimensions relative to the wavelength of operation. In general, the contrast ratio improves farther from the band edge, but the maximum absorption is smaller there, so the modulator must be longer, which increases its capacitance, decreases its speed, and increases its loss. Contrast ratios may reach 10/1 or more with <2 V applied for optimized electroabsorption modulators. The contrast ratio does not depend on the filling factor of the QW in the waveguide, but the required length L does. Since high-speed modulators require small capacitance and small length, the filling factor should be as high as possible.

Waveguide modulators are used at wavelengths where the absorption is not too large, well below the band edge. In this wavelength region, electroabsorption at a fixed wavelength can be modeled by a pure quadratic dependence on field. Thus:

$$\alpha(E) \approx \alpha_o + \alpha_2 E^2 \tag{86}$$

where α_2 will typically depend on the wavelength, the QW and barrier dimensions and composition, and the waveguide filling factor. Intimately connected with this change in absorption is a change in refractive index with a similar field dependence:

$$\delta n(E) \approx n_2 E^2 \tag{87}$$

where n_2 is also strongly dependent on wavelength. Both electroabsorption and electro-refraction are about an order of magnitude larger in QWs than in bulk material. Specific numerical values depend on the detailed design, but typical values are on the order of $\alpha_o \sim 100$ cm^{-1}, $\delta\alpha \sim 1000$ cm^{-1}, $L \sim 200$ μm for 2 V applied across an i region 2.5 μm thick, for a field of $\sim$10 kV/cm. This means $\alpha_2 \sim 2 \times 10^{-5}$ cm/V^2. Also, $n_2 \sim 2 \times 10^{-11}$ V^{-2}.

Applying a Field in a Semiconductor

The electric field is usually applied by reverse biasing a *pin* junction. The electric field is supported by the semiconductor depletion region that exists within a *pin* junction, or at a metal-semiconductor junction (Schottky barrier). Charge carrier depletion in the *n* and *p* regions may play a role in determining the electric field across thin intrinsic regions. Taking this into account while assuming an undoped *i* region, the electric field across the *i* region of an ideally abrupt *pin* junction is given by[51]:

$$E = \frac{eN_d}{\varepsilon} \frac{-d_i + \sqrt{d_i^2 + 2(\varepsilon V_{tot}/eN_o)(1 + N_d/N_a)}}{1 + N_d/N_a} \tag{88}$$

where N_d is the (donor) doping density in the *n* region, N_a is the (acceptor) doping density in the *p* region, *e* is the elementary charge, ε is the dielectric constant, d_i is the thickness of the intrinsic region, and V_{tot} is the sum of the applied and built-in field (defined positive for reverse bias). When the *n* and *p* regions are highly doped and the *i* region is undoped, most of the voltage is dropped across the *i* region. When d_i is sufficiently large, the square root can be approximated, the d_i terms in the numerator cancel, and Eq. (88) becomes:

$$E = \frac{V_{tot}}{d_i} \left[1 - \frac{\varepsilon V_{tot}(1 + N_d/N_a)}{2\, eN_d d_i^2} \right] \tag{89}$$

which, to lowest order, is just the field across a capacitor of thickness d_i. For a typical applied voltage of 5 V and $d_i = 0.25$ μm, with $N_a = 10^{18}$ cm^{-3} and $N_d = 10^{18}$ cm^{-3}, $E = 2 \times 10^5$ V/cm. How much absorption and refractive index change this results in depends on wavelength and, of course, material design.

Integrating the Modulator

Stripe-geometry modulators can be cleaved from a wafer, antireflection coated, and butt-coupled to either a laser or a fiber pigtail. Typical insertion losses may be ~10 dB. Or, the modulator may be monolithically integrated with the laser. A portion of the same epitaxial layer grown for the laser active region can be used as an electroabsorption modulator by providing a separate contact and applying a reverse bias. When such a modulator is placed inside the laser cavity, a multielement laser results that can have interesting switching properties, including wavelength tunability. When the electroabsorption modulator is placed outside the laser cavity, it is necessary to operate an electroabsorption modulator at wavelengths well below the band edge. Then, the modulator region must have a higher energy bandgap than the laser medium. Otherwise, the incident light will be absorbed, creating free electron-hole pairs that will move to screen the applied field and ruin the modulator.

The integration of an electroabsorption modulator, therefore, usually requires that the light traverse some portion of the sample that has a different bandgap from that of the laser region. Four techniques have been developed: etching and regrowth, vertical coupling between layers, selective area epitaxy, and postgrowth well and barrier intermixing.

Etching and Regrowth.   Typically, a first set of epitaxial layers is grown everywhere, which includes the laser structure up through the QW layer. Then the QW layer is etched away from the regions where it is not needed. The structure is then overgrown everywhere with the same upper cladding layers. This typically results in a bulk electroabsorption modulator, consisting of laser cladding material. A more complex fabrication process might mask the laser region during the regrowth process and grow a different QW composition that would provide an integrated butt-coupled modulator for the DFB (or DBR) QW laser.

Vertical Coupling Between Layers. This approach makes it possible to use a QW modulator as well as a QW laser, with a different QW composition in each. Two sets of QWs can be grown one on top of the other and the structures can be designed so that light couples vertically from one layer to the other, using, for example, grating assisted coupling. This may involve photolithographically defining a grating followed by a regrowth of cladding layers, depending on the design.

Selective Area Epitaxy. Growth on a patterned substrate allows the width of the QWs to be varied across the wafer during a single growth. The substrate is usually coated with a SiO_2 mask in which slots are opened. Under a precise set of growth conditions no growth takes place on top of the dielectric, but surface migration of the group III species (indium) can take place for some distance across the mask to the nearest opening. The growth rate in the opened area depends on the width of the opening and the patterning on the mask. Another approach is epitaxial growth on faceted mesas, making use of the different surface diffusion lengths of deposited atomic species on different crystal facets.

Well and Barrier Intermixing. The bandgap of a QW structure can be modified after growth by intermixing the well and barrier materials to form an alloy. This causes a rounding of the initially square QW bandgap profile and, in general, results in an increase of the bandgap energy. This provides a way to fabricate lasers and bandgap-shifted QCSE modulators using only one epitaxial step. Intermixing is greatly enhanced by the presence of impurities or defects in the vicinity of the QW interfaces. Then the bandgap is modified using impurity induced disordering, laser beam induced disordering, impurity-free vacancy diffusion, or ion implantation enhanced interdiffusion. The challenge is to ensure that the electrical quality of the *pin* junction remains after interdiffusion; sometimes regrowth of a top *p* layer helps.[52]

Operating Characteristics

In addition to contrast ratio, insertion loss, and required voltage, the performance of electroabsorption modulators depends on speed, chirp, polarization dependence, optical power-handling capabilities, and linearity. These factors all depend on the wavelength of operation, the materials, the presence of strain, the QW and waveguide geometry, and the device design. There will be extensive trade-offs that must be considered to achieve the best possible operation for a given application. Modulators will differ, depending on the laser and the proposed applications.

Chirp. Because a change in refractive index is simultaneous with any absorption change, electroabsorption modulators, in general, exhibit *chirp* (frequency broadening due to the time-varying refractive index, also observed in modulated lasers), which can seriously limit their usefulness. As with semiconductor lasers, the figure of merit is:

$$\beta = k_o \frac{\delta n}{\delta \alpha} \tag{90}$$

Unlike with lasers, however, there are particular wavelengths of sizable absorption change at which $\delta n = 0$. Studies have shown that these nulls in index change can be positioned where $\delta \alpha$ is large by using coupled quantum wells (CQWs).[53] These structures provide two, three, or more wells so closely spaced that the electron wave functions overlap between them. If desired, several sets of these CQWs may be used in a single waveguide, if they are separated by large enough barriers that they do not interact. Chirp-free design is an important aspect of electroabsorption modulators.

On the other hand, since the chirp can be controlled in electroabsorption modulators, there are conditions under which it is advantageous to provide a negative chirp to cancel out the positive chirp introduced by fibers. This allows 1.55-μm laser pulses to travel down normal dispensive fiber (zero material dispersion at 1.3 μm wavelength) without the pulses unduly spreading.

Polarization Dependence.   In general, the quantum-confined Stark effect is strongly polarization dependent, although there may be specific wavelengths at which TE and TM polarized light experience the same values of electroabsorption (and/or electrorefraction). It turns out that polarization-independent modulation is more readily achieved by using strained QWs. In addition, the contrast ratio of electroabsorption change at long wavelengths can be improved by using strained QWs.

Optical Power Dependence.   During the process of electroabsorption, the modulators can absorb some of the incident light. This will create electron-hole pairs. If these electron-hole pairs remain in the QWs, at high optical powers they will introduce a free carrier plasma field that can screen the exciton resonance. This broadens the absorption spectrum and reduces the contrast ratio. In some cases, electroabsorption modulators operating at the band edge of bulk semiconductors (the *Franz-Keldysh effect*) may be able to operate with higher laser power. A common approach is to use shallow QWs, so that the electrons and holes may escape easily.

Even when the electron-hole pairs created by absorption escape the QWs, they will move across the junction to screen the applied fields. This will tend to reduce the applied field, and the performance will depend on the magnitude of absorbed light. Photogenerated carriers must also be removed, or they will slow down the modulator's response time. Carriers may be removed by leakage currents in the electrodes or by recombination.

Built-in Bias.   Because *pin* junctions have built-in fields, even at zero applied voltage, electroabsorption modulators have a prebias. Some applications use a small forward bias to achieve even larger modulation depths. However, the large forward current resulting from the forward bias limits the usefulness of this approach. There are, at present, some research approaches to remove the internal fields using an internal strain-induced piezoelectric effect to offset the *pn* junction intrinsic field.

Advanced Concepts for Electroabsorption in QWs

Coupled QWs offer the possibility of chirp control, and strained QWs offer the possibility of polarization independence, as previously explained. Adding these degrees of freedom to electroabsorption modulator design has been crucial in obtaining the highest performance devices.

High-performance Discrete Electroabsorption Modulators.   A discrete modulator at 1.3 μm uses compressively strained InAsP wells grown on InP with InGaP barriers that are under tensile stress for strain compensation. High-speed operation with 3-dB bandwidth of >10 GHz and operating voltage of <2 V has been reported with a 20-dB on-off ratio.[54] The electroabsorptive layer contained five QWs, each 11 nm thick. The waveguide was an etched high-mesa structure 3 μm wide and 200 μm long. A modulator had a 10-dB insertion loss, and the measured electroabsorptive figure of merit was $\delta\alpha/\alpha_- = 10/1$.

A discrete modulator at 1.55 μm planned for polarization insensitivity and capable of handling high optical powers was designed with two strongly coupled tensily strained QWs.[55] An 8-dB extinction ratio at 1.5-V drive voltage with a 3-dB bandwidth of 20 GHz was reported for average optical powers as high as 20 mW. Two InGaAs wells 5 nm thick with a 0.5-nm

InGaAlAs barrier between them were grown in pairs, with 9-nm barriers separating each pair. A total of 13 pairs of wells were grown and etched to form ridge waveguides 3 μm wide.

High-Performance Integrated Electroabsorption Modulators.   When an electroabsorption modulator is integrated with a DFB laser, strain is not required because polarization insensitivity is not needed. Selective epitaxy has been used to grow a 200-μm-long modulator region consisting of lattice-matched quaternary wells ~5 nm in width. The reported extinction ratio was >13 dB at 1.5 V, with output powers in the on state of >4 mW, at a current of 100 mA.[56]

A two-step growth procedure provided a butt joint between a modulator and a DFB laser. Compressively strained wells were used to reduce the potential well that the hole sees, speeding the device.[57] Providing 3 V to a 200-μm-long modulator reduced the DFB laser output from 25 mW to 1 mW, for a 25/1 extinction ratio. The 3-dB bandwidth was 15 GHz. While there was a condition of zero chirp at −2 V, biasing to a regime of negative chirp allowed cancellation of the chromatic dispersion of fiber at the 1.55-μm laser wavelength. As a result, 10 Gb/s non-return-to-zero (NRZ) transmission was demonstrated over 60 km of standard fiber.

Wannier Stark Localization.   A variation on the quantum-confined Stark effect uses an array of closely coupled quantum wells, which exhibit *Wannier Stark localization* (WSL). Because of the close spacing of the QWs, in the absence of a field, the electron wave function is free to travel across all wells, creating a miniband. When a field is applied, the wells decouple, and the electrons localize within individual wells. This removes the miniband, sharpening the absorption spectrum and creating a *decrease* in absorption below the band edge.[58]

Electron Transfer Modulators.   A large absorption change can be created by filling the states near the edge of the semiconductor bands with electrons (or holes). This filling requires free carriers to be injected into the optical modulator. Quantum wells enhance the magnitude of this absorption change. Using applied voltage to transfer electrons from a reservoir across a barrier into a QW produces an effective long-wavelength modulator, termed a *barrier, reservoir, and quantum well electron transfer* (BRAQWET) modulator.[59] By changing the bias across the device, the bound states of the QW arc moved above and below the Fermi level fixed by the electron reservoir. These states are then emptied or filled by a transfer of electrons to or from the reservoir region. Optical modulation is achieved due to state filling and by carrier screening of the coulombic interaction between the electrons and holes in the QW. The combined effects reduce the absorption as the QW fills with electrons. Since electron transfer across the spacer is a very fast process, these modulators can have high modulation speeds, demonstrated at almost 6 GHz.

4.12   ELECTRO-OPTIC AND ELECTROREFRACTIVE SEMICONDUCTOR MODULATORS

Some semiconductor modulators are based on phase modulation that is converted to amplitude modulation by using a Mach-Zehnder interferometer, in the same manner as discussed in Sec. 4.10. Such modulators can be integrated on the same substrate as the laser, but do not have the chirp issues that electroabsorption modulators exhibit.

Electro-Optic Effect in Semiconductors

The III-V semiconductors are electro-optic. Although they are not initially anisotropic, they become so when an electric field is applied, and so they can be used as phase modulators. Referring to the discussion of the electro-optic effect in Sec. 4.10 for definitions, the GaAs electro-optic coefficients have only one nonzero term: $r_{41} = r_{52} = r_{63} = 1.4 \times 10^{-12}$ m/V. Crystals are typically grown on the (001) face, with the Z axis normal to the surface. This means that the field is usually applied along Z. The only electro-optically induced index change will be $\Delta(1/n^2)_4 = r_{41}E_z$. Inserting this into the equation for the index ellipsoid, the electric field causes a rotation of the index ellipsoid around Z. Performing the diagonalization shows that the new values of the index ellipsoid are: $1/n_{x'}^2 = 1/n_o^2 + r_{41}E_z$ and $1/n_{y'}^2 = 1/n_o^2 - r_{41}E_z$. These axes are at 45° to the crystal axes.

Performing the differential gives the refractive index changes at 45° to crystal axes:

$$n_{x'} = n_o + \frac{n_o^3}{2} r_{41}E_z \quad \text{and} \quad n_{y'} = n_o - \frac{n_o^3}{2} r_{41}E_z. \tag{91}$$

The direction of these new optic axes (45° to the crystal axes) turns out to be in the direction that the zincblende material cleaves. Thus, TE-polarized light traveling down a waveguide normal to a cleave experiences the index change shown here. Light polarized along Z will not see any index change. Depending on whether light is polarized along X' or Y', the index will increase or decrease.[60]

With an electro-optic coefficient of $r_{41} = 1.4 \times 10^{-10}$ cm/V, in a field of 10 kV/cm (2 V across 2 μm), and since $n_o = 3.3$, the index change for the TE polarization in GaAs will be 2.5×10^{-5}. The index change in InP-based materials is comparable. The phase shift in a sample of length L is ΔnkL. At 1-μm wavelength, this will require a sample of length 1 cm to achieve a π phase shift, so that the voltage-length product for electro-optic GaAs (or other semiconductor) will be ~20 V·mm. Practical devices require larger refractive index changes, which can be achieved by using quantum wells and choosing the exciton resonance at a shorter wavelength than that of the light to be modulated. These wells have an electrorefractive effect.

Electrorefraction in Semiconductors

Near the band edge in semiconductors, the change in refractive index with applied field can be particularly large, especially in quantum wells, and is termed *electrorefraction,* or the *electrorefractive effect.* Electrorefraction is calculated from the spectrum of electroabsorption using the Kramers-Kronig relations. Enhanced electroabsorption means enhanced electrorefraction at wavelengths below the band edge. Electrorefraction allows significant reductions in length and drive voltages required for phase modulation in waveguides. The voltage-length product depends on how close to the absorption resonance the modulator is operated. It also depends on device design. As with electroabsorption modulators, the field is usually applied across a *pin* junction. Some reported π voltage-length products are 2.3 V·mm in GaAs/AlGaAs QW (at 25-V bias), 1.8 V·mm in InGaAs/InAlAs QW and 1.8 V·mm in GaAs/AlGaAs double heterostructures.[61] These voltage-length products depend on wavelength detuning from the exciton resonance and therefore on insertion and electroabsorption losses. The larger the voltage-length product, the greater the loss.

Typical Performance.   Electrorefraction is polarization dependent, because the quantum-confined Stark effect is polarization dependent. In addition, the TE polarization experiences the electro-optic effect, which may add to or subtract from the electrorefractive effect, depending on the crystal orientation. Typically, Δn for TE polarization (in an orientation that sums these effects) will be 8×10^{-4} at 82 kV/cm (7 V across a waveguide with an *i* layer 0.85 μm thick). Of this, the contribution from the electro-optic effect is 2×10^{-4}. Thus, electrorefraction is about 4 times larger than the electro-optic effect. The voltage-length product will thus be enhanced by a factor of four, reducing to 5 V·mm. Of course, this ratio depends on the field, since the electro-optic effect is linear in field and electrorefraction is quadratic in the field. This ratio also depends on wavelength; electrorefraction can be larger closer to the exciton resonance, but the residual losses go up. The TM polarization, which experiences electrorefraction alone, will be 5×10^{-4} at the same field, slightly smaller than the TE electrorefraction, because QCSE is smaller for TM than TE polarization.[62]

Advanced QW Concepts.   Compressive strain increases electrorefraction, as it does QCSE. Measurements at the same 82 kV/cm show an increase from 2.5×10^{-4} to 7.5×10^{-4} by increasing compressive strain.[63] Strained QWs also make it possible to achieve polarization-independent electrorefractive modulators (although when integrated with a semiconductor laser, which typically has a well-defined polarization, this should not be a necessity).

Advanced QW designs have the potential to increase the refractive index change below the exciton resonance. One example analyzes asymmetric coupled QWs and finds more than 10 times enhancement in Δn below the band edge, at least at small biases. However, when fabricated and incorporated into Mach-Zehnder modulators, the complex three-well structure lowered V_{π} by only a factor of 3, attributed to the growth challenges of these structures.[64]

Nipi *Modulators.*   One way to obtain a particularly low voltage-length product is to use MQWs in a hetero-*nipi* waveguide. These structures incorporate multiple *pin* junctions (alternating *n-i-p-i-n-i-p*) and include QWs in each *i* layer. Selective contacts to each electrode are required, which limits how fast the modulator can be switched. A voltage-length product of 0.8 V·mm was observed at a wavelength 115 meV below the exciton resonance. The lowest voltage InGaAs modulator had $V_{\pi} = 0.5$ V, at speeds up to 110 MHz. Faster speeds require shorter devices and higher voltages.[61]

Band-Filling Modulators.   When one operates sufficiently far from the band edge that the absorption is not large, then the electrorefractive effect is only 2 to 3 times larger than the bulk electro-optic effect. This is because oscillations in the change in absorption with wavelength tend to cancel out their contributions to the change in refractive index at long wavelengths. By contrast, the long-wavelength refractive index change during band filling is large because band filling decreases the absorption at all wavelengths. However, because band filling relies on real carriers, it lasts as long as the carriers do, and it is important to find ways to remove these carriers to achieve high-speed operation.

Voltage-controlled transfer of electrons into and out of QWs (BRAQWET modulator) can yield large electrorefraction by band filling (Sec. 4.11 discusses electroabsorption in this structure under "Electron Transfer Modulators"). The refractive index change at 1.55 μm can be as large as $\Delta n = 0.02$ for 6 V. One structure consists of 12 repeating elements,[65] with the single QW replaced by three closely spaced strongly coupled QWs, demonstrating $V_{\pi}L = 3.2$ V·mm with negligible loss.

Semiconductor Interferometric Modulators

The issues for Mach-Zehnder modulators fabricated in semiconductors are similar to those for modulators in lithium niobate, but the design and fabrication processes in semiconductors are by no means as finalized. Fabrication tolerances, polarization dependence, interaction with lasers, and operation at high optical input powers are just some of the issues that need to be addressed. The interferometer can be composed of Y branches, fabricated by etching to form ridge waveguides. Alternatively, 3-dB couplers are often formed by a *multimode interferometer* (MMI), composed of two parallel waveguides placed very close together with a bridging region that introduces coupling between them. Proper choice of this coupling region yields a 3-dB coupler.

One example reports a Mach-Zehnder interferometer at 1.55 μm in InGaAlAs QWs with InAlAs barriers.[66] A polarization-independent extinction ratio of 30 dB was reported, over a

20-nm wavelength range without degradation at input powers of 18 dBm (63 mW). The interferometer phase-shifting region was 1000 µm long, and each MMI was 200 µm long. The insertion loss of 13 dB was due to the mismatch between the mode of the single-mode optical fiber and of the semiconductor waveguide, which was 2 µm wide and 3.5 µm high. Various semiconductor structures to convert spot size should bring this coupling loss down.

4.13 PIN *DIODES*

The detectors used for fiber-optic communication systems are usually *pin photodiodes*. In high-sensitivity applications, such as long-distance systems operating at 1.55-µm wavelength, *avalanche photodiodes* are sometimes used because they have internal gain. Occasionally, *metal-semiconductor-metal* (MSM) *photoconductive detectors* with interdigitated electrode geometry are used because of ease of fabrication and integration. For the highest speed applications, *Schottky photodiodes* may be chosen. This section reviews properties of *pin* photodiodes. The next section outlines the other photodetectors.

The material of choice for these photodiodes depends on the wavelength at which they will be operated. The short-wavelength *pin* silicon photodiode is perfectly suited for GaAs wavelengths (850 nm); these inexpensive photodetectors are paired with GaAs/AlGaAs LEDs for low-cost data communications applications. They are also used in the shorter-wavelength plastic fiber applications at 650 nm. The longer-wavelength telecommunication systems at 1.3 and 1.55 µm require longer-wavelength detectors. These are typically *pin* diodes composed of lattice-matched ternary $In_{0.47}Ga_{0.53}As$ grown on InP. Silicon is an indirect bandgap semiconductor while InGaAs is a direct band material; this affects each material's absorption and therefore its photodiode design. In particular, silicon photodiodes tend to be slower than those made of GaAs or InGaAs, because silicon intrinsic regions must be thicker. Speeds are also determined by carrier mobilities, which are higher in the III-V materials.

Previous volumes in this handbook have outlined the concepts behind the photodetectors discussed here. Volume I, Chap. 15 places *pin* photodetectors in context with other detectors, and gives specific characteristics of some commercially available detectors, allowing direct comparison of silicon, InGaAsP, and germanium detectors. Volume I, Chap. 16, describes the principles by which the *pin* and the avalanche photodiodes operate (Figure 6 of that chapter contains a misprint, in that the central portion of the top layer should be labeled p^+ InP, rather than n^+ InP). The properties of greatest interest to fiber communications are repeated here. Volume I, Chap. 17 concentrates on high-speed photodetectors and provides particularly useful information on their design for high-speed applications. Finally, some of the key issues for photodetectors in fiber-optic communication systems were outlined in Vol. II, Chap. 10. This chapter considers these issues in much more detail.

The *pin* junction consists of a thin, heavily doped *n* region, a near-intrinsic *n* region (the *i* region), and a heavily doped *p* region. When an incident photon has energy greater than or equal to the bandgap of the semiconductor, electron-hole pairs are generated. In a well-designed photodiode, this generation takes place in the space-charge region of the *pn* junction. As a result of the electric field in this region, the electrons and holes separate and drift in opposite directions, causing current to flow in the external circuit. This current is monitored as a change in voltage across a load resistor. The *pin* photo-diode is the workhorse of fiber communication systems.

Typical Geometry

Typically, the electric field is applied across the *pn* junction and photocarriers are collected across the diode. A typical geometry for a silicon photodiode is shown in Fig. 30a. A *pn* junction is formed by a thin p^+ diffusion into a lightly doped n^- layer (also called the *i* layer since

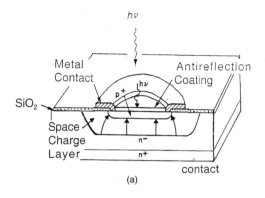

(a)

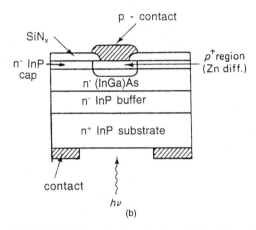

(b)

FIGURE 30 Geometry for *pin* photodiodes: (*a*) cutaway of silicon, illuminated from the top, showing the ring electrode and static electric field lines in the space-charge region; (*b*) cross-section of InGaAs/InP, illuminated from the bottom. The p^+ region is formed by diffusion. The low-doped n^- layer is the *i* or nearly intrinsic layer.

it is almost *intrinsic*) through a window in a protective SiO_2 film. The n^- region between the p^+ and n^+ regions supports a space-charge region, which, in the dark, is depleted of free carriers and supports the voltage drop that results from the *pn* junction. When light is absorbed in this space-charge region, the absorption process creates electron-hole pairs that separate in the electric field (field lines are shown in Fig. 30a), the electrons falling down the potential hill to the *n* region and the holes moving to the *p* region. This separation of charge produces a current in the external circuit, which is read out as a measure of the light level. Free carriers generated within a diffusion length of the junction may diffuse into the junction, adding to the measured current.

Long-wavelength detectors utilize n^- or *i* layers that are grown with a composition that will absorb efficiently in the wavelength region of interest. The ternary $In_{0.47}Ga_{0.53}As$ can be grown lattice-matched to InP and has a spectral response that is suitable for both the 1.3- and 1.55-μm wavelength regions. Thus, this ternary is usually the material of choice, rather than the more difficult to grow quaternary InGaAsP, although the latter provides more opportunity to tune the wavelength response. Figure 30b shows a typical geometry. Epitaxial growth is used to provide lightly doped n^- layers on a heavily doped n^+ substrate. The InP buffer layers are grown to keep the dopants from diffusing into the lightly doped absorbing InGaAs layer. The required thin p region is formed by diffusion through a silicon nitride insulating window. Because InP is transparent to 1.3 and 1.55 μm, the photodiode can be back-illuminated, which makes electrical contacting convenient. In some embodiments, a well is etched in the substrate and the fiber is glued in place just below the photosensitive region.

Carriers generated outside the depletion region may enter into the junction by diffusion, and can introduce considerable time delay. In silicon, the diffusion length is as long as 1 cm, so any photocarriers generated anywhere within the silicon photodiode can contribute to the photocurrent. Because the diffusion velocity is much slower than the transit time across the space-charge region, diffusion currents slow down silicon photodiodes. This is particularly true in pn diodes. Thus, high-speed applications typically use pin diodes with absorption only in the i layer.

To minimize diffusion from the p^+ entrance region, the junction should be formed very close to the surface. The space-charge region should be sufficiently thick that most of the light will be absorbed (thickness $\approx 1/\alpha$). With sufficient reverse bias, carriers will drift at their scattering-limited velocity. The space-charge layer must not be too thick, however, or transit-time effects will limit the frequency response. Neither should it be too thin, or excessive capacitance will result in a large RC time constant. The optimum compromise occurs when the modulation period is on the order of twice the transit time. For example, for a modulation frequency of 10 GHz, the optimum space-charge layer thickness in silicon is about 5 μm. However, this is not enough thickness to absorb more than ~50 percent of the light at 850 nm. Thus, there is a trade-off between sensitivity and speed. If the wavelength drops to 980 nm, only 10 percent of the light is absorbed in a 10-μm thickness of silicon space-charge layer.

The doping must be sufficiently small that the low n^- doped region can support the voltage drop of the built-in voltage V_{bi} plus the applied voltage. When the doping density of the p^+ region is much higher than the doping density of the n^- layer, the thickness of the space-charge layer is:

$$W_s = \sqrt{\frac{\dfrac{2\varepsilon_s}{e}(V_{bi} - V)}{N_D}} \tag{92}$$

To achieve $W_s = 10$ μm in a silicon photodiode with 10 V applied requires $N_D \approx 10^{14}$ cm^{-3}. If the doping is not this low, the voltage drops more rapidly, and the field will not extend fully across the low-doped region.

In InGaAs photodiodes (also GaAs/AlGaAs photodiodes), the n^+ and p^+ layers are transparent, and no photocarriers are generated in them. Thus, no photocarriers will enter from the n^+ and p^+ regions, even though the diffusion length is ~100 μm. The thickness of the i layer is chosen thin enough to achieve the desired speed (trading off transit time and capacitance), with a possible sacrifice of sensitivity.

Typically, light makes a single pass through the active layer. In silicon photodiodes, the light usually enters through the p contact at the surface of the diode (Fig. 30a); the top metal contact must have a window for light to enter (or be a transparent contact). The InGaAs photodiodes may receive light from the p side or the n side, because neither is absorbing. In addition, some back-illuminated devices use a double pass, reflecting off a mirrored top surface, to double the absorbing length. Some more advanced detectors, *resonant photodiodes*, use integrally grown Fabry-Perot cavities (using DBR mirrors, as in VCSELs) that resonantly reflect the light back and forth across the active region, enhancing the quantum efficiency. These are typically used only at the highest bandwidths (>20 GHz) or for wavelength division multiplexing (WDM) applications, where wavelength-selective photodetection is required. In addition, photodiodes designed for integration with other components are illuminated through a waveguide in the plane of the pn junction. The reader is directed to Vol. I, Chap. 17 to obtain more information on these advanced geometries.

Sensitivity (Responsivity)

To operate a *pin* photodiode, it is sufficient to place a load resistor between ground and the *n* side and apply reverse voltage to the *p* side ($V < 0$). The photocurrent is monitored as a voltage drop across this load resistor. The photodiode current in the presence of an optical signal of power P_s is negative, with a magnitude given by:

$$I = \eta_D \left(\frac{e}{h\nu} \right) P_s + I_D \qquad (93)$$

where I_D is the magnitude of the (negative) current measured in the dark. The detector *quantum efficiency* η_D (electron-hole pairs detected per photon) is determined by how much light is lost before reaching the space-charge region, how much light is absorbed (which depends on the absorption coefficient), and how much light is reflected from the surface of the photodiode (a loss which can be reduced by adding antireflective coatings). Finally, depending on design, there may be some loss from metal electrodes. These factors are contained in the following expression for the quantum efficiency:

$$\eta_D = (1 - R)T[1 - \exp(-\alpha W)] \qquad (94)$$

where R is the surface reflectivity, T is the transmission of any lossy electrode layers, W is the thickness of the absorbing layer, and α is its absorption coefficient.

The *sensitivity* (or *responsivity* $\Re$) of a detector is the ratio of milliamps of current out per milliwatt of light in. Thus, the responsivity is:

$$\Re = \frac{I_{PD}}{P_S} = \eta_D \frac{e}{h\nu} \qquad (95)$$

For detection of a given wavelength, the photodiode material must be chosen with a bandgap sufficient to provide suitable sensitivity. The absorption spectra of candidate detector materials are shown in Fig. 31. Silicon photodiodes provide low-cost detectors for most data communications applications, with acceptable sensitivity at 850 nm (absorption coefficient ~500 cm^{-1}). These detectors work well with the GaAs lasers and LEDs that are used in the inexpensive datacom systems and for short-distance or low-bandwidth local area network (LAN) applications. GaAs detectors are faster, both because their absorption can be larger and because their electron mobility is higher, but they are more expensive. Systems that require longer-wavelength InGaAsP/InP lasers typically use InGaAs photodiodes. Germanium has a larger dark current, so it is not usually employed for optical communications applications. Essentially all commercial photodetectors use bulk material, not quantum wells, as these are simpler, are less wavelength sensitive, and have comparable performance.

The spectral response of typical photodetectors is shown in Fig. 32. The detailed response depends on the detector design and on applied voltage, so these are only representative examples. Important communication wavelengths are marked.

Table 1 gives the sensitivity of typical detectors of interest in fiber communications, measured in units of amps per watt, along with speed and relative dark current.

Speed

Contributions to the speed of a *pin* diode come from the transit time across the space-charge region and from the RC time constant of the diode circuit in the presence of a load resistor R_L. Finally, in silicon there may be a contribution from the diffusion of carriers generated in undepleted regions.

In a properly designed *pin* photodiode, light should be absorbed in the space-charge region that extends from the p^+ junction across the low *n*-doped layer (the *i* layer). Equation

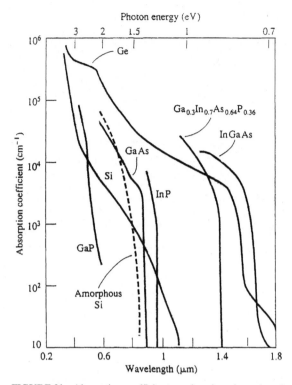

FIGURE 31 Absorption coefficient as a function of wavelength for several semiconductors used in *pin* diode detectors.

(92) gives the thickness of the space charge region W_s, as long as it is less than the thickness of the *i* layer W_i. Define V_i as that voltage at which $W_s = W_i$. Then

$$-V_i = W_i^2 \frac{eN_D}{2\varepsilon_s} - V_{bi}.$$

For any voltage larger than this, the space-charge width is essentially W_i, since the space charge extends a negligible distance into highly doped regions.

If the electric field across the space-charge region is high enough for the carriers to reach their saturation velocity v_s and high enough to fully deplete the *i* region, then the carrier transit time will be $\tau_i = W_i/v_s$. For $v_s = 10^7$ cm/s and $W_i = 4$ µm, the transit time $\tau_i = 40$ ps. It can be shown that a finite transit time τ_i reduces the response at modulation frequency ω[67]:

$$\Re(\omega) = \Re_o \frac{\sin(\omega\tau_i/2)}{\omega\tau_i/2} \tag{96}$$

Defining the 3-dB bandwidth as that modulation frequency at which the electrical power decreases by 50 percent, it can be shown that the transit-limited 3-dB bandwidth is $\delta\omega_i = 2.8/\tau_i = 2.8\,v_s/W_i$. (Electrical power is proportional to I^2 and $\Re^2$, so the half-power point is achieved when the current is reduced by $1/\sqrt{2}$.) There is a trade-off between diode sensitivity and diode transit time, since, for thin layers, from Eq. (94), $\eta_D \approx (1 - R)T\alpha W_i$. Thus, the quantum efficiency–bandwidth product is:

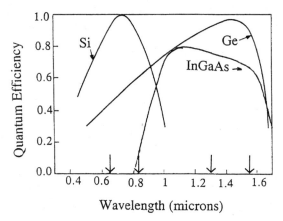

FIGURE 32   Spectral response of typical photodetectors.

$$\eta_D \, \delta\omega_i \approx 2.8\alpha v_s(1 - R)T \tag{97}$$

The speed of a *pin* photodiode is also limited by its capacitance, through the *RC* of the load resistor. Sandwiching a space-charge layer, which is depleted of carriers, between conductive *n* and *p* layers causes a diode capacitance proportional to the detector area *A*:

$$C_D = \frac{\varepsilon_s A}{W_i} \tag{98}$$

For a given load resistance, the smaller the area, the smaller the *RC* time constant, and the higher the speed. We will see also that the dark current I_s decreases as the detector area decreases. The detector should be as small as possible, as long as all the light from the fiber can be collected onto the detector. Multimode fibers easily butt-couple to detectors whose area matches the fiber core size. High-speed detectors compatible with single-mode fibers can be extremely small, but this increases the alignment difficulty; high-speed photodetectors can be obtained already pigtailed to single-mode fiber. A low load resistance may be needed to keep the *RC* time constant small, but this may result in a small signal that needs amplification. Speeds in excess of 1 GHz are straightforward to achieve, and speeds of 50 GHz are not uncommon.

Thicker space-charge regions provide smaller capacitance, but too thick a space charge region causes the speed to be limited by the carrier transit time. The bandwidth with a load resistor R_L is:

$$\omega_{3\,dB} = \frac{2.8}{\tau_i} + \frac{1}{R_L C} = \frac{2.8 v_s}{W_i} + \frac{W_i}{\varepsilon_s A R_L} \tag{99}$$

TABLE 1   Characteristics of Typical Photodiodes

	Wavelength, μm	Sensitivity $\mathfrak{R}$, As/W	Speed τ, ns	Dark current, normalized units
Sil	0.85	0.55	3	1
	0.65	0.4	3	
GaInAs	1.3–1.6	0.95	0.2	3
Ge (*pn*)	1.55	0.9	3	66

This shows that there is an optimum thickness W_i for high-speed operation. Any additional series resistance R_s or parasitic capacitance C_P must be added by using $R \rightarrow R_L + R_S$ and $C \rightarrow C + C_P$. The external connections to the photodetector can also limit speed. The gold bonding wire may provide additional series inductance. It is important to realize that the photodiode is a high impedance load, with very high electrical reflection, so that an appropriate load resistor must be used. As pointed out in Vol. I, Chap. 17, it is possible to integrate a matching load resistor inside the photodiode device, with a reduction in sensitivity of a factor of two (since half the photocurrent goes through the load resistor), but double the speed (since the RC time constant is halved). A second challenge is to build external bias circuits without high-frequency electrical resonances. Innovative design of the photodetector may integrate the necessary bias capacitor and load resistor, ensuring smooth electrical response.

Silicon photodetectors are inherently not as fast. Because their highly doped p and n regions are also absorbing, and because they are indirect bandgap materials and do not have as high an absorption coefficient, there will be a substantial contribution from carriers generated in undepleted regions. These carriers have to diffuse into the space charge region before they can be collected. Therefore, the photoresponse of the diode has a component of a slower response time governed by the carrier diffusion time:

$$T_D = \frac{W_D^2}{2D} \tag{100}$$

where W_D is the width of the absorbing undepleted region, and D is the diffusion constant for whichever carrier is dominant (usually holes in the n region). For silicon, $D = 12$ cm²/s, so that when $W_D = 10$ μm, $\tau_D = 40$ ns.

Dark Current

Semiconductor diodes can pass current even in the dark, giving rise to *dark current* that provides a background present in any measurement. This current comes primarily from the thermally generated diffusion of minority carriers out of the n and p regions into the depleted junction region, where they recombine. The current-voltage equation for a pn diode (in the dark) is:

$$I = I_S \left[\exp\left(\frac{eV}{\beta kT}\right) - 1 \right] \tag{101}$$

where I_S is the *saturation current* that flows at large back bias (V large and negative). This equation represents the current that passes through any biased pn junction. Photodiodes use pn junctions reverse biased ($V < 0$) to avoid large leakage current.

Here β is the *ideality factor*, which varies from 1 to 2, depending on the diode structure. In a metal-semiconductor junction (Schottky barrier) or an ideal pn junction in which the only current in the dark is due to minority carriers that diffuse from the p and n regions, then $\beta = 1$. However, if there is thermal generation and recombination of carriers in the space-charge region, then β tends toward the value 2. This is more likely to occur in long-wavelength detectors.

The saturation current I_S is proportional to the area A of the diode in an ideal junction:

$$I_S = e\, \frac{D_p p_{n0}}{L_p} + \frac{D_n n_{p0}}{L_n}\, A \tag{102}$$

where D_n, D_p are diffusion constants, L_n, L_p are diffusion lengths, and n_{p0}, p_{n0} are equilibrium minority carrier densities, all of electrons and holes, respectively. The saturation current I_S can be related to the diode resistance measured in the dark when $V = 0$. Defining

$$\frac{1}{R_0} = -\frac{\partial I}{\partial V}\bigg|_{V=0}$$

then:

$$R_0 = \frac{\beta kT}{eI_S} \tag{103}$$

The dark resistance is inversely proportional to the saturation current, and therefore to the area of the diode.

The diffusion current in Eq. (101) has two components that are of opposite sign in a forward-biased diode: a forward current $I_S \exp(eV/\beta kT)$ and a backward current $-I_S$. Each of these components is statistically independent, coming from diffusive contributions to the forward current and backward current, respectively. This fact is important in understanding the noise properties of photodiodes.

In photodiodes, $V \leq 0$. For clarity, write $V = -V'$ and use V' as a *positive* quantity in the equations that follow. For a reverse-biased diode in the dark, diffusion current flows as a negative *dark current,* with a magnitude given by

$$I_D = I_S \left[1 - \exp\left(\frac{-eV'}{\beta kT}\right) \right] \tag{104}$$

The negative dark current flows opposite to the current flow in a forward-biased diode. Holes move toward the p region and electrons move toward the n region; both currents are negative and add. This dark current adds to the negative photocurrent. The hole current must be thermally generated because there are no free holes in the n region to feed into the p region. By the same token, the electron current must be thermally generated since there are no free electrons in the p region to move toward the n region. The dark current at large reverse bias voltage is due to thermally generated currents.

Using Eq. (104) and assuming $eV' \gg kT$, the negative dark current equals the saturation current:

$$I_D = I_S \approx \left(\frac{\beta kT}{eR_0}\right) \tag{105}$$

It can be seen that the dark current increases linearly with temperature and is independent of (large enough) reverse bias. *Trap-assisted* thermal generation current increases β; in this process, carriers trapped in impurity levels can be thermally elevated to the conduction band. The temperature of photodiodes should be kept moderate in order to avoid excess dark current.

When light is present in a photodiode, the photocurrent is negative, in the direction of the applied voltage, and adds to the negative dark current. The net effect of carrier motion will be to tend to screen the internal field. Defining the magnitude of the photocurrent as $I_{PC} = \eta_D(e/h\nu)P_S$, then the total current is negative:

$$I = -[I_D + I_{PC}] = -I_S \left[1 - \exp\left(\frac{-eV'}{\beta kT}\right) \right] - I_{PC} \tag{106}$$

Noise in Photodiodes

Successful fiber-optic communication systems depend on a large signal-to-noise ratio. This requires photodiodes with high sensitivity and low noise. Background noise comes from shot noise due to the discrete process of photon detection, from thermal processes in the load resistor (Johnson noise), and from generation-recombination noise due to carriers within the semiconductor. When used with a field-effect transistor (FET) amplifier, there will also be shot noise from the amplifier and $1/f$ noise in the drain current.

Shot Noise.   *Shot noise* is fundamental to all photodiodes and is due to the discrete nature of the conversion of photons to free carriers. The shot noise current is a statistical process. If N photons are detected in a time interval Δt, Poisson noise statistics cause the uncertainty in N to be $\sqrt{N}$. Using the fact that N electron-hole pairs create a current I through $I = eN/\Delta t$, then the signal-to-noise ratio (SNR) is $N/\sqrt{N} = \sqrt{N} = \sqrt{(I\Delta t/e)}$. Writing the frequency bandwidth Δf in terms of the time interval through $\Delta f = 1/(2\Delta t)$ gives:

$$\text{SNR} = \sqrt{\frac{I}{2e\Delta f}}$$

The root mean square (rms) photon noise, given by $\sqrt{N}$, creates an rms shot noise current of:

$$i_{\text{SH}} = e\,\frac{\sqrt{N}}{\Delta t} = \sqrt{\frac{eI}{\Delta t}} = \sqrt{2eI\Delta f} \tag{107}$$

Shot noise depends on the average current I; therefore, for a given photodiode, it depends on the details of the current voltage characteristic. Expressed in terms of P_S, the optical signal power (when the dark current is small enough to be neglected), the rms shot noise current is

$$i_{\text{SH}} = \sqrt{2eI_{\text{PC}}\Delta f} = \sqrt{2e\Re P_s\Delta f} \tag{108}$$

where $\Re$ is the responsivity (or sensitivity), given in units of amps per watt.

The shot noise can be expressed directly in terms of the properties of the diode when all sources of noise are included. Since they are statistically independent, the contributions to the noise current will be additive. Noise currents can exist in both the forward and backward directions, and these contributions must add, along with the photocurrent contribution. The entire noise current squared becomes:

$$i_N^2 = 2e\left\{I_{\text{PC}} + \left(\frac{\beta kT}{eR_0}\right)\left[1 + \exp\left(\frac{-eV'}{\beta kT}\right)\right]\right\}\Delta f \tag{109}$$

Clearly, noise is reduced by increasing the reverse bias. When the voltage is large, the shot noise current squared becomes:

$$i_N^2 = 2e\left[I_{\text{PC}} + I_D\right]\Delta f \tag{110}$$

The dark current adds linearly to the photocurrent in calculating the shot noise.

In addition to shot noise due to the random variations in the detection process, the random thermal motion of charge carriers contributes to a *thermal noise* current, often called *Johnson* or *Nyquist noise*. It can be calculated by assuming thermal equilibrium with $V = 0$, $\beta = 1$, so that Eq. (109) becomes:

$$i_{\text{th}}^2 = 4\left(\frac{kT}{R_0}\right)\Delta f \tag{111}$$

This is just Johnson noise in the resistance of the diode. The noise appears as a fluctuating voltage, independent of bias level.

Johnson Noise from External Circuit.   An additional noise component will be from the load resistor R_L and resistance from the input to the preamplifier, R_i:

$$i_{NJ}^2 = 4kT \left(\frac{1}{R_L} + \frac{1}{R_i} \right) \Delta f \tag{112}$$

Note that the resistances add in parallel as they contribute to noise current.

Noise Equivalent Power.   The ability to detect a signal requires having a photocurrent equal to or higher than the noise current. The amount of noise that detectors produce is often characterized by the *noise equivalent power* (NEP), which is the amount of optical power required to produce a photocurrent just equal to the noise current. Define the noise equivalent photocurrent I_{NE}, which is set equal to the noise current i_{SH}. When the dark current is negligible,

$$i_{SH} = \sqrt{2eI_{NE}\Delta f} = I_{NE}$$

Thus, the noise equivalent current is $I_{NE} = 2e\Delta f$, and depends only on the bandwidth Δf. The noise equivalent power can now be expressed in terms of the noise equivalent current:

$$NEP = \frac{I_{NE}}{\eta} \frac{h\nu}{e} = 2 \frac{h\nu}{\eta} \Delta f \tag{113}$$

The second equality assumes the absence of dark current. In this case, the NEP can be decreased only by increasing the quantum efficiency (for a fixed bandwidth). In terms of sensitivity (amps per watt):

$$NEP = 2 \frac{e}{\Re} \Delta f = I_{NE} \Delta f \tag{114}$$

This expression is usually valid for photodetectors used in optical communication systems, which have small dark currents.

When dark current is dominant, $i_N = \sqrt{2e\, I_D\, \Delta f}$, so that:

$$NEP = \frac{I_{NE}}{\eta} \frac{h\nu}{e} = \sqrt{\frac{2I_D \Delta f}{e}} \frac{h\nu}{\eta} \tag{115}$$

This is often the case in infrared detectors such as germanium. Note that the dark-current-limited noise equivalent power is proportional to the square root of the area of the detector because the dark current is proportional to the detector area. The NEP is also proportional to the square root of the bandwidth Δf. Thus, in photodetectors whose noise is dominated by dark current, NEP divided by the square root of area times bandwidth should be a constant. The inverse of this quantity has been called the *detectivity D** and is often used to describe infrared detectors. In photodiodes used for communications, dark current usually does not dominate and it is better to use Eq. (114), an expression which is independent of area, but depends linearly on bandwidth.

4.14   *AVALANCHE PHOTODIODES, MSM DETECTORS, AND SCHOTTKY DIODES*

Avalanche Detectors

When large voltages are applied to photodiodes, the avalanche process produces gain, but at the cost of excess noise and slower speed. In fiber telecommunications applications, where speed and signal-to-noise are of the essence, avalanche photodiodes (APDs) are frequently at a disadvantage. Nonetheless, in long-haul systems at 2488 Mb/s, APDs may provide up to 10 dB greater sensitivity in receivers limited by amplifier noise. While APDs are inherently complex and costly to manufacture, they are less expensive than optical amplifiers and may be used when signals are weak.

Gain (Multiplication).   When a diode is subject to a high reverse-bias field, the process of impact ionization makes it possible for a single electron to gain sufficient kinetic energy to knock another electron from the valence to the conduction band, creating another electron-hole pair. This enables the quantum efficiency to be >1. This internal multiplication of photocurrent could be compared to the gain in photomultiplier tubes. The *gain* (or *multiplication*) *M* of an APD is the ratio of the photocurrent divided by that which would give unity quantum efficiency. Multiplication comes with a penalty of an excess noise factor, which multiplies shot noise. This excess noise is function of both the gain and the ratio of impact ionization rates between electrons and holes.

Phenomenologically, the low-frequency multiplication factor is:

$$M_{DC} = \frac{1}{1 - (V/V_B)^n} \tag{116}$$

where the parameter n varies between 3 and 6, depending on the semiconductor, and V_B is the breakdown voltage. Gains of $M > 100$ can be achieved in silicon APDs, while they are more typically 10 to 20 for longer-wavelength detectors, before multiplied noise begins to exceed multiplied signal. A typical voltage will be 75 V in InGaAs APDs, while in silicon it can be 400 V.

The avalanche process involves using an electric field high enough to cause carriers to gain enough energy to accelerate them into ionizing collisions with the lattice, producing electron-hole pairs. Then, both the original carriers and the newly generated carriers can be accelerated to produce further ionizing collisions. The result is an avalanche process.

In an *i* layer (where the electric field is uniform) of width W_i, the gain relates to the fundamental avalanche process through $M = 1/(1 - aW_i)$, where a is the *impact ionization coefficient*, which is the number of ionizing collisions per unit length. When $aW_i \rightarrow 1$, the gain becomes infinity and the diode breaks down. This means that avalanche multiplication appears in the regime before the probability of an ionizing collision is 100 percent. The gain is a strong function of voltage, and these diodes must be used very carefully. The total current will be the sum of avalanching electron current and avalanching hole current.

In most *pin* diodes the *i* region is really low *n*-doped. This means that the field is not exactly constant, and an integration of the avalanche process across the layer must be performed to determine a. The result depends on the relative ionization coefficients; in III-V materials they are approximately equal. In this case, aW_i is just the integral of the ionizing coefficient that varies rapidly with electric field.

Separate Absorber and Multiplication (SAM) APDs.   In this design the long-wavelength infrared light is absorbed in an intrinsic narrow-bandgap InGaAs layer and photocarriers move to a separate, more highly *n*-doped InP layer that supports a much higher field. This layer is designed to provide avalanche gain in a separate region without excessive dark currents from tunneling processes. This layer typically contains the *pn* junction, which traditionally has been diffused. Fabrication procedures such as etching a mesa, burying it, and introducing a guard ring electrode are all required to reduce noise and dark current. All-epitaxial structures provide low-cost batch-processed devices with high performance characteristics.[68]

Speed.   When the gain is low, the speed is limited by the *RC* time constant. As the gain increases, the avalanche buildup time limits the speed, and for modulated signals the multiplication factor decreases. The multiplication factor as a function of modulation frequency is:

$$M(\omega) = \frac{M_{DC}}{\sqrt{1 + M_{DC}^2 \omega^2 \tau_1^2}} \qquad (117)$$

where $\tau_1 = p\tau$, where τ is the multiplication-region transit time and p is a number that changes from 2 to ⅓ as the gain changes from 1 to 1000. The gain decreases from its low-frequency value when $M_{DC}\omega = 1/\tau_1$. It can be seen that it is the gain-bandwidth product that describes the characteristics of an avalanche photodiode in a communication system.

Noise. The shot noise in an APD is that of a *pin* diode multiplied by M^2 times an excess noise factor F_e:

$$i_S^2 = 2e \, I_{PC} \, \Delta f \, M^2 \, F_e \qquad (118)$$

where

$$F_e(M) = \beta M + (1 - \beta)\left(2 - \frac{1}{M}\right)$$

In this expression, β is the ratio of the ionization coefficient of the opposite type divided by the ionization coefficient of the carrier type that initiates multiplication. In the limit of equal ionization coefficients of electrons and holes (usually the case in III-V semiconductors), $F_e = M$ and $F_h = 1$. Typical numerical values for enhanced APD sensitivity are given in Vol. I, Chap. 17, Fig. 15.

Dark Current. In an APD, dark current is the sum of the unmultiplied current I_{du}, mainly due to surface leakage, and the bulk dark current experiencing multiplication I_{dm}, multiplied by the gain:

$$I_d = I_{du} + MI_{dm} \qquad (119)$$

The shot noise from dark (leakage) current i_d:

$$i_d^2 = 2e \, [i_{du} + I_{dm}M^2 \, F_e(M)] \, \Delta f \qquad (120)$$

The proper use of APDs requires choosing the proper design, carefully controlling the voltage, and using the APD in a suitably designed system, since the noise is so large.

MSM Detectors

Volume I, Chap. 17, Fig. 1 of this handbook shows that interdigitated electrodes on top of a semiconductor can provide a planar configuration for electrical contacts. Either a *pn* junction or bulk semiconductor material can reside under the interdigitated fingers. The MSM geometry has the advantage of lower capacitance for a given cross-sectional area, but the transit times may be longer, limited by the lithographic ability to produce very fine lines. Typically, MSM detectors are photoconductive. Volume I, Chap. 17, Fig. 17 shows the geometry of high-speed interdigitated photoconductors. These are simple to fabricate and can be integrated in a straightforward way onto MESFET preamplifiers.

Consider parallel electrodes deposited on the surface of a photoconductive semiconductor with a distance L between them. Under illumination, the photocarriers will travel laterally to the electrodes. The photocurrent in the presence of P_s input optical flux at photon energy $h\nu$ is:

$$I_{ph} = q\eta GP \, h\nu \qquad (121)$$

The photoconductive gain G is the ratio of the carrier lifetime τ to the carrier transit time τ_{tr}:

$$G = \frac{\tau}{\tau_{tr}}$$

Decreasing the carrier lifetime increases the speed but decreases the sensitivity.

The output signal is due to the time-varying resistance that results from the time-varying photoinduced carrier density $N(t)$:

$$R_s(t) = \frac{L}{eN(t)\,\mu w d_e} \tag{122}$$

where μ is the sum of the electron and hole mobilities, w is the length along the electrodes excited by light, and d_e is the effective absorption depth into the semiconductor.

Usually, MSM detectors are not the design of choice for high-quality communication systems. Nonetheless, their ease of fabrication and integration with other components makes them desirable for some low-cost applications—for example, when there are a number of parallel channels and dense integration is required.

Schottky Photodiodes

A Schottky photodiode uses a metal-semiconductor junction rather than a *pin* junction. An abrupt contact between metal and semiconductor can produce a space-charge region. Absorption of light in this region causes photocurrent that can be detected in an external circuit. Because metal-semiconductor diodes are majority carrier devices they may be faster than *pin* diodes (they rely on drift currents only, there is no minority carrier diffusion). Up to 100 GHz modulation has been reported in a 5- $\times$ 5-μm area detector with a 0.3-μm thin drift region using a semitransparent platinum film 10 nm thick to provide the abrupt Schottky contact. Resonance enhancement of the light has been used to improve sensitivity.

4.15   REFERENCES

1. D. Botez, *IEEE J. Quantum Electron.* **17**:178 (1981).
2. See, for example, E. Garmire and M. Tavis, *IEEE J. Quantum Electron.* **20**:1277 (1984).
3. B. B. Elenkrig, S. Smetona, J. G. Simmons, T. Making, and J. D. Evans, *J. Appl. Phys.* **85**:2367 (1999).
4. M. Yamada, T. Anan, K. Tokutome, and S. Sugou, *IEEE Photon. Technol. Lett.* **11**:164 (1999).
5. T. C. Hasenberg and E. Garmire, *IEEE J. Quantum Electron.* **23**:948 (1987).
6. D. Botez and M. Ettenberg, *IEEE J. Quantum Electron.* **14**:827 (1978).
7. G. P. Agrawal and N. K. Dutta, *Semiconductor Lasers,* 2d ed., Van Nostrand Reinhold, New York, 1993, Sec. 6.4.
8. G. H. B. Thompson, *Physics of Semiconductor Laser Devices,* John Wiley & Sons, New York, 1980, Fig. 7.8
9. K. Tatah and E. Garmire, *IEEE J. Quantum Electron.* **25**:1800 (1989).
10. Agrawal and Dutta, Sec. 6.4.3.
11. W. H. Cheng, A. Mar, J. E. Bowers, R. T. Huang, and C. B. Su, *IEEE J. Quantum Electron.* **29**:1650 (1993).
12. J. T. Verdeyen, *Laser Electronics,* 3d ed., Prentice Hall, Englewood Cliffs, N.J., 1995, p. 490.
13. Agrawal and Dutta, Eq. 6.6.32.
14. N. K. Dutta, N. A. Olsson, L. A. Koszi, P. Besomi, and R. B. Wilson, *J. Appl. Phys.* **56**:2167 (1984).
15. Agrawal and Dutta, Sec. 6.6.2.

16. Agrawal and Dutta, Sec. 6.5.2.

17. L. A. Coldren and S. W. Corizine, *Diode Lasers and Photonic Integrated Circuits,* John Wiley & Sons, New York, 1995, Sec. 5.5.

18. M. C. Tatham, I. F. Lealman, C. P. Seltzer, L. D. Westbrook, and D. M. Cooper, *IEEE J. Quantum Electron.* **28**:408 (1992).

19. H. Jackel and G. Guekos, *Opt. Quantum Electron.* **9**:223 (1977).

20. M. K. Aoki, K. Uomi, T. Tsuchiya, S. Sasaki, M. Okai, and N. Chinone, *IEEE J. Quantum Electron.* **27**:1782 (1991).

21. R. W. Tkach and A. R. Chaplyvy, *J. Lightwave Technol.* **LT-4**:1655 (1986).

22. K. Petermann, *IEEE J. Sel. Topics in Qu. Electr.* **1**:480 (1995).

23. T. Hirono, T. Kurosaki, and M. Fukuda, *IEEE J. Quantum Electron.* **32**:829 (1996).

24. Y. Kitaoka, *IEEE J. Quantum Electron.* **32**:822 (1996), Fig. 2.

25. M. Kawase, E. Garmire, H. C. Lee, and P. D. Dapkus, *IEEE J. Quantum Electron.* **30**:981 (1994).

26. Coldren and Corizine, Sec. 4.3.

27. S. L. Chuang, *Physics of Optoelectronic Devices,* John Wiley & Sons, New York, 1995, Fig. 10.33.

28. T. L. Koch and U. Koren, *IEEE J. Quantum Electron.* **27**:641 (1991).

29. B. G. Kim and E. Garmire, *J. Opt. Soc. Am.* **A9**:132 (1992).

30. A. Yariv, *Optical Electronics,* 4th ed., (Saunders, Philadelphia, Pa., 1991) Eq. 13.6-19.

31. Information from www.telecomdevices.com.

32. C. L. Jiang and B. H. Reysen, *Proc. SPIE* **3002**:168 (1997), Fig. 7.

33. See, for example, Pallab Bhattacharya, *Semiconductor Optoelectronic Devices,* Prentice-Hall, New York, 1998, App. 6.

34. C. H. Chen, M. Hargis, J. M. Woodall, M. R. Melloch, J. S. Reynolds, E. Yablonovitch, and W. Wang, *Appl. Phys. Lett.* **74**:3140 (1999).

35. Y. A. Wu, G. S. Li, W. Yuen, C. Caneau, and C. J. Chang-Hasnain, *IEEE J. Sel. Topics in Quantum Electron.* **3**:429 (1997).

36. See, for example, F. L. Pedrotti and L. S. Pedrotti, *Introduction to Optics,* Prentice-Hall, Englewood Cliffs, N.J., 1987.

37. B. Weigl, M. Grabherr, C. Jung, R. Jager, G. Reinėr, R. Michalzik, D. Sowada, and K. J. Ebeling, *IEEE J. Sel. Top. in Quantum Electron.* **3**:409 (1997).

38. J. W. Law and G. P. Agrawal, *IEEE J. Sel. Top. Quantum Electron.* **3**:353 (1997).

39. J. Boucart et al., *IEEE J. Sel. Top. Quantum Electron.* **5**:520 (1999).

40. S. K. Korotky and R. C. Alferness, "Ti:LiNbO$_3$ Integrated Optic Technology," in (ed.), L. D. Hutcheson, *Integrated Optical Circuits and Components,* Dekker, New York, 1987.

41. G. Y. Wang and E. Garmire, *Opt. Lett.* **21**:42 (1996).

42. Yariv, Table 9.2.

43. G. D. Boyd, R. C. Miller, K. Nassau, W. L. Bond, and A. Savage, *Appl. Phys. Lett.* **5**:234 (1964).

44. Information from www.utocorp.com, June 1999.

45. F. P. Leonberger and J. P. Donnelly, "Semiconductor Integrated Optic Devices," in T. Tamir (ed.), *Guided Wave Optoelectronics,* Springer-Verlag, 1990, p. 340.

46. C. C. Chen, H. Porte, A. Carenco, J. P. Goedgebuer, and V. Armbruster, *IEEE Photon. Technol. Lett.* **9**:1361 (1997).

47. C. T. Mueller and E. Garmire, *Appl. Opt.* **23**:4348 (1984).

48. H. Kogelnik, "Theory of Dielectric Waveguides," in T. Tamir (ed.), *Integrated Optics,* Springer-Verlag, 1975.

49. H. Q. Hou, A. N. Cheng, H. H. Wieder, W. S. C. Chang, and C. W. Tu, *Appl. Phys. Lett.* **63**:1833 (1993).

50. A. Ramdane, F. Devaux, N. Souli, D. Dalprat, and A. Ougazzaden, *IEEE J. Sel. Top. in Quantum Electron.* **2**:326 (1996).

51. S. D. Koehler and E. M. Garmire, in T. Tamir, H. Bertoni, and G. Griffel (eds.), *Guided-Wave Optoelectronics: Device Characterization, Analysis and Design,* Plenum Press, New York, 1995.

52. See, for example, S. Carbonneau, E. S. Koteles, P. J. Poole, J. J. He, G. C. Aers, J. Haysom, M. Buchanan, Y. Feng, A. Delage, F. Yang, M. Davies, R. D. Goldberg, P. G. Piva, and I. V. Mitchell, *IEEE J. Sel. Top. in Quantum Electron.* **4**:772 (1998).

53. J. A. Trezza, J. S. Powell, and J. S. Harris, *IEEE Photon. Technol. Lett.* **9**:330 (1997).

54. K. Wakita, I. Kotaka, T. Amano, and H. Sugiura, *Electron. Lett.* **31**:1339 (1995).

55. F. Devaux, J. C. Harmand, I. F. L. Dias, T. Guetler, O. Krebs, and P. Voisin, *Electron. Lett.* **33**:161 (1997).

56. H. Yamazaki, Y. Sakata, M. Yamaguchi, Y. Inomoto, and K. Komatsu, *Electron. Lett.* **32**:109 (1996).

57. K. Morito, K. Sato, Y. Kotaki, H. Soda, and R. Sahara, *Electron. Lett.* **31**:975 (1995).

58. F. Devaux, E. Bigan, M. Allovon, et al., *Appl. Phys. Lett.* **61**:2773 (1992).

59. J. E. Zucker, K. L. Jones, M. Wegener, T. Y. Chang, N. J. Sauer, M. D. Divino, and D. S. Chemla, *Appl. Phys. Lett.* **59**:201 (1991).

60. M. Jupina, E. Garmire, M. Zembutsu, and N. Shibata, *IEEE J. Quantum Electron.* **28**:663 (1992).

61. S. D. Koehler, E. M. Garmire, A. R. Kost, D. Yap, D. P. Doctor, and T. C. Hasenberg, *IEEE Photon. Technol. Lett.* **7**:878 (1995).

62. A. Sneh, J. E. Zucker, B. I. Miller, and L. W. Stultz, *IEEE Photon. Technol. Lett.* **9**:1589 (1997).

63. J. Pamulapati, et al., *J. Appl. Phys.* **69**:4071 (1991).

64. H. Feng, J. P. Pang, M. Sugiyama, K. Tada, and Y. Nakano, *IEEE J. Quantum Electron.* **34**:1197 (1998).

65. J. Wang, J. E. Zucker, J. P. Leburton, T. Y. Chang, and N. J. Sauer, *Appl. Phys. Lett.* **65**:2196 (1994).

66. N. Yoshimoto, Y. Shibata, S. Oku, S. Kondo, and Y. Noguchi, *IEEE Photon. Technol. Lett.* **10**:531 (1998).

67. Yariv, Sec. 11.7.

68. E. Hasnain et al., *IEEE J. Quantum Electron.* **34**:2321 (1998).

CHAPTER 5
OPTICAL FIBER AMPLIFIERS

John A. Buck
School of Electrical and Computer Engineering
Georgia Institute of Technology
Atlanta, Georgia

5.1 INTRODUCTION

The development of rare-earth-doped fiber amplifiers has led to dramatic increases in the channel capacities of fiber communication systems, and has provided the key components in many new forms of optical sources and signal processing devices. The most widely used fiber amplifiers are formed by doping the glass fiber host with erbium ions, from which gain by stimulated emission occurs at wavelengths in the vicinity of 1.55 µm. The amplifiers are optically pumped using light at either 1.48-µm or 0.98-µm wavelengths. Other rare-earth dopants include praseodymium, which provides gain at 1.3 µm and which is pumped at 1.02 µm,[1] ytterbium, which amplifies from 975 to 1150 nm using pump wavelengths between 910 and 1064 nm,[2] and erbium-ytterbium codoping, which enables use of pump light at 1.06 µm while providing gain at 1.55 µm.[3] Additionally, thulium- and thulium/terbium-doped fluoride fibers have been constructed for amplification at 0.8, 1.4, and 1.65 µm.[4] Aside from systems applications, much development has occurred in fiber ring lasers based on erbium-doped-fiber amplifiers (EDFAs),[5] in addition to optical storage loops[6] and nonlinear switching devices.[7]

The original intent in fiber amplifier development was to provide a simpler alternative to the electronic repeater by allowing the signal to remain in optical form throughout a link or network. Fiber amplifiers as repeaters offer additional advantages, which include the ability to change system data rates as needed, or to simultaneously transmit multiple rates—all without the need to modify the transmission channel. A further advantage is that signal power at multiple wavelengths can be simultaneously boosted by a single amplifier—a task that would otherwise require a separate electronic repeater for each wavelength. This latter feature contributed to the realization of dense wavelength-division multiplexed (DWDM) systems, in which terabit/sec data rates have been demonstrated.[8] The usable gain in an EDFA occupies a wavelength range spanning 1.53 to 1.56 µm. In DWDM systems this allows, for example, the use of some 40 channels having 100-GHz spacing. A fundamental disadvantage of the fiber amplifier as a repeater is that dispersion is not reset. This requires additional network design efforts in dispersion management,[9] which may include optical equalization methods.[10] The deployment of fiber amplifiers in commercial networks demonstrates the move toward *transparent* fiber systems, in which signals are maintained in optical form, and in which multiple wavelengths, data rates, and modulation formats are supported.

Aside from rare-earth-doped glass fibers, which provide gain through stimulated emission, there has been renewed interest in fiber Raman amplifiers, in which gain at the signal wavelength occurs as a result of glass-mediated coupling to a shorter-wavelength optical pump.[11] Raman amplifiers have recently been demonstrated in DWDM systems that operate in the vicinity of 1.3 μm.[12] This chapter emphasizes the rare-earth systems—particularly erbium-doped fiber amplifiers, since these are the most important ones in practical use.

5.2 RARE-EARTH-DOPED AMPLIFIER CONFIGURATION AND OPERATION

Pump Configuation and Optimum Amplifier Length

A typical fiber amplifier configuration consists of the doped fiber positioned between polarization-independent optical isolators. Pump light is input by way of a wavelength-selective coupler which can be configured for forward, backward, or bidirectional pumping (see Fig. 1). Pump absorption throughout the amplifier length results in a population inversion that varies with position along the fiber; this reaches a minimum at the fiber end opposite the pump laser for unidirectional pumping, or minimizes at midlength for bidirectional pumping using equal pump powers. To achieve the highest overall gain, the length is chosen so that the fiber is transparent to the signal at the point of minimum pump power. For example, using forward pumping, the optimum fiber length is determined by requiring transparency to occur at the output end. If longer fiber lengths are used, some reabsorption of the signal will occur beyond the transparency point. With lengths shorter than the optimum, full use is not made of the available pump energy. Other factors may modify the optimum length, particularly if substantial gain saturation occurs, or if amplified spontaneous emission (ASE) is present, which can result in additional gain saturation and noise.[13]

Isolators maintain unidirectional light propagation so that, for example, backscattered or reflected light from further down the link cannot reenter the amplifier and cause gain quenching, noise enhancement, or possibly lasing. Double-pass and segmented configurations are also used; in the latter, isolators are positioned between two or more lengths of amplifying fiber that

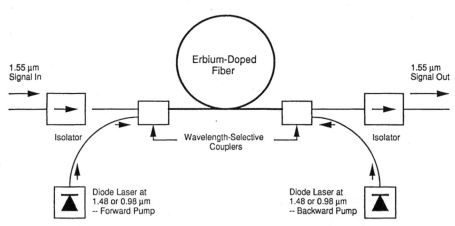

FIGURE 1   General erbium-doped fiber configuration, showing bidirectional pumping.

are separately pumped. The result is that gain quenching and noise arising from back-scattered light or from ASE can be lower than those of a single fiber amplifier of the combined lengths.

Regimes of Operation

There are roughly three operating regimes, the choice between which is determined by the use intended for the amplifier.[14, 15] These are (1) small-signal, or linear, (2) saturation, and (3) deep saturation regimes. In the linear regime, low input signal levels (<1 μW) are amplified with negligible gain saturation, assuming the amplifier length has been optimized. Amplifier gain in decibels is defined in terms of the input and output signal powers as $G(\text{dB}) = 10 \log_{10} (P_s^{\text{out}}/P_s^{\text{in}})$. EDFA small signal gains range between 25 and 35 dB.[15]

In the saturation regime, the input signal level is high enough to cause a measurable reduction in the net gain. A useful figure of merit is the *input saturation power,* defined as the input signal power required to reduce the net amplifier gain by 3 dB, assuming an optimized fiber length. Specifically, the gain in this case is $G = G_{\text{max}} - 3$ dB, where G_{max} is the small-signal gain. A related parameter is the *saturation output power,* $P_{\text{sat}}^{\text{out}}$, defined as the amplifier output that is achieved when the overall gain is reduced by 3 dB. The two quantities are thus related through $G_{\text{max}} - 3\text{dB} = 10 \log_{10} (P_{\text{sat}}^{\text{out}}/P_{\text{sat}}^{\text{in}})$. Using these parameters, the *dynamic range* of the amplifier is defined through $P_s^{\text{in}} \leq P_{\text{sat}}^{\text{in}}$, or equivalently $P_s^{\text{out}} \leq P_{\text{sat}}^{\text{out}}$. For an N-channel wavelength-division multiplexed signal, the dynamic range is reduced accordingly by a factor of 1/N, assuming a flat gain spectrum.[15]

With the amplifier operating in deep saturation, gain reductions on the order of 20 to 40 dB occur.[14] This is typical of *power amplifier* applications, in which input signal levels are high, and where the maximum output power is desired. In this application, the concept of *power conversion efficiency* (PCE) between pump and signal becomes important. It is defined as $\text{PCE} = (P_s^{\text{out}} - P_s^{\text{in}})/P_p^{\text{in}}$, where P_p^{in} is the input pump power. Another important quantity that is pertinent to the deep saturation regime is the *saturated output power,* P_s^{out} (max), not to be confused with the saturation output power previously described. P_s^{out} (max) is the maximum output signal power that can be achieved for a given input signal level and available pump power. This quantity would maximize when the amplifier, having previously been fully inverted, is then completely saturated by the signal. Maximum saturation, however, requires the input signal power to be extremely high, such that ultimately, P_s^{out} (max) $\approx P_s^{\text{in}}$, representing a net gain of nearly 0 dB. Clearly the more important situations are those in which moderate signal powers are to be amplified; in these cases the choice of pump power and pumping configuration can substantially influence P_s^{out} (max).

5.3 EDFA PHYSICAL STRUCTURE AND LIGHT INTERACTIONS

Energy Levels in the EDFA

Gain in the erbium-doped fiber system occurs when an inverted population exists between parts of the $^4I_{13/2}$ and $^4I_{15/2}$ states, as shown in Fig. 2a. This notation uses the standard form, $^{(2S+1)}L_J$, where L, S, and J are the orbital, spin, and total angular momenta, respectively. EDFAs are manufactured by incorporating erbium ions into the glass matrix that forms the fiber core. Interactions between the ions and the host matrix induce Stark splitting of the ion energy levels, as shown in Fig. 2a. This produces an average spacing between adjacent Stark levels of 50 cm^{-1}, and an overall spread of 300 to 400 cm^{-1} within each state. A broader emission spectrum results, since more deexcitation pathways are produced, which occur at different transition wavelengths.

Other mechanisms further broaden the emission spectrum. First, the extent to which ions interact with the glass varies from site to site, as a result of the nonuniform structure of the

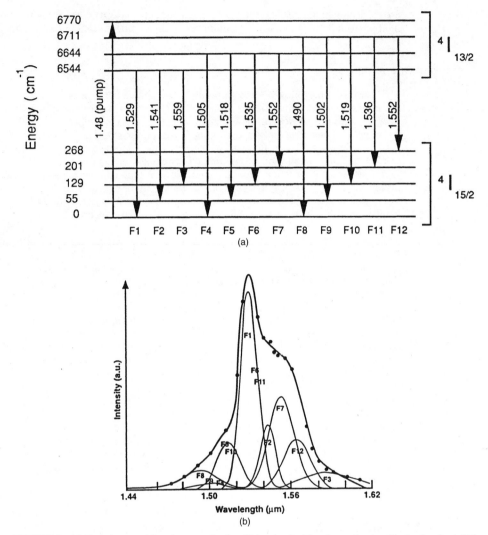

FIGURE 2 (*a*) Emissive transitions between Stark-split levels of erbium in an aluminosilicate glass host. Values on transition arrows indicate wavelengths in micrometers. *(Adapted from Ref. 19).* (*b*) EDFA fluorescence spectrum associated with the transitions in Figure 2a. *(Reprinted with permission from Optical Fiber Amplifiers: Materials, Devices, and Applications, by S. Sudo. Artech House Publishers, Norwood, MA, USA. www.artech-house.com).*

amorphous glass matrix. This produces some degree of inhomogeneous broadening in the emission spectrum, the extent of which varies with the type of glass host used.[16] Second, thermal fluctuations in the material lead to homogeneous broadening of the individual Stark transitions. The magnitudes of the two broadening mechanisms are 27 to 60 cm^{-1} for inhomogeneous, and 8 to 49 cm^{-1} for homogeneous.[16] The choice of host material strongly affects the shape of the emission spectrum, owing to the character of the ion-host interactions. For example, in pure silica (SiO$_2$), the spectrum of the Er-doped system is narrowest and has the least smoothness. Use of an aluminosilicate host (SiO_2-Al_2O_3) produces slight broadening and smoothing.[17] The broadest spectra, however, occur when using fluoride-based glass, such as ZBLAN.[18]

Gain Formation

Fig. 2b shows how the net emission (fluorescence) spectrum is constructed from the super-position of the individual Stark spectra; the latter are associated with the transitions shown in Fig. 2a. Similar diagrams can be constructed for the upward (absorptive) transitions, from which the absorption spectrum can be developed.[19] The shapes of both spectra are further influenced by the populations within the Stark split levels, which assume a Maxwell-Boltzman distribution. The sequence of events in the population dynamics is (1) pump light boosts population from the ground state, $^4I_{15/2}$, to the upper Stark levels in the first excited state, $^4I_{13/2}$; (2) the upper state Stark level populations thermalize; and (3) deexcitation from $^4I_{13/2}$ to $^4I_{15/2}$ occurs through either spontaneous or stimulated emission.

The system can be treated using a simple two-level (1.48-μm pump) or three-level model (0.98-μm pump), from which rate equations can be constructed that incorporate the actual wavelength- and temperature-dependent absorption and emission cross sections. These models have been formulated with and without inhomogeneous broadening, but in most cases excellent agreement with experiment has been achieved by assuming only homogeneous broadening.[20, 21, 22]

Pump Wavelength Options in EDFAs

The 1.48-μm pump wavelength corresponds to the energy difference between the two most widely spaced Stark levels, as shown in Fig. 2a. A better alternative is to pump with light at 0.98 μm, which boosts the ground state population to the second excited state, $^4I_{11/2}$, which lies above $^4I_{13/2}$. This is followed by rapid nonradiative decay into $^4I_{13/2}$ and gain is formed as before. The pumping efficiency suffers slightly at 0.98 μm, owing to some excited state absorption (ESA) from $^4I_{11/2}$ to the higher-lying $^4F_{7/2}$.[23] Use of 0.98-μm pump light as opposed to 1.48 μm will nevertheless yield a more efficient system, since the 0.98-μm pump will not contribute to the deexcitation process, which occurs when 1.48 μm is used.

The *gain efficiency* of a rare-earth-doped fiber is defined as the ratio of the maximum small signal gain to the input pump power, using the optimized fiber length. EDFA efficiencies are typically on the order of 10 dB/mW for pumping at .98 μm. For pumping at 1.48 μm, efficiencies are about half the values obtainable at .98 μm, and require about twice the fiber length. Other pump wavelengths can be used,[24] but with some penalty to be paid in the form of excited state absorption from the $^4I_{13/2}$ state into various upper levels, thus depleting the gain that would otherwise be available. This problem is minimized when using either 0.98 or 1.48 μm, and so these two wavelengths are almost exclusively used in erbium-doped fibers.

Noise

Performance is degraded by the presence of noise from two fundamental sources. These are (1) amplified spontaneous emission (ASE) and (2) Rayleigh scattering. Both processes lead to additional light that propagates in the forward and backward directions, and which can encounter considerable gain over long amplifier lengths. The more serious of the two noise sources is ASE. In severe cases, involving high-gain amplifiers of long lengths, ASE can be of high enough intensity to partially saturate the gain (self-saturation), thus reducing the available gain for signal amplification. Backward pumping has been found to reduce this effect.[25]

In general, ASE can be reduced by (1) assuring that the population inversion is as high as possible (ideally, completely inverted); (2) operating the amplifier in the deep saturation regime; or (3) using two or more amplifier stages rather than one continuous length of fiber, and positioning bandpass filters and isolators between stages. Rayleigh scattering noise can be minimized by using multistage configurations, in addition to placing adequate controls on dopant concentration and confinement during the manufacturing stage.[26]

The *noise figure* of a rare-earth-doped fiber amplifier is stated in a manner consistent with the IEEE standard definition for a general amplifier. This is the signal-to-noise ratio of the fiber amplifier input divided by the signal-to-noise ratio of the output, expressed in decibels, where the input signal is shot noise limited. Although this definition is widely used, it has recently come under some criticism, owing to the physical nature of ASE noise, and the resulting awkwardness in applying the definition to cascaded amplifier systems.[27] The best noise figures for EDFAs are obtained by using whatever pump configurations produce the highest population inversions. Again, use of .98-µm pumping is preferred, with noise figures of about 3 dB obtainable; using 1.48 µm yields best results of about 4 dB.[14]

Gain Flattening

The use of multiple wavelength channels motivates the need to effectively flatten the emission spectrum, and thus equalize the gain for all wavelengths. Flattening techniques can be classified into roughly three categories. First, intrinsic methods can be used; these involve choices of fiber host materials such as fluoride glass[28] that yield smoother and broader gain spectra. In addition, by carefully choosing pump power levels, a degree of population inversion can be obtained which will allow some cancellation to occur between the slopes of the absorption and emission spectra.[29] Second, spectral filtering at the output of a single amplifier or between cascaded amplifiers can be employed; this effectively produces higher loss for wavelengths that have achieved higher gain. Examples of successful filtering devices include long-period fiber gratings[30] and Mach-Zehnder filters.[31] Third, hybrid amplifiers that use cascaded configurations of different gain media can be used to produce an overall gain spectrum that is reasonably flat. Flattened gain spectra have been obtained having approximate widths that range from 12 to 85 nm. Reference 32 is recommended for an excellent discussion and comparison of the methods.

5.4 GAIN FORMATION IN OTHER RARE-EARTH SYSTEMS

Praseodymium-Doped Fiber Amplifiers (PDFAs)

In the praseodymium-doped fluoride system, the strongest gain occurs in the vicinity of 1.3 µm, with the pump wavelength at 1.02 µm. Gain formation is described by a basic three-level model, in which pump light excites the system from the ground state, $^3H_4$, to the metastable excited state, $^1G_4$. Gain for 1.3-µm light is associated with the downward $^1G_4 \rightarrow ^3H_5$ transition, which peaks in the vicinity of 1.32 to 1.34 µm. Gain diminishes at longer wavelengths, principally as a result of ground state absorption from $^3H_4 \rightarrow ^3F_3$.[23]

The main problem with the PDFA system is reduction of the available gain through the competing $^1G_4 \rightarrow ^3F_4$ transition (2900 cm^{-1} spacing), occurring through multiphonon relaxation.[23] The result is that the *radiative quantum efficiency* (defined as the ratio of the desired transition rate to itself plus all competing transition rates) can be low enough to make the system impractical. The multiphonon relaxation rate is reduced when using hosts having low phonon energies, such as fluoride or chalcogenide glasses. Using these produces radiative quantum efficiencies on the order of 2 percent. For comparison, erbium systems exhibit quantum efficiencies of nearly 100 percent for the 1.5-μm transition. Nevertheless, PDFAs have found practical use in 1.3-μm transmission systems, and have yielded net gains that are comparable to EDFAs, but of course with substantially higher pump power requirements. Other considerations such as broadening mechanisms and excited state absorption are analogous to the erbium system. References 1, 32, and 33 are recommended for further reading.

Erbium/Ytterbium-Doped Fiber Amplifiers (EYDFAs)

Erbium/ytterbium codoping offers special advantages in fiber amplifier performance. Ytterbium ions absorb very efficiently over the wavelength range of 800 to 1100 nm. Once excited, they transfer their energy to the erbium ions, and gain between 1.53 and 1.56 μm is formed as before.[3] Advantages of such a system include the following: With high pump absorption, side-pumping is possible, thus allowing the use of large-area diode lasers as pumps. In addition, high gain can be established over a shorter propagation distance in the fiber than is possible in a conventional EDFA. As a result, shorter-length amplifiers having lower ASE noise can be constructed. An added benefit is that the absorption band allows pumping by high-power lasers such as Nd:YAG (at 1.06 μm) or Nd:YLF (at 1.05 μm), and there is no excited state absorption. Currently, Yb-sensitized fibers are primarily attractive for use as power amplifiers, and in the construction of fiber lasers, in which a short-length, high-gain medium is needed.[35]

5.5   *REFERENCES*

1. Y. Ohishi et al., "Pr^{3+}-Doped Fluoride Fiber Amplifier Operation at 1.3 μm," *Optics Letters* **16**: 1747–1749 (1991).

2. R. Paschotta et al., "Ytterbium-Doped Fiber Amplifiers," *IEEE Journal of Quantum Electronics* **33**: 1049–1056 (1997).

3. J. E. Townsend et al., "Yb^{3+}-Sensitized Er^{3+}-Doped Silica Optical Fiber with Ultra High Efficiency and Gain," *Electronics Letters* **27**:1958–1959 (1991).

4. S. Sudo, "Progress in Optical Fiber Amplifiers," *Current Trends in Optical Amplifiers and their Applications,* T. P. Lee (ed.), World Scientific, New Jersey, 1996, 19–21.

5. I. N. Duling III, "Subpicosecond All-Fiber Erbium Laser," *Electronics Letters* **27**:544–545 (1991).

6. B. Moslehi and J. W. Goodman, "Novel Amplified Fiber Optic Recirculating Delay Line Processor," *IEEE Journal of Lightwave Technology* **10**:1142–1147 (1992).

7. R. H. Pantell, R. W. Sadowski, M. J. F. Digonnet, and H. J. Shaw, "Laser Diode-Pumped Nonlinear Switch in Erbium-Doped Fiber," *Optics Letters* **17**:1026–1028 (1992).

8. A. R. Chraplyvy and R. W. Tkach, "Terabit/Second Transmission Experiments," *IEEE Journal of Quantum Electronics* **34**:2103–2108 (1998).

9. A. R. Chraplyvy, A. H. Gnauck, R. W. Tkach, and R. M. Derosier, "8 × 10 Gb/s Transmission through 280 km of Dispersion-Managed Fiber," *IEEE Photonics Technology Letters* **5**:1233–1235 (1993).

10. *IEEE Journal of Lightwave Technology*—Special Miniissue on Dispersion Compensation **12**:1706–1765 (1994).

11. J. Auyeung and A. Yariv, "Spontaneous and Stimulated Raman Scattering in Long Low-Loss Fibers," *IEEE Journal of Quantum Electronics* **14**:347–352 (1978).

12. T. N. Nielsen, P. B. Hansen, A. J. Stentz, V. M. Aguaro, J. R. Pedrazzani, A. A. Abramov, and R. P. Espindola, "8 × 10 Gb/s 1.3-μm Unrepeatered Transmission over a Distance of 141 km with Raman Post- and Pre-Amplifiers," *IEEE Photonics Technology Letters,* **10**:1492–1494 (1998).

13. P. C. Becker, N. A. Olsson, and J. R. Simpson, *Erbium-Doped Fiber Amplifiers, Fundamentals and Technology,* Academic Press, San Diego, 1999, 139–140.

14. J.-M. P. Delavaux and J. A. Nagel, "Multi-Stage Erbium-Doped Fiber Amplifier Design," *IEEE Journal of Lightwave Technology* **13**:703–720 (1995).

15. E. Desurvire, *Erbium-Doped Fiber Amplifiers, Principles and Applications,* Wiley-Interscience, New York, 1994, 337–340.

16. S. Sudo, "Outline of Optical Fiber Amplifiers," *Optical Fiber Amplifiers: Materials, Devices, and Applications,* S. Sudo (ed.), Artech House, Boston, 1997, 81–83 and references therein.

17. W. J. Miniscalco, "Erbium-Doped Glasses for Fiber Amplifiers at 1500nm," *IEEE Journal of Lightwave Technology* **9**:234–250 (1991).

18. S. T. Davey and P. W. France, "Rare-Earth-Doped Fluorozirconate Glass for Fibre Devices," *British Telecom Technical Journal* **7**:58 (1989).

19. E. Desurvire, *Erbium-Doped Fiber Amplifiers, Principles and Applications,* Wiley-Interscience, New York, 1994, 238.

20. C. R. Giles and E. Desurvire, "Modeling Erbium-Doped Fiber Amplifiers," *IEEE Journal of Lightwave Technology* **9**:271–283 (1991).

21. E. Desurvire, "Study of the Complex Atomic Susceptibility of Erbium-Doped Fiber Amplifiers," *IEEE Journal of Lightwave Technology* **8**:1517–1527 (1990).

22. Y. Sun, J. L. Zyskind, and A. K. Srivastava, "Average Inversion Level, Modeling, and Physics of Erbium-Doped Fiber Amplifiers," *IEEE Journal of Selected Topics in Quantum Electronics* **3**:991–1007 (1997).

23. S. Sudo, "Outline of Optical Fiber Amplifiers," *Optical Fiber Amplifiers: Materials, Devices, and Applications,* S. Sudo (ed.), Artech House, Boston, 1997, 73.

24. M. Horiguchi et al., "Erbium-Doped Fiber Amplifiers Pumped in the 660- and 820-nm Bands," *IEEE Journal of Lightwave Technology,* **12**:810–820 (1994).

25. E. Desurvire, "Analysis of Gain Difference Between Forward- and Backward-Pumped Erbium-Doped Fibers in the Saturation Regime," *IEEE Photonics Technology Letters,* **4**:711–713 (1992).

26. M. N. Zervas and R. I. Laming, "Rayleigh Scattering Effect on the Gain Efficiency and Noise of Erbium-Doped Fiber Amplifiers," *IEEE Journal of Quantum Electronics* **31**:469–471 (1995).

27. H. A. Haus, "The Noise Figure of Optical Amplifiers," *IEEE Photonics Technology Letters* **10**:1602–1604 (1998).

28. D. Bayart, B. Clesca, L. Hamon, and J. L. Beylat, "Experimental Investigation of the Gain Flatness Characteristics for 1.55μm Erbium-Doped Fluoride Fiber Amplifiers," *IEEE Photonics Technology Letters* **6**:613–615 (1994).

29. E. L. Goldstein, L. Eskildsen, C. Lin, and R. E. Tench, "Multiwavelength Propagation in Lightwave Systems with Strongly-Inverted Fiber Amplifiers," *IEEE Photonics Technology Letters* **6**:266–269 (1994).

30. C. R. Giles, "Lightwave Applications of Fiber Bragg Gratings," *IEEE Journal of Lightwave Technology* **15**:1391–1404 (1997).

31. J.-Y. Pan, M. A. Ah, A. F. Elrefaie, and R. E. Wagner, "Multiwavelength Fiber Amplifier Cascades with Equalization Employing Mach-Zehnder Optical Filters," *IEEE Photonics Technology Letters* **7**:1501–1503 (1995).

32. P. C. Becker, N. A. Olsson, and J. R. Simpson, op. cit., 285–295.

33. T. J. Whitley et al., "Quarter-Watt Output from a Praseodymium-Doped Fluoride Fibre Amplifier with a Diode-Pumped Nd:YLF Laser," *IEEE Photonics Technology Letters* **5**:399–401 (1993).

34. T. J. Whitley, "A Review of Recent System Demonstrations Incorporating Praseodymium-Doped Fluoride Fiber Amplifiers," *IEEE Journal of Lightwave Technology* **13**:744–760 (1995).

35. G. G. Vienne et al., "Fabrication and Characterization of $Yb^{3+}:Er^{3+}$ Phosphosilicate Fibers for Lasers," *IEEE Journal of Lightwave Technology* **16**:1990–2001 (1998).

CHAPTER 6
FIBER-OPTIC COMMUNICATION LINKS (TELECOM, DATACOM, AND ANALOG)

Casimer DeCusatis
IBM Corporation
Poughkeepsie, New York

Guifang Li
School of Optics/The Center for Research
and Education in Optics and Lasers (CREOL)
University of Central Florida
Orlando, Florida

6.1 INTRODUCTION

There are many different applications for fiber-optic communication systems, and each has its own unique performance requirements. For example, analog communication systems may be subject to different types of noise and interference than digital systems, and consequently require different figures of merit to characterize their behavior. At first glance, telecommunication and data communication systems appear to have much in common, as both use digital encoding of data streams; in fact, both types can share a common network infrastructure. Upon closer examination, however, we find important differences between them. First, datacom systems must maintain a much lower bit error rate (BER), defined as the number of transmission errors per second in the communication link. (We will discuss BER in more detail in the following sections.) For telecom (voice) communications, the ultimate receiver is the human ear, and voice signals have a bandwidth of only about 4 kHz; transmission errors often manifest as excessive static noise such as encountered on a mobile phone, and most users can tolerate this level of fidelity. In contrast, the consequences of even a single bit error to a datacom system can be very serious; critical data such as medical or financial records could be corrupted, or large computer systems could be shut down. Typical telecom systems operate at a BER of about 10e-9, compared with about 10e-12 to 10e-15 for datacom systems. Another unique requirement of datacom systems is eye safety versus distance trade-offs. Most telecommunications equipment is maintained in a restricted environment and accessible only to personnel trained in the proper handling of high-power optical sources. Datacom equipment is maintained in a computer center and must comply with international regulations for inherent eye safety; this limits the amount of optical power that can safely be launched into the fiber, and consequently limits the maximum transmission distances that can be achieved without using repeaters or regenerators. For the same reason, datacom equipment must be rugged enough to withstand casual use, while telecom equipment is more often handled by specially trained service personnel. Telecom systems also make more extensive use of multiplexing techniques, which are only now being introduced into the data center, and more extensive use of optical repeaters. For example, commercial phone lines require repeaters spaced about every 12 km; optical links have increased this distance to around 40 km, and some recently installed systems (1997) extend the unrepeated distance up to 120 km or more.

In the following sections, we will examine the technical requirements for designing fiber optic communication systems suitable for these different environments. We begin by defining some figures of merit to characterize the system performance. Then, concentrating on digital optical communication systems, we will describe how to design an optical link loss budget and how to account for various types of noise sources in the link.

6.2 FIGURES OF MERIT: SNR, BER, MER, AND SFDR

Several possible figures of merit may be used to characterize the performance of an optical communication system. Furthermore, different figures of merit may be more suitable for different applications, such as analog or digital transmission. In this section we will describe some of the measurements used to characterize the performance of optical communication systems. Even if we ignore the practical considerations of laser eye safety standards, an optical transmitter is capable of launching only a limited amount of optical power into a fiber; similarly, there is a limit as to how weak a signal can be detected by the receiver in the presence of noise and interference. Thus, a fundamental consideration in optical communication systems design is the optical link power budget, or the difference between the transmitted and received optical power levels. Some power will be lost due to connections, splices, and bulk attenuation in the fiber. There may also be optical power penalties due to dispersion, modal noise, or other effects in the fiber and electronics. The optical power levels define the signal-to-noise ratio (SNR) at the receiver, which is often used to characterize the performance of analog communication systems. For digital transmission, the most common figure of merit is the bit error rate (BER), defined as the ratio of received bit errors to the total number of transmitted bits. Signal-to-noise ratio is related to the bit error rate by the Gaussian integral

$$\text{BER} = \frac{1}{\sqrt{2\pi}} \int_Q^\infty e^{-Q^2/2} \, dQ \cong \frac{1}{Q\sqrt{2\pi}} e^{-Q^2/2} \tag{1}$$

where Q represents the SNR for simplicity of notation.[1-4] From Eq. (1), we see that a plot of BER versus received optical power yields a straight line on semilog scale, as illustrated in Fig. 1. Nominally, the slope is about 1.8 dB/decade; deviations from a straight line may indicate the presence of nonlinear or non-Gaussian noise sources. Some effects, such as fiber attenuation, are linear noise sources; they can be overcome by increasing the received optical power, as seen from Fig. 1, subject to constraints on maximum optical power (laser safety) and the limits of receiver sensitivity. There are other types of noise sources, such as mode partition noise or relative intensity noise (RIN), which are independent of signal strength. When such noise is present, no amount of increase in transmitted signal strength will affect the BER; a noise floor is produced, as shown by curve B in Fig. 1. This type of noise can be a serious limitation on link performance. If we plot BER versus receiver sensitivity for increasing optical power, we obtain a curve similar to that in Fig. 2, which shows that for very high power levels

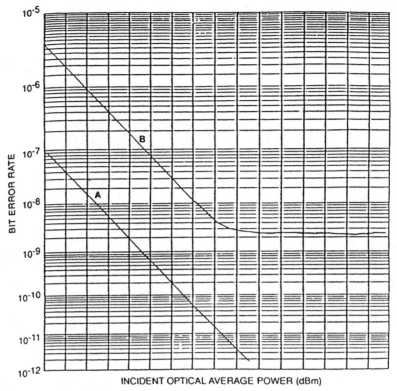

FIGURE 1   Bit error rate as a function of received optical power. Curve A shows typical performance, whereas curve B shows a BER floor [5].

the receiver will go into saturation. The characteristic bathtub-shaped curve illustrates a window of operation with both upper and lower limits on the received power. There may also be an upper limit on optical power due to eye safety considerations.

We can see from Fig. 1 that receiver sensitivity is specified at a given BER, which is often too low to measure directly in a reasonable amount of time (for example, a 200-Mbit/s link operating at a BER of 10e-15 will only take one error every 57 days on average, and several hundred errors are recommended for a reasonable BER measurement). For practical reasons, the BER is typically measured at much higher error rates (such as 10e-4 to 10e-8), where the data can be collected more quickly and then extrapolated to find the sensitivity at low BER. This assumes the absence of nonlinear noise floors, as cautioned previously. The relationship between optical input power, in watts, and the BER is the complimentary Gaussian error function

$$BER = 1/2 \ erfc \ (Pout - Psignal \ / \ RMS \ noise) \tag{2}$$

where the error function is an open integral that cannot be solved directly. Several approximations have been developed for this integral, which can be developed into transformation functions that yield a linear least squares fit to the data.[1] The same curve-fitting equations can also be used to characterize the eye window performance of optical receivers. Clock position/ phase versus BER data are collected for each edge of the eye window; these data sets are then

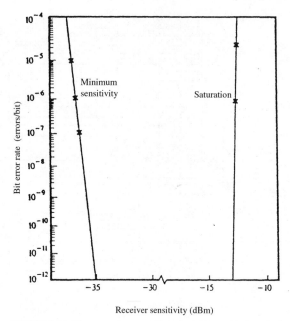

FIGURE 2 Bit error rate as a function of received optical power illustrating range of operation from minimum sensitivity to saturation.

curve-fitted with the previously noted expressions to determine the clock position at the desired BER. The difference in the two resulting clock positions, on either side of the window, gives the clear eye opening.[1–4]

In describing Figs. 1 and 2, we have also made some assumptions about the receiver circuit. Most data links are asynchronous and do not transmit a clock pulse along with the data; instead, a clock is extracted from the incoming data and used to retime the received data stream. We have made the assumption that the BER is measured with the clock at the center of the received data bit; ideally, this is when we compare the signal with a preset threshold to determine if a logical 1 or 0 was sent. When the clock is recovered from a receiver circuit such as a phase lock loop, there is always some uncertainty about the clock position; even if it is centered on the data bit, the relative clock position may drift over time. The region of the bit interval in the time domain where the BER is acceptable is called the *eyewidth;* if the clock timing is swept over the data bit using a delay generator, the BER will degrade near the edges of the eye window. Eyewidth measurements are an important parameter in link design, which will be discussed further in the section on jitter and link budget modeling.

In the designs of some analog optical communication systems, as well as of some digital television systems (for example, those based on 64-bit Quadrature Amplitude Modulation), another possible figure of merit is the modulation error ratio (MER). To understand this metric, we will consider the standard definition of the Digital Video Broadcasters (DVB) Measurements Group.[5] First, the video receiver captures a time record of N received signal coordinate pairs, representing the position of information on a two-dimensional screen. The ideal position coordinates are given by the vector (X_j, Y_j). For each received symbol, a decision is made as to which symbol was transmitted, and an error vector $(\Delta X_j, \Delta Y_j)$ is defined as the distance from the ideal position to the actual position of the received symbol. The MER is then defined as the sum of the squares of the magnitudes of the ideal symbol vector divided by the sum of the squares of the magnitudes of the symbol error vectors:

$$\text{MER} = 10 \log \frac{\sum_{j=1}^{N} (X_j^2 + Y_j^2)}{\sum_{j=1}^{N} (\Delta X_j^2 + \Delta Y_j^2)} \quad \text{dB} \tag{3}$$

When the signal vectors are corrupted by noise, they can be treated as random variables. The denominator in Eq. (3) becomes an estimate of the average power of the error vector (in other words, its second moment) and contains all signal degradation due to noise, reflections, transmitter quadrature errors, and so forth. If the only significant source of signal degradation is additive white Gaussian noise, then MER and SNR are equivalent. For communication systems that contain other noise sources, MER offers some advantages; in particular, for some digital transmission systems there may be a very sharp change in BER as a function of SNR (a so-called *cliff effect*), which means that BER alone cannot be used as an early predictor of system failures. MER, on the other hand, can be used to measure signal-to-interference ratios accurately for such systems. Because MER is a statistical measurement, its accuracy is directly related to the number of vectors, N, used in the computation; an accuracy of 0.14 dB can be obtained with $N = 10,000$, which would require about 2 ms to accumulate at the industry standard digital video rate of 5.057 Msymbols/s.

In order to design a proper optical data link, the contribution of different types of noise sources should be assessed when developing a link budget. There are two basic approaches to link-budget modeling. One method is to design the link to operate at the desired BER when all the individual link components assume their worst-case performance. This conservative approach is desirable when very high performance is required, or when it is difficult or inconvenient to replace failing components near the end of their useful lifetimes. The resulting design has a high safety margin; in some cases it may be overdesigned for the required level of performance. Since it is very unlikely that all the elements of the link will assume their worst-case performance at the same time, an alternative is to model the link budget statistically. For this method, distributions of transmitter power output, receiver sensitivity, and other parameters are either measured or estimated. They are then combined statistically using an approach such as the Monte Carlo method, in which many possible link combinations are simulated to generate an overall distribution of the available link optical power. A typical approach is the 3-sigma design, in which the combined variations of all link components are not allowed to extend more than three standard deviations from the average performance target in either direction. The statistical approach results in greater design flexibility and generally increased distance compared with a worst-case model at the same BER.

Harmonic Distortions, Intermodulation Distortions, and Dynamics Range

Fiber-optic analog links are in general nonlinear. That is, if the input electrical information is a harmonic signal of frequency f_0, the output electrical signal will contain the fundamental frequency f_0 as well as high-order harmonics of frequencies nf_0 ($n > 2$). These high-order harmonics comprise the harmonic distortions of analog fiber-optic links.[6] The nonlinear behavior is caused by nonlinearities in the transmitter, the fiber, and the receiver.

The same sources of nonlinearities in the fiber-optic links lead to intermodulation distortions (IMD), which can best be illustrated in a two-tone transmission scenario. If the input electrical information is a superposition of two harmonic signals of frequencies f_1 and f_2, the output electrical signal will contain second-order intermodulation at frequencies $f_1 + f_2$ and $f_1 - f_2$ as well as third-order intermodulation at frequencies $2f_1 - f_2$ and $2f_2 - f_1$.

Most analog fiber-optic links require bandwidth of less than one octave ($f_{max} < 2f_{min}$). As a result, harmonic distortions as well as second-order IMD products are not important as they can be filtered out electronically. However, third-order IMD products are in the same frequency range (between f_{min} and f_{max}) as the signal itself and therefore appear in the output signal as the spurious response. Thus the linearity of analog fiber-optic links is determined by the level of third-order IMD products. In the case of analog links where third-order IMD is eliminated through linearization circuitry, the lowest odd-order IMD determines the linearity of the link.

To quantify IMD distortions, a two-tone experiment (or simulation) is usually conducted, where the input RF powers of the two tones are equal. The linear and nonlinear power transfer functions—the output RF power of each of two input tones and the second- or third-order IMD product as a function of the input RF power of each input harmonic signal—are schematically presented in Fig. 3. When plotted on a log-log scale, the fundamental power transfer function should be a line with a slope of unity. The second- (third-) order power transfer function should be a line with a slope of two (three). The intersections of the power transfer functions are called *second-* and *third-order* intercept points, respectively. Because of the fixed slopes of the power transfer functions, the intercept points can be calculated from measurements obtained at a single input power level. Suppose that at a certain input level, the output power of each of the two fundamental tones, the second-order IMD product, and third-order IMD products are P_1, P_2, and P_3, respectively. When the power levels are in units of dB or dBm, the second-order and third-order intercept points are

$$IP_2 = 2P_1 - P_2 \tag{4}$$

and

$$IP_3 = (3P_1 - P_3)/2 \tag{5}$$

The dynamic range is a measure of an analog fiber-optic link's ability to faithfully transmit signals at various power levels. At the low input power end, the analog link can fail due to insufficient power level, so that the output power is below the noise level. At the high input power

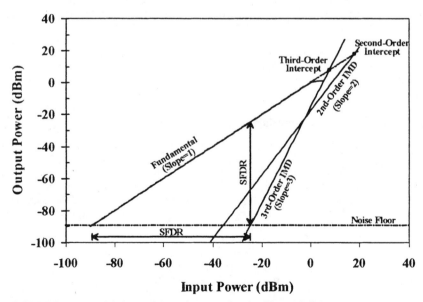

FIGURE 3   Intermodulation and dynamic range of analog fiber-optic links.

end, the analog link can fail due to the fact that the IMD products become the dominant source of signal degradation. In terms of the output power, the dynamic range (of the output power) is defined as the ratio of the fundamental output to the noise power. However, it should be noted that the third-order IMD products increase three times faster than the fundamental signal. After the third-order IMD products exceeds the noise floor, the ratio of the fundamental output to the noise power is meaningless, as the dominant degradation of the output signal comes from IMD products. So a more meaningful definition of the dynamic range is the so-called spurious-free dynamic range (SFDR),[6,7] which is the ratio of the fundamental output to the noise power at the point where the IMD products is at the noise level. The spurious-free dynamic range is then practically the maximum dynamic range. Since the noise floor depends on the bandwidth of interest, the unit for SFDR should be (dB $Hz^{2/3}$). The dynamic range decreases as the bandwidth of the system is increased. The spurious-free dynamic range is also often defined with reference to the input power, which corresponds to SFDR with reference to the output power if there is no gain compression.

6.3 LINK BUDGET ANALYSIS: INSTALLATION LOSS

It is convenient to break down the link budget into two areas: *installation loss* and *available power*. Installation or DC loss refers to optical losses associated with the fiber cable plant, such as connector loss, splice loss, and bandwidth considerations. Available optical power is the difference between the transmitter output and receiver input powers, minus additional losses due to optical noise sources on the link (also known as *AC losses*). With this approach, the installation loss budget may be treated statistically and the available power budget as worst case. First, we consider the installation loss budget, which can be broken down into three areas, namely transmission loss, fiber attenuation as a function of wavelength, and connector or splice losses.

Transmission Loss

Transmission loss is perhaps the most important property of an optical fiber; it affects the link budget and maximum unrepeated distance. Since the maximum optical power launched into an optical fiber is determined by international laser eye safety standards,[8] the number and separation between optical repeaters and regenerators is largely determined by this loss. The mechanisms responsible for this loss include material absorption as well as both linear and nonlinear scattering of light from impurities in the fiber.[1-5] Typical loss for single-mode optical fiber is about 2 to 3 dB/km near 800 nm wavelength, 0.5 dB/km near 1300 nm, and 0.25 dB/km near 1550 nm. Multimode fiber loss is slightly higher, and bending loss will only increase the link attenuation further.

Attenuation versus Wavelength

Since fiber loss varies with wavelength, changes in the source wavelength or use of sources with a spectrum of wavelengths will produce additional loss. Transmission loss is minimized near the 1550-nm wavelength band, which unfortunately does not correspond with the dispersion minimum at around 1310 nm. An accurate model for fiber loss as a function of wavelength has been developed by Walker[9]; this model accounts for the effects of linear scattering, macrobending, and material absorption due to ultraviolet and infrared band edges, hydroxide (OH) absorption, and absorption from common impurities such as phosphorus. Using this model, it is possible to calculate the fiber loss as a function of wavelength for different impurity levels; the fiber properties can be specified along with the acceptable wavelength limits of the source to limit the fiber loss over the entire operating wavelength range. Design tradeoffs are possible between center wavelength and fiber composition to achieve the desired result. Typical loss due to wavelength-dependent attenuation for laser sources on single-mode fiber can be held below 0.1 dB/km.

Connector and Splice Losses

There are also installation losses associated with fiber-optic connectors and splices; both of these are inherently statistical in nature and can be characterized by a Gaussian distribution. There are many different kinds of standardized optical connectors, some of which have been discussed previously; some industry standards also specify the type of optical fiber and connectors suitable for a given application.[10] There are also different models which have been published for estimating connection loss due to fiber misalignment[11, 12]; most of these treat loss due to misalignment of fiber cores, offset of fibers on either side of the connector, and angular misalignment of fibers. The loss due to these effects is then combined into an overall estimate of the connector performance. There is no general model available to treat all types of connectors, but typical connector loss values average about 0.5 dB worst case for multimode, and slightly higher for single mode (see Table 1).

TABLE 1 Typical Cable Plant Optical Losses [5]

Component	Description	Size (μm)	Mean loss	Variance (dB^2)
Connector[a]	Physical contact	62.5–62.5	0.40 dB	0.02
		50.0–50.0	0.40 dB	0.02
		9.0–9.0[b]	0.35 dB	0.06
		62.5–50.0	2.10 dB	0.12
		50.0–62.5	0.00 dB	0.01
Connector[a]	Nonphysical contact (multimode only)	62.5–62.5	0.70 dB	0.04
		50.0–50.0	0.70 dB	0.04
		62.5–50.0	2.40 dB	0.12
		50.0–62.5	0.30 dB	0.01
Splice	Mechanical	62.5–62.5	0.15 dB	0.01
		50.0–50.0	0.15 dB	0.01
		9.0–9.0[b]	0.15 dB	0.01
Splice	Fusion	62.5–62.5	0.40 dB	0.01
		50.0–50.0	0.40 dB	0.01
		9.0–9.0[b]	0.40 dB	0.01
Cable	IBM multimode jumper	62.5	1.75 dB/km	NA
	IBM multimode jumper	50.0	3.00 dB/km at 850 nm	NA
	IBM single-mode jumper	9.0	0.8 dB/km	NA
	Trunk	62.5	1.00 dB/km	NA
	Trunk	50.0	0.90 dB/km	NA
	Trunk	9.0	0.50 dB/km	NA

[a] The connector loss value is typical when attaching identical connectors. The loss can vary significantly if attaching different connector types.

[b] Single-mode connectors and splices must meet a minimum return loss specification of 28 dB.

Optical splices are required for longer links, since fiber is usually available in spools of 1 to 5 km, or to repair broken fibers. There are two basic types, *mechanical splices* (which involve placing the two fiber ends in a receptacle that holds them close together, usually with epoxy) and the more commonly used *fusion splices* (in which the fibers are aligned, then heated sufficiently to fuse the two ends together). Typical splice loss values are given in Table 1.

6.4 LINK BUDGET ANALYSIS: OPTICAL POWER PENALTIES

Next, we will consider the *assembly loss budget,* which is the difference between the transmitter output and receiver input powers, allowing for optical power penalties due to noise sources in the link. We will follow the standard convention in the literature of assuming a digital optical communication link which is best characterized by its BER. Contributing factors to link performance include the following:

- Dispersion (modal and chromatic) or intersymbol interference
- Mode partition noise
- Mode hopping
- Extinction ratio
- Multipath interference
- Relative intensity noise (RIN)
- Timing jitter
- Radiation-induced darkening
- Modal noise

Higher order, nonlinear effects, including Stimulated Raman and Brillouin scattering and frequency chirping, will be discussed elsewhere.

Dispersion

The most important fiber characteristic after transmission loss is *dispersion,* or *intersymbol interference.* This refers to the broadening of optical pulses as they propagate along the fiber. As pulses broaden, they tend to interfere with adjacent pulses; this limits the maximum achievable data rate. In multimode fibers, there are two dominant kinds of dispersion, *modal* and *chromatic.* Modal dispersion refers to the fact that different modes will travel at different velocities and cause pulse broadening. The fiber's modal bandwidth, in units of MHz-km, is specified according to the expression

$$BW_{\mathrm{modal}} = BW_1/L^\gamma \qquad (6)$$

where BW_{modal} is the modal bandwidth for a length L of fiber, BW_1 is the manufacturer-specified modal bandwidth of a 1-km section of fiber, and γ is a constant known as the *modal bandwidth concatenation length scaling factor.* The term γ usually assumes a value between 0.5 and 1, depending on details of the fiber manufacturing and design as well as the operating wavelength; it is conservative to take $\gamma = 1.0$. Modal bandwidth can be increased by mode mixing, which promotes the interchange of energy between modes to average out the effects of modal dispersion. Fiber splices tend to increase the modal bandwidth, although it is conservative to discard this effect when designing a link.

The other major contribution is chromatic dispersion, BW_{chrom}, which occurs because different wavelengths of light propagate at different velocities in the fiber. For multimode fiber, this is given by an empirical model of the form

$$BW_{chrom} = \frac{L^{\gamma c}}{\sqrt{\lambda_w}\,(a_0 + a_1|\lambda_c - \lambda_{eff}|)} \tag{7}$$

where L is the fiber length in km; λ_c is the center wavelength of the source in nm; λ_w is the source FWHM spectral width in nm; γ_c is the chromatic bandwidth length scaling coefficient, a constant; λ_{eff} is the effective wavelength, which combines the effects of the fiber zero dispersion wavelength and spectral loss signature; and the constants a_1 and a_0 are determined by a regression fit of measured data. From Ref. (13), the chromatic bandwidth for 62.5/125-micron fiber is empirically given by

$$BW_{chrom} = \frac{10^4 L^{-0.69}}{\sqrt{\lambda_w}\,(1.1 + 0.0189|\lambda_c - 1370|)} \tag{8}$$

For this expression, the center wavelength was 1335 nm and λ_{eff} was chosen midway between λ_c and the water absorption peak at 1390 nm; although λ_{eff} was estimated in this case, the expression still provides a good fit to the data. For 50/125-micron fiber, the expression becomes

$$BW_{chrom} = \frac{10^4 L^{-0.65}}{\sqrt{\lambda_w}\,(1.01 + 0.0177|\lambda_c - 1330|)} \tag{9}$$

For this case, λ_c was 1313 nm and the chromatic bandwidth peaked at $\lambda_{eff} = 1330$ nm. Recall that this is only one possible model for fiber bandwidth.[1] The total bandwidth capacity of multimode fiber BW_t is obtained by combining the modal and chromatic dispersion contributions, according to

$$\frac{1}{BW_t^2} = \frac{1}{BW_{chrom}^2} + \frac{1}{BW_{modal}^2} \tag{10}$$

Once the total bandwidth is known, the dispersion penalty can be calculated for a given data rate. One expression for the dispersion penalty in dB is

$$P_d = 1.22 \left[\frac{\text{Bit Rate } (Mb/s)}{BW_t(\text{MHz})} \right]^2 \tag{11}$$

For typical telecommunication grade fiber, the dispersion penalty for a 20-km link is about 0.5 dB.

Dispersion is usually minimized at wavelengths near 1310 nm; special types of fiber have been developed which manipulate the index profile across the core to achieve minimal dispersion near 1550 nm, which is also the wavelength region of minimal transmission loss. Unfortunately, this dispersion-shifted fiber suffers from some practical drawbacks, including susceptibility to certain kinds of nonlinear noise and increased interference between adjacent channels in a wavelength multiplexing environment. There is a new type of fiber, called *dispersion-optimized fiber,* that minimizes dispersion while reducing the unwanted crosstalk effects. By using a very sophisticated fiber profile, it is possible to minimize dispersion over the entire wavelength range from 1300 to 1550 nm, at the expense of very high loss (around 2 dB/km); this is known as *dispersion-flattened fiber.* Yet another approach is called *dispersion-compensating fiber;* this fiber is designed with negative dispersion characteristics, so that when used in series with conventional fiber it will "undisperse" the signal. Dispersion-compensating fiber has a much narrower core than standard single-mode fiber, which makes it susceptible to nonlinear effects; it is also birefringent and suffers from polarization mode dispersion, in which different states of polarized light propagate with very different group velocities. Note that standard single-mode fiber does not preserve the polarization state of the incident light; there is yet another type of specialty fiber, with asymmetric core profiles, capable of preserving the polarization of incident light over long distances.

By definition, single-mode fiber does not suffer modal dispersion. Chromatic dispersion is an important effect, though, even given the relatively narrow spectral width of most laser diodes. The dispersion of single-mode fiber corresponds to the first derivative of group velocity τ_g with respect to wavelength, and is given by

$$D = \frac{d\tau_g}{d\lambda} = \frac{S_0}{4}\left(\lambda_c - \frac{\lambda_0^4}{\lambda_c^3}\right) \tag{12}$$

where D is the dispersion in ps/(km-nm) and λ_c is the laser center wavelength. The fiber is characterized by its zero dispersion wavelength, λ_0, and zero dispersion slope, S_0. Usually, both center wavelength and zero dispersion wavelength are specified over a range of values; it is necessary to consider both upper and lower bounds in order to determine the worst-case dispersion penalty. This can be seen from Fig. 4, which plots D versus wavelength for some typical values of λ_0 and λ_c; the largest absolute value of D occurs at the extremes of this region. Once the dispersion is determined, the intersymbol interference penalty as a function of link length L can be determined to a good approximation from a model proposed by Agrawal[14]:

$$P_d = 5 \log\left[1 + 2\pi(BD\,\Delta\lambda)^2\, L^2\right] \tag{13}$$

where B is the bit rate and $\Delta\lambda$ is the root mean square (RMS) spectral width of the source. By maintaining a close match between the operating and zero dispersion wavelengths, this penalty can be kept to a tolerable 0.5 to 1.0 dB in most cases.

Mode Partition Noise

Group velocity dispersion contributes to other optical penalties that remain the subject of continuing research—*mode partition noise* and *mode hopping*. These penalties are related to

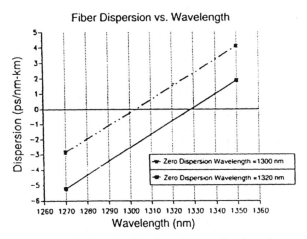

FIGURE 4 Single-mode fiber dispersion as a function of wavelength [5].

the properties of a Fabry-Perot type laser diode cavity; although the total optical power output from the laser may remain constant, the optical power distribution among the laser's longitudinal modes will fluctuate. This is illustrated by the model depicted in Fig. 5; when a laser diode is directly modulated with injection current, the total output power stays constant from pulse to pulse; however, the power distribution among several longitudinal modes will vary between pulses. We must be careful to distinguish this behavior of the instantaneous laser spectrum, which varies with time, from the time-averaged spectrum that is normally observed experimentally. The light propagates through a fiber with wavelength-dependent dispersion or attenuation, which deforms the pulse shape. Each mode is delayed by a different amount due to group velocity dispersion in the fiber; this leads to additional signal degradation at the receiver, in addition to the intersymbol interference caused by chromatic dispersion alone, discussed earlier. This is known as *mode partition noise;* it is capable of generating bit error rate floors such that additional optical power into the receiver will not improve the link BER. This is because mode partition noise is a function of the laser spectral fluctuations and wavelength-dependent dispersion of the fiber, so the signal-to-noise ratio due to this effect is independent of the signal power. The power penalty due to mode partition noise was first calculated by Ogawa[15] as

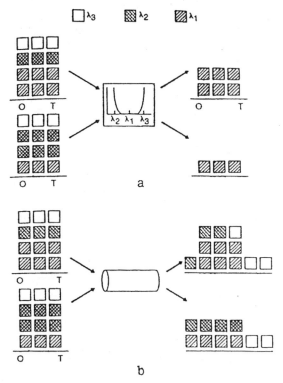

FIGURE 5 Model for mode partition noise; an optical source emits a combination of wavelengths, illustrated by different color blocks: (a) wavelength-dependent loss; (b) chromatic dispersion.

$$P_{mp} = 5 \log (1 - Q^2 \sigma_{mp}^2) \tag{14}$$

where

$$\sigma_{mp}^2 = \frac{1}{2} k^2 (\pi B)^4 [A_1^4 \Delta\lambda^4 + 42 A_1^2 A_2^2 \Delta\lambda^6 + 48 A_2^4 \Delta\lambda^8] \tag{15}$$

$$A_1 = DL \tag{16}$$

and

$$A_2 = \frac{A_1}{2(\lambda_c - \lambda_0)} \tag{17}$$

The mode partition coefficient k is a number between 0 and 1 that describes how much of the optical power is randomly shared between modes; it summarizes the statistical nature of mode partition noise. According to Ogawa, k depends on the number of interacting modes and rms spectral width of the source, the exact dependence being complex. However, subsequent work has shown[16] that Ogawa's model tends to underestimate the power penalty due to mode partition noise because it does not consider the variation of longitudinal mode power between successive baud periods, and because it assumes a linear model of chromatic dispersion rather than the nonlinear model given in the just-cited equation. A more detailed model has been proposed by Campbell,[17] which is general enough to include effects of the laser diode spectrum, pulse shaping, transmitter extinction ratio, and statistics of the data stream. While Ogawa's model assumed an equiprobable distribution of zeros and ones in the data stream, Campbell showed that mode partition noise is data dependent as well. Recent work based on this model[18] has rederived the signal variance:

$$\sigma_{mp}^2 = E_{av}(\sigma_0^2 + \sigma_{+1}^2 + \sigma_{-1}^2) \tag{18}$$

where the mode partition noise contributed by adjacent baud periods is defined by

$$\sigma_{+1}^2 + \sigma_{-1}^2 = \frac{1}{2} k^2 (\pi B)^4 (1.25 A_1^4 \Delta\lambda^4 + 40.95 A_1^2 A_2^2 \Delta\lambda^6 + 50.25 A_2^4 \Delta\lambda^8) \tag{19}$$

and the time-average extinction ratio $E_{av} = 10 \log (P_1/P_0)$, where P_1, P_0 represent the optical power by a 1 and 0, respectively. If the operating wavelength is far away from the zero dispersion wavelength, the noise variance simplifies to

$$\sigma_{mp}^2 = 2.25 \frac{k^2}{2} E_{av} (1 - e^{-\beta L^2})^2 \tag{20}$$

which is valid provided that

$$\beta = (\pi B D \Delta\lambda)^2 \ll 1 \tag{21}$$

Many diode lasers exhibit mode hopping or mode splitting, in which the spectrum appears to split optical power between 2 or 3 modes for brief periods of time. The exact mechanism is not fully understood, but stable Gaussian spectra are generally only observed for CW operation and temperature-stabilized lasers. During these mode hops the previously cited theory does not apply, since the spectrum is non-Gaussian, and the model will overpredict the power penalty; hence, it is not possible to model mode hops as mode partitioning with $k = 1$. There is no currently published model describing a treatment of mode-hopping noise, although recent papers[19] suggest approximate calculations based on the statistical properties of the laser cavity. In a practical link, some amount of mode hopping is probably unavoidable as a contributor to burst noise; empirical testing of link hardware remains the only reliable way to reduce this effect. A practical rule of thumb is to keep the mode partition noise penalty less than 1.0 dB maximum, provided that this penalty is far away from any noise floors.

Extinction Ratio

The receiver extinction ratio also contributes directly to the link penalties. The receiver BER is a function of the modulated AC signal power; if the laser transmitter has a small extinction ratio, the DC component of total optical power is significant. Gain or loss can be introduced in the link budget if the extinction ratio at which the receiver sensitivity is measured differs from the worst-case transmitter extinction ratio. If the extinction ratio E_t at the transmitter is defined as the ratio of optical power when a 1 is transmitted versus when a 0 is transmitted,

$$E_t = \frac{\text{Power}(1)}{\text{Power}(0)} \tag{22}$$

then we can define a modulation index at the transmitter M_t according to

$$M_t = \frac{E_t - 1}{E_t + 1} \tag{23}$$

Similarly, we can measure the linear extinction ratio at the optical receiver input and define a modulation index M_r. The extinction ratio penalty is given by

$$P_{er} = -10 \log\left(\frac{M_t}{M_r}\right) \tag{24}$$

where the subscripts t and r refer to specifications for the transmitter and receiver, respectively. Usually, the extinction ratio is specified to be the same at the transmitter and receiver, and is large enough that there is no power penalty due to extinction ratio effects.

Multipath Interference

Another important property of the optical link is the amount of light reflected from the fiber endfaces that returns up the link and back into the transmitter. Whenever there is a connection or splice in the link, some fraction of the light is reflected back; each connection is thus a potential noise generator, since the reflected fields can interfere with one another to create noise in the detected optical signal. The phenomenon is analogous to the noise caused by multiple atmospheric reflections of radio waves, and is known as *multipath interference noise*. To limit this noise, connectors and splices are specified with a minimum return loss. If there is a total of N reflection points in a link and the geometric mean of the connector reflections is alpha, then based on the model of Duff et al.[20] the power penalty due to multipath interference (adjusted for bit error rate and bandwidth) is closely approximated by

$$P_{mpi} = 10 \log (1 - 0.7Na) \tag{25}$$

Multipath noise can usually be reduced well below 0.5 dB with available connectors, whose return loss is often better than 25 dB.

Relative Intensity Noise (RIN)

Stray light reflected back into a Fabry-Perot type laser diode gives rise to intensity fluctuations in the laser output. This is a complicated phenomenon, strongly dependent on the type of laser; it is called either *reflection-induced intensity noise* or *relative intensity noise* (RIN). This effect is important, since it can also generate BER floors. The power penalty due to RIN is the subject of ongoing research; since the reflected light is measured at a specified signal level, RIN is data dependent, although it is independent of link length. Since many laser diodes are packaged in windowed containers, it is difficult to correlate the RIN measurements on an unpackaged laser with those of a commercial product. There have been several detailed attempts to characterize RIN[21, 22]; typically, the RIN noise is assumed Gaussian in amplitude and uniform in frequency over the receiver bandwidth of interest. The RIN value is specified for a given laser by measuring changes in the optical power when a controlled amount of light is fed back into the laser; it is signal dependent, and is also influenced by temperature, bias voltage, laser structure, and other factors which typically influence laser output power.[22] If we assume that the effect of RIN is to produce an equivalent noise current at the receiver, then the additional receiver noise σ_r may be modeled as

$$\sigma_r = \gamma^2 \, S^{2g} \, B \tag{26}$$

where S is the signal level during a bit period, B is the bit rate, and g is a noise exponent that defines the amount of signal-dependent noise. If $g = 0$, noise power is independent of the signal, while for $g = 1$ noise power is proportional to the square of the signal strength. The coefficient γ is given by

$$\gamma^2 = S_i^{2(1-g)} \, 10^{(\mathrm{RIN}_i/10)} \tag{27}$$

where RIN_i is the measured RIN value at the average signal level S_i, including worst-case back-reflection conditions and operating temperatures. The Gaussian BER probability due to the additional RIN noise current is given by

$$P_{\mathrm{error}} = \frac{1}{2} \left[P_e^1 \left(\frac{S_1 - S_0}{2\sigma_1} \right) + P_e^0 \left(\frac{S_1 - S_0}{2\sigma_0} \right) \right] \tag{28}$$

where σ_1 and σ_0 represent the total noise current during transmission of a digital 1 and 0, respectively and P_e^1 and P_e^0 are the probabilities of error during transmission of a 1 or 0, respectively. The power penalty due to RIN may then be calculated by determining the additional signal power required to achieve the same BER with RIN noise present as without the RIN contribution. One approximation for the RIN power penalty is given by

$$P_{\mathrm{rin}} = -5 \log \left[1 - Q^2(BW)(1 + M_r)^{2g}(10^{\mathrm{RIN}/10})\left(\frac{1}{M_r} \right)^2 \right] \tag{29}$$

where the RIN value is specified in dB/Hz, BW is the receiver bandwidth, M_r is the receiver modulation index, and the exponent g is a constant varying between 0 and 1 which relates the magnitude of RIN noise to the optical power level. The maximum RIN noise penalty in a link can usually be kept to below 0.5 dB.

Jitter

Although it is not strictly an optical phenomenon, another important area in link design deals with the effects of timing jitter on the optical signal. In a typical optical link, a clock is extracted from the incoming data signal which is used to retime and reshape the received digital pulse; the received pulse is then compared with a threshold to determine if a digital 1 or 0 was transmitted. So far, we have discussed BER testing with the implicit assumption that the measurement was made in the center of the received data bit; to achieve this, a clock transition at the center of the bit is required. When the clock is generated from a receiver timing recovery circuit, it will have some variation in time, and the exact location of the clock edge will be uncertain. Even if the clock is positioned at the center of the bit, its position may drift over time. There will be a region of the bit interval, or eye, in the time domain where the BER is acceptable; this region is defined as the eyewidth.[1-3] Eyewidth measurements are an important parameter for evaluation of fiber-optic links; they are intimately related to the BER, as well as to the acceptable clock drift, pulse width distortion, and optical power. At low optical power levels, the receiver signal-to-noise ratio is reduced; increased noise causes amplitude variations in the received signal. These amplitude variations are translated into time domain variations in the receiver decision circuitry, which narrows the eyewidth. At the other extreme, an optical receiver may become saturated at high optical power, reducing the eyewidth and making the system more sensitive to timing jitter. This behavior results in the typical bathtub curve shown in Fig. 2; for this measurement, the clock is delayed from one end of the bit cell to the other, with the BER calculated at each position. Near the ends of the cell, a large number of errors occur; toward the center of the cell, the BER decreases to its true value. The eye opening may be defined as the portion of the eye for which the BER remains constant; pulse width distortion occurs near the edges of the eye, which denotes the limits of the valid clock timing. Uncertainty in the data pulse arrival times causes errors to occur by closing the eye window and causing the eye pattern to be sampled away from the center. This is one of the fundamental problems of optical and digital signal processing, and a large body of work has been done in this area.[23, 24] In general, multiple jitter sources will be present in a link; these will tend to be uncorrelated. However, jitter on digital signals, especially resulting from a cascade of repeaters, may be coherent.

International standards on jitter were first published by the CCITT (Central Commission for International Telephony and Telegraphy, now known as the International Telecommunications Union, or ITU). This standards body has adopted a definition of jitter[24] as short-term variations of the significant instants (rising or falling edges) of a digital signal from their ideal position in time. Longer-term variations are described as *wander;* in terms of frequency, the distinction between jitter and wander is somewhat unclear. The predominant sources of jitter include the following:

- Phase noise in receiver clock recovery circuits, particularly crystal-controlled oscillator circuits; this may be aggravated by filters or other components which do not have a linear phase response. Noise in digital logic resulting from restricted rise and fall times may also contribute to jitter.

- Imperfect timing recovery in digital regenerative repeaters, which is usually dependent on the data pattern.

- Different data patterns may contribute to jitter when the clock recovery circuit of a repeater attempts to recover the receive clock from inbound data. Data pattern sensitivity can produce as much as 0.5-dB penalty in receiver sensitivity. Higher data rates are more susceptible (>1 Gbit/s); data patterns with long run lengths of 1s or 0s, or with abrupt phase transitions between consecutive blocks of 1s and 0s, tend to produce worst-case jitter.

- At low optical power levels, the receiver signal-to-noise ratio, Q, is reduced; increased noise causes amplitude variations in the signal, which may be translated into time domain variations by the receiver circuitry.

- Low frequency jitter, also called *wander,* resulting from instabilities in clock sources and modulation of transmitters.

- Very low frequency jitter caused by variations in the propagation delay of fibers, connectors, etc., typically resulting from small temperature variations (this can make it especially difficult to perform long-term jitter measurements).

In general, jitter from each of these sources will be uncorrelated; jitter related to modulation components of the digital signal may be coherent, and cumulative jitter from a series of repeaters or regenerators may also contain some well-correlated components.

There are several parameters of interest in characterizing jitter performance. Jitter may be classified as either random or deterministic, depending on whether it is associated with pattern-dependent effects; these are distinct from the duty cycle distortion that often accompanies imperfect signal timing. Each component of the optical link (data source, serializer, transmitter, encoder, fiber, receiver, retiming/clock recovery/deserialization, decision circuit) will contribute some fraction of the total system jitter. If we consider the link to be a "black box" (but not necessarily a linear system), then we can measure the level of output jitter in the absence of input jitter; this is known as the *intrinsic jitter* of the link. The relative importance of jitter from different sources may be evaluated by measuring the spectral density of the jitter. Another approach is the maximum tolerable input jitter (MTIJ) for the link. Finally, since jitter is essentially a stochastic process, we may attempt to characterize the jitter transfer function (JTF) of the link, or estimate the probability density function of the jitter. When multiple traces occur at the edges of the eye, this can indicate the presence of data-dependent jitter or duty cycle distortion; a histogram of the edge location will show several distinct peaks. This type of jitter can indicate a design flaw in the transmitter or receiver. By contrast, random jitter typically has a more Gaussian profile and is present to some degree in all data links.

The problem of jitter accumulation in a chain of repeaters becomes increasingly complex; however, we can state some general rules of thumb. It has been shown[25] that jitter can be generally divided into two components, one due to repetitive patterns and one due to random data. In receivers with phase-lock loop timing recovery circuits, repetitive data patterns will tend to cause jitter accumulation, especially for long run lengths. This effect is commonly modeled as a second-order receiver transfer function. Jitter will also accumulate when the link is transferring random data; jitter due to random data is of two types, *systematic* and *random*. The classic model for systematic jitter accumulation in cascaded repeaters was published by Byrne.[26] The Byrne model assumes cascaded identical timing recovery circuits, and then the systematic and random jitter can be combined as rms quantities so that total jitter due to random jitter may be obtained. This model has been generalized to networks consisting of different components,[27] and to nonidentical repeaters.[28] Despite these considerations, for well-designed practical networks the basic results of the Byrne model remain valid for N nominally identical repeaters transmitting random data; systematic jitter accumulates in proportion to $N^{1/2}$; and random jitter accumulates in proportion to $N^{1/4}$. For most applications, the maximum timing jitter should be kept below about 30 percent of the maximum receiver eye opening.

Modal Noise

An additional effect of lossy connectors and splices is modal noise. Because high-capacity optical links tend to use highly coherent laser transmitters, random coupling between fiber modes causes fluctuations in the optical power coupled through splices and connectors; this phenomena is known as *modal noise*.[29] As one might expect, modal noise is worst when using laser sources in conjunction with multimode fiber; recent industry standards have allowed the use of short-wave lasers (750 to 850 nm) on 50-micron fiber, which may experience this problem. Modal noise is usually considered to be nonexistent in single-mode systems. However, modal noise in single-mode fibers can arise when higher-order modes are generated at imperfect connections or splices. If the lossy mode is not completely attenuated before it reaches the next connection, interference with the dominant mode may occur. The effects of modal noise have been modeled previously,[29] assuming that the only significant interaction occurs between the LP01 and LP11 modes for a sufficiently coherent laser. For N sections of fiber, each of length L in a single-mode link, the worst-case sigma for modal noise can be given by

$$\sigma_m = \sqrt{2}\, N\eta(1-\eta)e^{-aL} \tag{30}$$

where a is the attenuation coefficient of the LP11 mode and η is the splice transmission efficiency, given by

$$\eta = 10^{-(\eta_0/10)} \tag{31}$$

where η_0 is the mean splice loss (typically, splice transmission efficiency will exceed 90 percent). The corresponding optical power penalty due to modal noise is given by

$$P = -5 \log (1 - Q^2 \sigma_m^2) \qquad (32)$$

where Q corresponds to the desired BER. This power penalty should be kept to less than 0.5 dB.

Radiation-Induced Loss

Another important environmental factor as mentioned earlier is exposure of the fiber to ionizing radiation damage. There is a large body of literature concerning the effects of ionizing radiation on fiber links.[30, 31] There are many factors that can affect the radiation susceptibility of optical fiber, including the type of fiber, type of radiation (gamma radiation is usually assumed to be representative), total dose, dose rate (important only for higher exposure levels), prior irradiation history of the fiber, temperature, wavelength, and data rate. Optical fiber with a pure silica core is least susceptible to radiation damage; however, almost all commercial fiber is intentionally doped to control the refractive index of the core and cladding, as well as dispersion properties. Trace impurities are also introduced which become important only under irradiation; among the most important are Ge dopants in the core of graded index (GRIN) fibers, in addition to F, Cl, P, B, OH content, and the alkali metals. In general, radiation sensitivity is worst at lower temperatures, and is also made worse by hydrogen diffusion from materials in the fiber cladding. Because of the many factors involved, a comprehensive theory does not exist to model radiation damage in optical fibers. The basic physics of the interaction have been described[30, 31]; there are two dominant mechanisms, *radiation-induced darkening* and *scintillation*. First, high-energy radiation can interact with dopants, impurities, or defects in the glass structure to produce color centers which absorb strongly at the operating wavelength. Carriers can also be freed by radiolytic or photochemical processes; some of these become trapped at defect sites, which modifies the band structure of the fiber and causes strong absorption at infrared wavelengths. This radiation-induced darkening increases the fiber attenuation; in some cases it is partially reversible when the radiation is removed, although high levels or prolonged exposure will permanently damage the fiber. A second effect is caused if the radiation interacts with impurities to produce stray light, or scintillation. This light is generally broadband, but will tend to degrade the BER at the receiver; scintillation is a weaker effect than radiation-induced darkening. These effects will degrade the BER of a link; they can be prevented by shielding the fiber, or partially overcome by a third mechanism, *photobleaching*. The presence of intense light at the proper wavelength can partially reverse the effects of darkening in a fiber. It is also possible to treat silica core fibers by briefly exposing them to controlled levels of radiation at controlled temperatures; this increases the fiber loss, but makes the fiber less susceptible to future irradiation. These so-called radiation-hardened fibers are often used in environments where radiation is anticipated to play an important role. Recently, several models have been advanced[31] for the performance of fiber under moderate radiation levels; the effect on BER is a power law model of the form

$$BER = BER_0 + A(\text{dose})^b \qquad (33)$$

where BER_0 is the link BER prior to irradiation, the dose is given in rads, and the constants A and b are empirically fitted. The loss due to normal background radiation exposure over a typical link lifetime can be held below about 0.5 dB.

6.5   REFERENCES

1. S. E. Miller and A. G. Chynoweth (eds.), *Optical Fiber Telecommunications*, Academic Press, New York, 1979.

2. J. Gowar, *Optical Communication Systems,* Prentice Hall, Englewood Cliffs, New Jersey, 1984.

3. C. DeCusatis, E. Maass, D. Clement, and R. Lasky (eds.), *Handbook of Fiber Optic Data Communication,* Academic Press, New York, 1998; see also *Optical Engineering* special issue on optical data communication (December 1998).

4. R. Lasky, U. Osterberg, and D. Stigliani (eds.), *Optoelectronics for Data Communication,* Academic Press, New York, 1995.

5. "Digital Video Broadcasting (DVB) Measurement Guidelines for DVB Systems," European Telecommunications Standards Institute ETSI Technical Report ETR 290, May 1997; "Digital Multi-Programme Systems for Television Sound and Data Services for Cable Distribution," International Telecommunications Union ITU-T Recommendation J.83, 1995; "Digital Broadcasting System for Television, Sound and Data Services; Framing Structure, Channel Coding and Modulation for Cable Systems," European Telecommunications Standards Institute ETSI 300 429, 1994.

6. W. E. Stephens and T. R. Hoseph, "System Characteristics of Direct Modulated and Externally Modulated RF Fiber-Optic Links," *IEEE J. Lightwave Technol.,* **LT-5(3)**:380–387 (1987).

7. C. H. Cox, III, and E. I. Ackerman, "Some Limits on the Performance of an Analog Optical Link," *Proceedings of the SPIE—The International Society for Optical Engineering* **3463**:2–7 (1999).

8. Laser safety standards in the United States are regulated by the Department of Health and Human Services (DHHS), Occupational Safety and Health Administration (OSHA), Food and Drug Administration (FDA) Code of Radiological Health (CDRH) 21 Code of Federal Regulations (CFR) subchapter J; the relevant standards are ANSI Z136.1, "Standard for the Safe Use of Lasers" (1993 revision) and ANSI Z136.2, "Standard for the Safe Use of Optical Fiber Communication Systems Utilizing Laser Diodes and LED Sources" (1996–1997 revision); elsewhere in the world, the relevant standard is International Electrotechnical Commission (IEC/CEI) 825 (1993 revision).

9. S. S. Walker, "Rapid Modeling and Estimation of Total Spectral Loss in Optical Fibers," *IEEE Journ. Lightwave Tech.* **4**:1125–1132 (1996).

10. Electronics Industry Association/Telecommunications Industry Association (EIA/TIA) Commercial Building Telecommunications Cabling Standard (EIA/TIA-568-A), Electronics Industry Association/Telecommunications Industry Association (EIA/TIA) Detail Specification for 62.5 Micron Core Diameter/125 Micron Cladding Diameter Class 1a Multimode Graded Index Optical Waveguide Fibers (EIA/TIA-492AAAA), Electronics Industry Association/Telecommunications Industry Association (EIA/TIA) Detail Specification for Class IV-a Dispersion Unshifted Single-Mode Optical Waveguide Fibers Used in Communications Systems (EIA/TIA-492BAAA), Electronics Industry Association, New York.

11. D. Gloge, "Propagation Effects in Optical Fibers," *IEEE Trans. Microwave Theory and Tech.* **MTT-23**: p. 106–120 (1975).

12. P. M. Shanker, "Effect of Modal Noise on Single-Mode Fiber Optic Network," *Opt. Comm.* **64**: 347–350 (1988).

13. J. J. Refi, "LED Bandwidth of Multimode Fiber as a Function of Source Bandwidth and LED Spectral Characteristics," *IEEE Journ. of Lightwave Tech.* **LT-14**:265–272 (1986).

14. G. P. Agrawal et al., "Dispersion Penalty for 1.3 Micron Lightwave Systems with Multimode Semiconductor Lasers," *IEEE Journ. Lightwave Tech.* **6**:620–625 (1988).

15. K. Ogawa, "Analysis of Mode Partition Noise in Laser Transmission Systems," *IEEE Journ. Quantum Elec.* **QE-18**:849–855 (1982).

16. K. Ogawa, "Semiconductor Laser Noise; Mode Partition Noise," in *Semiconductors and Semimetals,* Vol. 22C, R. K. Willardson and A. C. Beer (eds.), Academic Press, New York, 1985.

17. J. C. Campbell, "Calculation of the Dispersion Penalty of the Route Design of Single-Mode Systems," *IEEE Journ. Lightwave Tech.* **6**:564–573 (1988).

18. M. Ohtsu et al., "Mode Stability Analysis of Nearly Single-Mode Semiconductor Laser," *IEEE Journ. Quantum Elec.* **24**:716–723 (1988).

19. M. Ohtsu and Y. Teramachi, "Analysis of Mode Partition and Mode Hopping in Semiconductor Lasers," *IEEE Quantum Elec.* **25**:31–38 (1989).

20. D. Duff et al., "Measurements and Simulations of Multipath Interference for 1.7 Gbit/s Lightwave Systems Utilizing Single and Multifrequency Lasers," *Proc. OFC:* 128 (1989).

21. J. Radcliffe, "Fiber Optic Link Performance in the Presence of Internal Noise Sources," IBM Technical Report, Glendale Labs, Endicott, New York (1989).

22. L. L. Xiao, C. B. Su, and R. B. Lauer, "Increase in Laser RIN Due to Asymmetric Nonlinear Gain, Fiber Dispersion, and Modulation," *IEEE Photon. Tech. Lett.* **4**:774–777 (1992).

23. P. Trischitta and P. Sannuti, "The Accumulation of Pattern Dependent Jitter for a Chain of Fiber Optic Regenerators," *IEEE Trans. Comm.* **36**:761–765 (1988).

24. CCITT Recommendations G.824, G.823, O.171, and G.703 on Timing Jitter in Digital Systems (1984).

25. R. J. S. Bates, "A Model for Jitter Accumulation in Digital Networks," *IEEE Globecom Proc.:* 145–149 (1983).

26. C. J. Byrne, B. J. Karafin, and D. B. Robinson, Jr., "Systematic Jitter in a Chain of Digital Regenerators," *Bell System Tech. Journal* **43**:2679–2714 (1963).

27. R. J. S. Bates and L. A. Sauer, "Jitter Accumulation in Token Passing Ring LANs," *IBM Journal Research and Development* **29**:580–587 (1985).

28. C. Chamzas, "Accumulation of Jitter: A Stochastic Mode," *AT&T Tech. Journal:* 64 (1985).

29. D. Marcuse and H. M. Presby, "Mode Coupling in an Optical Fiber with Core Distortion," *Bell Sys. Tech. Journal.* **1**:3 (1975).

30. E. J. Frieble et al., "Effect of Low Dose Rate Irradiation on Doped Silica Core Optical Fibers," *App. Opt.* **23**:4202–4208 (1984).

31. J. B. Haber et al., "Assessment of Radiation Induced Loss for AT&T Fiber Optic Transmission Systems in the Terestrial Environment," *IEEE Journ. Lightwave Tech.* **6**:150–154 (1988).

CHAPTER 7

SOLITONS IN OPTICAL FIBER COMMUNICATION SYSTEMS

P. V. Mamyshev
Bell Laboratories—Lucent Technologies
Holmdel, New Jersey

7.1 INTRODUCTION

To understand why optical solitons are needed in optical fiber communication systems, we should consider the problems that limit the distance and/or capacity of optical data transmission. A fiber-optic transmission line consists of a transmitter and a receiver connected with each other by a transmission optical fiber. Optical fibers inevitably have chromatic dispersion, losses (attenuation of the signal), and nonlinearity. Dispersion and nonlinearity can lead to the distortion of the signal. Because the optical receiver has a finite sensitivity, the signal should have a high-enough level to achieve error-free performance of the system. On the other hand, by increasing the signal level, one also increases the nonlinear effects in the fiber. To compensate for the fiber losses in a long distance transmission, one has to periodically install optical amplifiers along the transmission line. By doing this, a new source of errors is introduced into the system—an amplifier spontaneous emission noise. (Note that even ideal optical amplifiers inevitably introduce spontaneous emission noise.) The amount of noise increases with the transmission distance (with the number of amplifiers). To keep the signal-to-noise ratio (SNR) high enough for the error-free system performance, one has to increase the signal level and hence the potential problems caused by the nonlinear effects. Note that the nonlinear effects are proportional to the product of the signal power, P, and the transmission distance, L, and both of these multipliers increase with the distance. Summarizing, we can say that all the problems—dispersion, noise, and nonlinearity—grow with the transmission distance. The problems also increase when the transmission bit rate (speed) increases. It is important to emphasize that it is very difficult to deal with the signal distortions when the nonlinearity is involved, because the nonlinearity can couple all the detrimental effects together [nonlinearity, dispersion, noise, polarization mode dispersion (i.e., random birefringence of the fiber), polarization-dependent loss/gain, etc]. That happens when the nonlinear effects are out of control. The idea of soliton transmission is to guide the nonlinearity to the desired direction and use it for your benefit. When soliton pulses are used as an information carrier, the effects of dispersion and nonlinearity balance (or compensate) each other and thus don't degrade the signal quality with the propagation distance. In such a regime, the pulses propagate through the fiber without changing their spectral and temporal shapes. This mutual compensation of dispersion and nonlinear effects takes place continuously with the distance in the case of "classical" solitons and periodically with the so-called dispersion map length in the case of dispersion-managed solitons. In addition, because of the unique features of optical solitons, soliton transmission can help to solve other problems of data transmission, like polarization mode dispersion. Also, when used with frequency guiding filters (sliding guiding filters in particular), the soliton systems provide continuous all-optical regeneration of the signal suppressing the detrimental effects of the noise and reducing the penalties associated with wavelength-division multiplexed (WDM) transmission. Because the soliton data looks essentially the same at different distances along the transmission, the soliton type of transmission is especially attractive for all-optical data networking. Moreover, because of the high quality of the pulses and return-to-zero (RZ) nature of the data, the soliton data is suitable for all-optical processing.

7.2   NATURE OF THE CLASSICAL SOLITON

Signal propagation in optical fibers is governed by the Nonlinear Schroedinger equation (NSE) for the complex envelope of the electric field of the signal.[1-3] This equation describes the combined action of the self-phase modulation and dispersion effects, which play the major role in the signal evolution in most practical cases. Additional linear and nonlinear effects can be added to the modified NSE.[4] Mathematically, one can say that solitons are stable solutions of NSE.[1,2] In this paper, however, we will give a qualitative physical description of the soliton regimes of pulse propagation, trying to avoid mathematics as much as possible.

Consider first the effect of dispersion. An optical pulse of width τ has a finite spectral bandwidth $BW \approx 1/\tau$. When the pulse is transform limited, or unchirped, all the spectral components have the same phase. In time domain, one can say that all the spectral components overlap in time, or sit on top of each other (see Fig. 1). Because of the dispersion, different spectral components propagate in the fiber with different group velocities, V_{gr}. As a result of the dispersion action alone, the initial unchirped pulse broadens and gets chirped (frequency modulated). The sign of the chirp depends on the sign of the fiber group velocity dispersion (see Fig. 1).

$$D = d\left(\frac{1}{V_{gr}}\right)/d\lambda \qquad (1)$$

(λ is the light wavelength). A characteristic fiber length called the *dispersion length*, at which the pulse broadens by a factor sqrt(2), is determined both by the fiber dispersion and the pulse width:

$$z_d = \frac{2\pi c\, 0.322\tau^2}{\lambda^2 D} \qquad (2)$$

(c is the speed of light). Note that the pulse spectral bandwidth remains unchanged because the dispersion is a linear effect.

Consider now the nonlinear effect of self-phase modulation (SPM).[5] Due to the Kerr effect, the fiber refractive index depends on the signal intensity, $n(I) = n_0 + n_2 I$, where n_2 is the nonlinear refractive index and intensity is $I = P/A$, P is the signal power and A is the fiber effective cross-section mode area. During a pulse propagation through the fiber, different parts of the pulse acquire different values of the nonlinear phase shift: $\phi(t) = 2\pi/\lambda\, n_2 I(t) L$. Here $I(t)$ is the intensity pulse shape in time domain and L is the transmission distance. This time-dependent nonlinear phase shift means that different parts of the pulse experience different frequency shifts:

$$\delta\omega(t) = \frac{d\phi}{dt} = -\frac{2\pi}{\lambda}\, n_2 L\, \frac{dI(t)}{dt} \qquad (3)$$

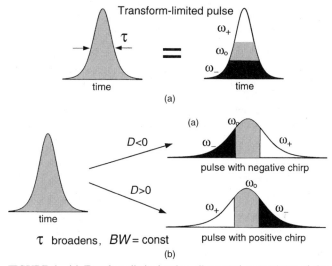

FIGURE 1 (*a*) Transform-limited pulse: all spectral components of the pulse "sit" on top of each other. (*b*) Effect of group velocity dispersion on a transform-limited pulse.

As one can see, the frequency shift is determined by the time derivative of the pulse shape. Because the nonlinear refractive index in silica-based fibers is positive, the self-phase modulation effect always shifts the front edge of the pulse to the "red" spectral region (downshift in frequency), and the trailing edge of the pulse to the "blue" spectral region (upshift in frequency). This means that an initially unchirped pulse spectrally broadens and gets negatively chirped (Fig. 2). A characteristic fiber length called the *nonlinear length,* at which the pulse spectrally broadens by a factor of two, is

$$z_{\mathrm{NL}} = \left(\frac{2\pi}{\lambda} \, n_2 I_0 \right)^{-1} \tag{4}$$

Note that, when acting alone, SPM does not change the temporal intensity profile of the pulse.

As it was mentioned earlier, when under no control, both SPM and dispersion may be very harmful for the data transmission distorting considerably the spectral and temporal characteristics of the signal. Consider now how to control these effects by achieving the soliton

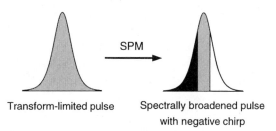

FIGURE 2   Effect of self-phase modulation on a transform-limited pulse.

regime of data transmission when the combined action of these effects results in a stable propagation of data pulses without changing their spectral and temporal envelopes.

In our qualitative consideration, consider the combined action of dispersion and nonlinearity (SPM) as an alternative sequence of actions of dispersion and nonlinearity. Assume that we start with a chirp-free pulse (see Fig. 3). The self-phase modulation broadens the pulse spectrum and produces a negative frequency chirp: The front edge of the pulse becomes red-shifted, and the trailing edge becomes blue-shifted. When positive GVD is then applied to this chirped pulse, the red spectral components are delayed in time with respect to the blue ones. If the right amount of dispersion is applied, the sign of the pulse chirp can be reversed to negative: The blue spectral components shift in time to the front pulse edge, while the red spectral components move to the trailing edge. When the nonlinearity is applied again, it shifts the frequency of the front edge to the red spectral region and upshifts the frequency of the trailing edge. That means that the blue front edge becomes green again, the red trailing edge also becomes green, and the pulse spectrum bandwidth narrows to its original width. The described regime of soliton propagation is achieved when the nonlinear and dispersion effect compensate each other exactly. In reality, the effects of dispersion and SPM act simultaneously, so that the pulse spectral and temporal widths stay constant with the distance, and the only net effect is a (constant within the entire pulse) phase shift of 0.5 rad per dispersion length of propagation.[6] The condition of the soliton regime is equality of the nonlinear and dispersion lengths: $z_d = z_{NL}$. One can rewrite this expression to find a relationship between the soliton peak power, pulse width, and fiber dispersion:

$$P_0 = \frac{\lambda^3 DA}{0.322\ 4\pi^2 c n_2 \tau^2} \tag{5}$$

Here, P_0 is the soliton peak power and τ is the soliton FWHM. Soliton pulses have a sech² form. Note that as it follows from our previous consideration, classical soliton propagation in fibers requires a positive sign of the fiber's dispersion, D (assuming that n_2 is positive). Consider a numerical example. For a pulse of width $\tau = 20$ ps propagating in a fiber with $D = 0.5$ ps nm⁻¹ km⁻¹, fiber cross-section mode area $A = 50$ μm², $\lambda = 1.55$ μm, and typical value of $n_2 = 2.6$ cm²/W, one can find the soliton peak power is 2.4 mW. The dispersion length is $z_d = 200$ km in this case.

7.3   PROPERTIES OF SOLITONS

The most important property of optical solitons is their robustness.[6-20] Consider what robustness means from a practical point of view. When a pulse is injected into the fiber, the pulse does not have to have the exact soliton shape and parameters (Eq. 5) to propagate as a soliton. As long as the input parameters are not too far from the optimum, during the nonlinear propagation the pulse "readjusts" itself, shaping into a soliton and shedding off nonsoliton components. For example, an unchirped pulse of width τ will be reshaped into a single soliton as long as its input

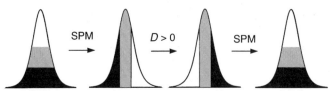

FIGURE 3   Qualitative explanation of classical soliton. Combined action of dispersion and nonlinearity (self-phase modulation) results in a stable pulse propagation with constant spectral and temporal widths. See text.

power, P, is greater than $P_0/4$ and less than $2.25P_0$. Here, P_0 is the soliton power determined by Eq. 5.[3]

Solitons are also robust with respect to the variations of the pulse energy and of the fiber parameters along the transmission line. As long as these variations are fast enough (period of perturbations is much smaller than the soliton dispersion length, z_d), the soliton "feels" only the average values of these parameters. This feature is extremely important for practical systems. In particular, it makes it possible to use solitons in long distance transmission systems where fiber losses are periodically compensated by lumped amplifiers. As long as the amplifier spacing is much less than the soliton dispersion length, $L_{amp} \ll z_d$, classical solitons work very well in these systems. Note that all soliton perturbations result in a loss of some part of the soliton energy, which is radiated into dispersive waves.

Consider now a case of slow variations of parameters along the transmission when a characteristic length at which a fiber parameter (or pulse energy) changes considerably is much longer than the soliton dispersion length. Soliton parameters follow adiabatically these changes. That means that all the parameters in Eq. 5 can be considered as distance dependent, and Eq. 5 remains valid. It can be rewritten in the following form:

$$\tau(z) = \text{const} \; \frac{D(z)A(z)}{P(z)\tau(z)} = \text{const} \; \frac{D(z)A(z)}{\text{Energy}(z)} \tag{6}$$

One can derive many important consequences from this equation.[13–22] One example would be the pulse broadening (and spectral narrowing) in a fiber with loss [assuming $D(z)$ and $A(z)$ are constant].[13–15] Note that the soliton broadening can be used in repeaterless data transmission systems when high-input signal power is required.[15] On the other hand, one can get a pulse compression in a fiber with adiabatic gain. Similar effects can be obtained by changing the fiber dispersion and/or mode area along the length. For example, adiabatic soliton compression can be obtained in a fiber with slowly decreasing dispersion (dispersion-tapered fiber).[16–22]

It is important to emphasize that the adiabatic soliton propagation does not necessarily require that each of these parameters—pulse energy, fiber dispersion, and mode area—changes adiabatically with the distance, as long as the whole expression, $[D(z) A(z)]/[\text{Energy}(z)]$ changes adiabatically with the distance. For example, soliton propagation in a dispersion-tapered fiber with losses is equivalent to transmission in a lossless, constant-dispersion fiber if the dispersion decreases with the same rate with the distance as the pulse energy [i.e., if $D(z)/\text{Energy}(z) = \text{const}$]. Note that this is true no matter what the fiber loss and the pulse width are.

So far, we've been discussing a single pulse propagation. In communication systems, one has to deal with streams of pulses. When two or more soliton pulses propagate in the fiber at the same wavelength, they can interact with each other: Tails from one soliton pulse may overlap with the other pulse. Due to the cross-phase modulation effect, this overlap leads to the frequency shifts of the interacting solitons. The signs of the frequency shifts are opposite for the two solitons. Through the fiber dispersion, the frequency changes result in the changes of the soliton group velocities. The strength of the interaction decreases very fast with the soliton separation and for most practical applications can be considered to be negligible when the separation is 4 to 5 times greater than the soliton pulse width, τ.[23,24] The character of interaction depends on the mutual optical phases of the solitons: When they are the same, the solitons attract to each other; when they are out of phase, the solitons repel from each other; when the phase difference is $\pi/2$, the solitons do not interact.

7.4 CLASSICAL SOLITON TRANSMISSION SYSTEMS

The soliton properties described earlier determine the engineering rules for designing the soliton-based transmission systems. First, to make sure that every individual pulse is stable in the transmission line with constant fiber dispersion and loss periodically compensated by lump amplifiers, the amplifier spacing, L_{amp}, should be much smaller than the soliton dispersion length, z_d. To avoid considerable pulse-to-pulse interaction, the minimum distance between adjacent pulses should be $T \geq 4\tau$, where $1/T$ is the transmission bit rate and τ is the soliton pulse width. The pulse power determined from Eq. 5 should be considered as a path-average power, P_{av}. If the signal energy decreases with the distance in the fiber spans between the amplifiers as $\exp(\gamma z)$ (here, γ is the loss rate), the path-average power is related to the pulse power at the output of each amplifier (input to the fiber span), P_{in}, as:

$$P_o = P_{in} \frac{1 - \exp(\gamma L_{amp})}{|\gamma| L_{amp}} \tag{7}$$

Here, L_{amp} is the amplifier spacing. As it was stated earlier, the dispersion and nonlinear effects "compensate" each other in the soliton regime of transmission, so that the pulses propagate practically without changing their temporal and spectral shapes. As long as the length scale of perturbations of the transmission parameters is much shorter than the soliton dispersion length, the pulses "feel" only the average parameters. Note, however, that perturbations may lead to shedding of dispersive waves by solitons.[12]

There are two main sources of errors in the soliton transmission systems: fluctuations of the pulse energies and fluctuations of the pulse arrival times.[25] The origin of the energy fluctuations is the same as in the other types of systems—spontaneous emission noise generated by the amplifiers. Each amplifier contributes a noise with a spectral density (power per unit bandwidth):

$$P_v = (G - 1) n_{sp} h\nu \tag{8}$$

Here, G is the power gain of the amplifier, $h\nu$ is the photon energy, and $n_{sp} \geq 1$ is the spontaneous emission factor that characterizes the quality of the amplifier. In the best case, when the amplifier is highly inverted, n_{sp} is close to unity. In a broadband transmission system (i.e., without in-line spectral filters), when the lumped amplifiers compensate exactly for the fiber loss, the noise grows linearly with the distance (with the number of amplifiers). At the output of a transmission line of length L, the path-averaged spectral density is:

$$P_{v\,av} = |\gamma| L n_{sp} h\nu F(G) \tag{9}$$

Here, function $F(G)$ describes the penalty one has to pay for having high-gain amplifiers (or long amplifier spacing):

$$F(G) = \frac{(G - 1)^2}{G \ln^2 G} \tag{10}$$

The penalty function has its minimum $[F(G) = 1]$ in the case of distributed amplification (when $G \to 1$) and grows with G. The SNR at the output of transmission should be high enough to have error-free transmission. Note that the noise spectral density, P_v, has units of energy. It is also the noise energy received in any time, T, in a spectral bandwidth, $1/T$. That is why P_v is also called the *equipartition energy*. To have the error probability less than 10^{-9} and 10^{-15}, the ratio of the pulse energy to the equipartition energy should be, correspondingly, 100 and 160. For example, consider a transmission system with the average loss of 0.21 dB/km, $n_{sp} = 1.5$, amplifier spacing of 50 km. The minimum pulse energy at the input of each fiber span to have the error probability less than 10^{-9} in such a system of length $L = 5000$ km is 20 fJ, and for $L = 10,000$ km, it is 40 fJ.

Another type of error in the soliton systems is the fluctuation in the pulse arrival times, or timing jitter. The timing jitter can be caused by several factors. The adjacent pulse-to-pulse interactions can cause the pulses to shift in time. As we have stated earlier, interaction problems can be practically eliminated by spacing the solitons in time by more than 4 or 5 of their width. A very important source of the timing jitter is the spontaneous emission noise. Every time the noise is added to the signal, it modulates the carrier frequencies of the solitons at random. The chromatic dispersion of the fiber then converts these frequency variations in a variation of the pulses' arrival times. This effect is known as the *Gordon-Haus effect*.[6,26] The variance of the timing jitter produced by the Gordon-Haus effect is:

$$\sigma_{GH}^2 \approx 0.2 n_2 h n_{sp} F(G) \frac{|\gamma|}{A} \frac{D}{\tau} L^3 \tag{11}$$

An error occurs when a pulse arrives outside of the acceptance time window, W, of the detection system (this window is usually slightly less than the bit slot, T). To have the error probability less than 10^{-9}, the acceptance window should be greater than 12 standard deviations of the timing jitter:

$$W \geq 12\sigma_{GH} \tag{12}$$

The Gordon-Haus jitter limits the maximum bit rate and transmission distance. As one can see from Eq. 11, the jitter increases very fast with the distance; it also increases when τ decreases. Another factor that limits the maximum transmission distance is that σ_{GH}^2 is proportional to the pulse energy [because the pulse energy is proportional to (D/τ)], and long-distance transmission systems should have high-enough pulse energies to keep the SNR high. Consider a numerical example, $L = 9,000$ km, $\tau = 20$ ps, $n_{sp} = 1.4$, $\gamma = -0.048$ km^{-1}, amplifier spacing $= 30$ km, $D = 0.5$ ps/(nm^{-1} km^{-1}), $A = 50$ µm^2. Equation 11 then gives the standard deviation of the Gordon-Haus timing jitter $\sigma = 11.7$ ps. As one can see, according to Eq. 12, this jitter is too high for 10 Gbit/s transmission ($1/T = 100$ ps) to be error-free, because $12\sigma_{GH} > 1/T$ in this case.

Another source of the timing jitter is the acoustic interaction of pulses.[27–30] Due to the electrostriction effect in the fiber, each propagating pulse generates an acoustic wave in the fiber. Other pulses experience the refractive index change caused by the acoustic wave. The resultant frequency changes of the pulses lead, through the effect of the fiber chromatic dispersion, to the fluctuation in the arrival times. The acoustic effect causes a "long-range" interaction: Pulses separated by a few nanoseconds can interact through this effect. One can estimate the acoustic timing jitter from the following simplified equation:

$$\sigma_a \approx 4.3 \frac{D^2}{\tau} (R - 0.99)^{1/2} L^2 \tag{13}$$

Here, standard deviation, σ_a, is in picoseconds; dispersion, D, is in picoseconds per nanometer per kilometer; the bit rate, $R = 1/T$, is in gigabits per second; and the distance, L, is in megameters. Equation 13 also assumes the fiber mode area of $A = 50$ µm^2. The acoustic jitter increases with the bit rate, and it has even stronger dependence on the distance than the Gordon-Haus jitter.

As it follows from the previous considerations, the timing jitter can impose severe limitations on the distance and capacity of the systems, and it has to be controlled.

7.5   FREQUENCY-GUIDING FILTERS

The Gordon-Haus and acoustic timing jitters originate from the frequency fluctuations of the pulses. That means that by controlling the frequency of the solitons, one can control the timing jitter as well. The frequency control can be done by periodically inserting narrowband filters (so-called frequency-guiding filters) along the transmission line, usually at the amplifier locations.[31,32] If, for some reason, the center frequency of a soliton is shifted from the filter peak, the filter-induced differential loss across the pulse spectrum "pushes" the pulse frequency back to the filter peak. As a result, the pulse spectrum returns back to the filter peak in a characteristic damping length, Δ. If the damping length is considerably less that the transmission distance, L, the guiding filters dramatically reduce the timing jitter. To calculate the timing jitter in a filtered system, one should replace L^3 by $3L\Delta^2$ in Eq. 11, and L^2 in Eq. 13 should be replaced by $2L\Delta$. Then, we get the following expression for the Gordon-Haus jitter:

$$\sigma^2_{\text{GH},f} \approx 0.6 n_2 h n_{\text{sp}} F(G) \frac{|\gamma|}{A} \frac{D}{\tau} L\Delta^2 \tag{14}$$

The damping properties of the guiding filters are determined mainly by the curvature of the filter response in the neighborhood of its peak. That means that shallow Fabry-Perot etalon filters can be used as the guiding filters. Fabry-Perot etalon filters have multiple peaks, and different peaks can be used for different WDM channels. The ability of the guiding filters to control the frequency jitter is determined both by the filter characteristics and by the soliton spectral bandwidth. In the case of Fabry-Perot filters with the intensity mirror reflectivity, R, and the free spectral range (FSR), the damping length is:

$$\Delta = 0.483(\tau \, \text{FSR})^2 \frac{(1-R)^2}{R} L_f \tag{15}$$

Here, L_f is the spacing between the guiding filters; usually, L_f equals the amplifier spacing L_{amp}.

Note that the Gordon-Haus and acoustic jitters are not specific for soliton transmission only. Any kind of transmission systems, including so-called linear transmission, are subject to these effects. However, the guiding filters can be used in the soliton systems only. Every time a pulse passes through a guiding filter, its spectrum narrows. Solitons can quickly recover their bandwidth through the fiber nonlinearity, whereas for a linear transmission the filter action continuously destroys the signal.

Note that even a more effective reduction of the timing jitter can be achieved if, in addition to the frequency-guiding filters, an amplitude and/or phase modulation at the bit rate is applied to the signal periodically with the distance. "Error-free" transmission over practically unlimited distances can be achieved in this case (1 million kilometers at 10 Gbit/s has been demonstrated).[33,34] Nevertheless, this technique is not passive, high-speed electronics is involved, and the clock recovery is required each time the modulation is applied. Also, in the case of WDM transmission, all WDM channels have to be demultiplexed before the modulation and then multiplexed back afterward; each channel has to have its own clock recovery and modulator. As one can see, this technique shares many drawbacks of the electronic regeneration schemes.

The frequency-guiding filters can dramatically reduce the timing jitter in the systems. At the same time, though, in some cases they can introduce additional problems. Every time a soliton passes through the filter, it loses some energy. To compensate for this loss, the amplifiers should provide an additional (excess) gain. Under this condition, the spontaneous emission noise and other nonsoliton components with the spectrum in the neighborhood of the filter peak experience exponential growth with the distance, which reduces the SNR and can lead to the soliton instabilities. As a result, one has to use weak-enough filters to reduce the excess gain. In practice, the filter strength is chosen to minimize the total penalty from the timing jitter and the excess gain.

7.6   *SLIDING FREQUENCY-GUIDING FILTERS*

As one can see, the excess gain prevents one from taking a full advantage of guiding filters. By using the sliding frequency-guiding filters,[35] one can essentially eliminate the problems associated with the excess gain. The trick is very simple: The transmission peak of each guiding filter is shifted in frequency with respect to the peak of the previous filter, so that the center frequency slides with the distance with the rate of $f' = df/dz$. Solitons, thanks to the nonlinearity, can follow the filters and slide in frequency with the distance. But all unwanted linear radiation (e.g., spontaneous emission noise, nonsoliton components shedded from the solitons, etc.) cannot slide and eventually is killed by the filters. The sliding allows one to use strong guiding filters and even to reduce the amount of noise at the output of transmission in comparison with the broadband (no guiding filters) case. The maximum filter strength[36] and maximum sliding rate[35] are determined by the soliton stability. The error-free transmission of 10 Gbit/s signal over 40,000 km and 20 Gbit/s over 14,000 km was demonstrated with the sliding frequency-guiding filters technique.[37,38]

It is important to emphasize that by introducing the sliding frequency-guiding filters into the transmission line, one converts this transmission line into an effective, all-optical passive regenerator (compatible with WDM). Solitons with only one energy (and pulse width) can propagate stably in such a transmission line. The parameters of the transmission line (the filter strength, excess gain, fiber dispersion, and mode area) determine the unique parameters of these stable solitons. The system is opaque for a low-intensity radiation (noise, for example). However, if the pulse parameters at the input of the transmission line are not too far from the optimum soliton parameters, the transmission line reshapes the pulse into the soliton of that line. Note, again, that the parameters of the resultant soliton do not depend on the input pulse parameters, but only on the parameters of the transmission line. Note also that all nonsoliton components generated during the pulse reshaping are absorbed by the filters. That means, in particular, that the transmission line removes the energy fluctuations from the input data signal.[6] Note that the damping length for the energy fluctuations is close to the frequency damping length of Eq. 15. A very impressive demonstration of regenerative properties of a transmission line with the frequency-guiding filters is the conversion of a nonreturn-to-zero (NRZ) data signal (frequency modulated at the bit rate) into a clean soliton data signal.[39] Another important consequence of the regenerative properties of a transmission line with the frequency-guiding filters is the ability to self-equalize the energies of different channels in WDM transmission.[40] Negative feedback provided by frequency-guiding filters locks the energies of individual soliton channels to values that do not change with distance, even in the face of considerable variation in amplifier gain among the different channels. The equilibrium values of the energies are independent of the input values. All these benefits of sliding frequency-guiding filters are extremely valuable for practical systems. Additional benefits of guiding filters for WDM systems will be discussed later.

7.7 WAVELENGTH DIVISION MULTIPLEXING

Due to the fiber chromatic dispersion, pulses from different WDM channels propagate with different group velocities and collide with each other.[41] Consider a collision of two solitons propagating at different wavelengths (different channels). When the pulses are initially separated and the fast soliton (the soliton at shorter wavelength, with higher group velocity) is behind the slow one, the fast soliton eventually overtakes and passes through the slow soliton. An important parameter of the soliton collision is the collision length, L_{coll}, the fiber length at which the solitons overlap with each other. If we let the collision begin and end with the overlap of the pulses at half power points, then the collision length is:

$$L_{coll} = \frac{2\tau}{D\Delta\lambda} \tag{16}$$

Here, $\Delta\lambda$ is the solitons wavelengths difference. Due to the effect of cross-phase modulation, the solitons shift each other's carrier frequency during the collision. The frequency shifts for the two solitons are equal in amplitudes (if the pulse widths are equal) and have opposite signs. During the first half of collision, the fast accelerates even faster (carrier frequency increases), while the slow soliton slows down. The maximum frequency excursion, δf_{max}, of the solitons is achieved in the middle of the collision, when the pulses completely overlap with each other:

$$\delta f_{max} = \pm \frac{1}{3\pi^2 0.322 \, \Delta f \tau^2} = \pm \frac{1.18 n_2 \varepsilon}{A \tau D \lambda \, \Delta \lambda} \tag{17}$$

Here, $\Delta f = -c \, \Delta \lambda / \lambda^2$ is the frequency separation between the solitons, and $\varepsilon = 1.13 P_0 \tau$ is the soliton energy. In the middle of collision, the accelerations of the solitons change their signs. As a result, the frequency shifts in the second half of collision undo the frequency shifts of the first half, so that the soliton frequency shifts go back to zero when the collision is complete. This is a very important and beneficial feature for practical applications. The only residual effect of complete collision in a lossless fiber is the time displacements of the solitons:

$$\delta t_{cc} = \pm \frac{0.1786}{\Delta f^2 \, \tau} = \pm \frac{2 \varepsilon n_2 \lambda}{c D A \, \Delta \lambda^2} \tag{18}$$

The symmetry of the collision can be broken if the collision takes place in a transmission line with loss and lumped amplification. For example, if the collision length, L_{coll}, is shorter than the amplifier spacing, L_{amp}, and the center of collision coincides with the amplifier location, the pulses intensities are low in the first half of collision and high in the second half. As a result, the first half of collision is practically linear. The soliton frequency shifts acquired in the first half of collision are very small and insufficient to compensate for the frequency shifts of opposite signs acquired by the pulses in the second half of collision. This results in nonzero residual frequency shifts. Note that similar effects take place when there is a discontinuity in the value of the fiber dispersion as a function of distance. In this case, if a discontinuity takes place in the middle of collision, one half of the collision is fast (where D is higher) and the other half is slow. The result is nonzero residual frequency shifts. Nonzero residual frequency shifts lead, through the dispersion of the rest of the transmission fiber, to variations in the pulses arrival time at the output of transmission. Nevertheless, if the collision length is much longer than the amplifier spacing and of the characteristic length of the dispersion variations in the fiber, the residual soliton frequency shifts are zero, just like in a lossless uniform fiber. In practice, the residual frequency shifts are essentially zero as long as the following condition is satisfied:[41]

$$L_{coll} \geq 2 L_{amp} \tag{19}$$

Another important case is so-called half-collisions (or partial collisions) at the input of the transmission.[42] These collisions take place if solitons from different channels overlap at the transmission input. These collisions result in residual frequency shifts of δf_{max} and the following pulse timing shifts, δt_{pc}, at the output of transmission of length L:

$$\delta t_{pc} \approx \delta f_{max} \, \frac{\lambda^2}{c} \, D(L - L_{coll}/4) = \pm \frac{1.18 \varepsilon n_2 \lambda}{c \tau A \, \Delta \lambda} \, (L - L_{coll}/4) \tag{20}$$

One can avoid half-collisions by staggering the pulse positions of the WDM channels at the transmission input.

Consider now the time shifts caused by all complete collisions. Consider a two-channel transmission, where each channel has a $1/T$ bit rate. The distance between subsequent collisions is:

$$l_{coll} = \frac{T}{D \, \Delta \lambda} \tag{21}$$

The maximum number of collisions that each pulse can experience is L/l_{coll}. This means that the maximum time shift caused by all complete collisions is:

$$\delta t_{\Sigma cc} \approx \delta t_{cc} L/l_{coll} = \pm \frac{2\varepsilon n_2 \lambda}{cTA\,\Delta\lambda}\,L \qquad (22)$$

It is interesting to note that $\delta t_{\Sigma cc}$ does not depend on the fiber dispersion. Note also that Eq. 22 describes the worst case when the pulse experiences the maximum number of possible collisions. Consider a numerical example. For a two-channel transmission, 10 Gbit/s each ($T = 100$ ps), pulse energy ($\varepsilon = 50$ fJ), channel wavelength separation ($\Delta\lambda = 0.6$ nm), fiber mode area ($A = 50$ μm^2 and $L = 10$ Mm), we find $\delta t_{\Sigma cc} = 45$ ps. Note that this timing shift can be reduced by increasing the channel separation. Another way to reduce the channel-to-channel interaction by a factor of two is to have these channels orthogonally polarized to each other. In WDM transmission, with many channels, one has to add timing shifts caused by all other channels. Note, however, that as one can see from Eq. 22, the maximum penalty comes from the nearest neighboring channels.

As one can see, soliton collisions introduce additional jitter to the pulse arrival time, which can lead to considerable transmission penalties. As we saw earlier, the frequency-guiding filters are very effective in suppressing the Gordon-Haus and acoustic jitters. They can also be very effective in suppressing the timing jitter induced by WDM collisions. In the ideal case of parabolical filters and the collision length being much longer than the filter spacing, $L_{coll} \gg L_f$, the filters make the residual time shift of a complete collision, δt_{cc}, exactly zero. They also considerably reduce the timing jitter associated with asymmetrical collisions and half-collisions. Note that for the guiding filters to work effectively in suppressing the collision penalties, the collision length should be at least a few times greater than the filter spacing. Note also that real filters, such as etalon filters, do not always perform as good as ideal parabolic filters. This is true especially when large-frequency excursions of solitons are involved, because the curvature of a shallow etalon filter response reduces with the deviation of the frequency from the filter peak. In any case, filters do a very good job in suppressing the timing jitter in WDM systems.

Consider now another potential problem in WDM transmission, which is the four-wave mixing. During the soliton collisions, the four-wave mixing spectral sidebands are generated. Nevertheless, in the case of a lossless, constant-dispersion fiber, these sidebands exist only during the collision, and when the collision is complete, the energy from the sidebands regenerates back into the solitons. That is why it was considered for a long time that the four-wave mixing should not be a problem in soliton systems. But this is true only in the case of a transmission in a lossless fiber. In the case of lossy fiber and periodical amplification, these perturbations can lead to the effect of the pseudo-phase-matched (or resonance) four-wave mixing.[43] The pseudo-phase-matched four-wave mixing lead to the soliton energy loss to the spectral sidebands and to a timing jitter (we called that effect an *extended Gordon-Haus effect*).[43] The effect can be so strong that even sliding frequency-guiding filters are not effective enough to suppress it. The solution to this problem is to use dispersion-tapered fiber spans. As we have discussed earlier, soliton propagation in the condition:

$$\frac{D(z)\,A(z)}{\text{Energy}(z)} = \text{const} \qquad (23)$$

is identical to the case of lossless, constant-dispersion fiber. That means that the fiber dispersion in the spans between the amplifiers should decrease with the same rate as the signal energy. In the case of lumped amplifiers, this is the exponential decay with the distance. Note that the dispersion-tapered spans solve not just the four-wave mixing problem. By making the soliton transmission perturbation-free, they lift the requirements to have the amplifier spacing much shorter than the soliton dispersion length. The collisions remain symmetrical even when the collision length is shorter than the amplifier spacing. (Note, however, that the dispersion-tapered fiber spans do not lift the requirement to have guiding filter spacing as short as possible in comparison with the collision length and with the dispersion length.) The dispersion-tapered fiber spans can be made with the present technology.[22] Stepwise approximation of the exact exponential taper made of fiber pieces of constant dispersion can also be used.[43] It was shown numerically and experimentally that by using fiber spans with only a few steps one can dramatically improve the quality of transmission.[44,45] In the experiment, each fiber span was dispersion tapered typically in three or four steps, the path-average dispersion value was 0.5 ± 0.05 ps nm^{-1} km^{-1} at 1557 nm. The use of dispersion-tapered fiber spans together with sliding frequency-guiding filters allowed transmission of eight 10-Gbit/s channels with the channel spacing, $\Delta\lambda = 0.6$ nm, over more than 9000 km. The maximum number of channels in this experiment was limited by the dispersion slope, $dD/d\lambda$, which was about 0.07 ps nm^{-2} km^{-1}. Because of the dispersion slope, different WDM channels experience different values of dispersion. As a result, not only the path average dispersion changes with the wavelength, but the dispersion tapering has exponential behavior only in a vicinity of one particular wavelength in the center of the transmission band. Wavelength-division multiplexed channels located far from that wavelength propagate in far from the optimal conditions. One solution to the problem is to use dispersion-flattened fibers (i.e., fibers with $dD/d\lambda = 0$). Unfortunately, these types of fibers are not commercially available at this time. This and some other problems of classical soliton transmission can be solved by using dispersion-managed soliton transmission.[46-63]

7.8 DISPERSION-MANAGED SOLITONS

In the dispersion-managed (DM) soliton transmission, the transmission line consists of the fiber spans with alternating signs of the dispersion. Let the positive and negative dispersion spans of the map have lengths and dispersions, L_+, D_+ and L_-, D_-, respectively. Then, the path-average dispersion, D_{av} is:

$$D_{av} = (D_+L_+ + L_-D_-)/L_{map} \qquad (24)$$

Here, L_{map}, is the length of the dispersion map:

$$L_{map} = L_+ + L_- \qquad (25)$$

Like in the case of classical soliton, during the DM soliton propagation, the dispersion and nonlinear effects cancel each other. The difference is that in the classical case, this cancellation takes place continuously, whereas in the DM case, it takes place periodically with the period of the dispersion map length, L_{map}. The strength of the DM is characterized by a parameter, S, which is determined as[47,50,52]

$$S = \frac{\lambda^2}{2\pi c} \frac{(D_+ - D_{av})L_+ - (D_- - D_{av})L_-}{\tau^2} \qquad (26)$$

The absolute values of the local dispersion are usually much greater than the path average dispersion: $|D_+|, |D_-| \gg |D_{av}|$. As one can see from Eq. 26, the strength of the map is proportional to the number of the local dispersion lengths of the pulse in the map length: $S \approx L_{map}/z_{d, local}$. The shape of the DM solitons are close to Gaussian. A very important feature of DM solitons is the so-called power enhancement. Depending on the strength of the map, the pulse energy of DM solitons, ε_{DM}, is greater than that of classical solitons, ε_0, propagating in a fiber with constant dispersion, $D = D_{av}$:[47,50]

$$\varepsilon_{DM} \approx \varepsilon_0 \, (1 + 0.7S^2) \qquad (27)$$

Note that this equation assumes lossless fiber. The power enhancement effect is very important for practical applications. It provides an extra degree of freedom in the system design by giving the possibility to change the pulse energy while keeping the path-average fiber dispersion constant. In particular, because DM solitons can have adequate pulse energy (to have a high-enough SNR) at or near zero path average dispersion, timing jitter from the Gordon-Haus and acoustic effects is greatly reduced (for example, the variance of the Gordon-Haus jitter, σ^2, scales almost as $1/\varepsilon_{DM}$).[49] Single-channel high-bit-rate DM soliton transmission over long distances with weak guiding filters and without guiding filters was experimentally demonstrated.[46,51]

Dispersion-managed soliton transmission is possible not only in transmission lines with positive dispersion, $D_{av} > 0$, but also in the case of $D_{av} = 0$ and even $D_{av} < 0$.[52] To understand this, consider qualitatively the DM soliton propagation (Fig. 4). Locally, the dispersive effects are always stronger than the nonlinear effect (i.e., the local dispersion length is much shorter than the nonlinear length). In the zero approximation, the pulse propagation in the map is almost linear. Let's call the middle of the positive D sections "point a," the middle of the negative sections "point c," transitions between positive and negative sections "point b," and transitions between negative and positive sections "point d." The chirp-free (minimum pulse width) positions of the pulse are in the middle of the positive- and negative-D sections (points a and c). The pulse chirp is positive between points a, b, and c (see Fig. 4). That means that the high-frequency (blue) spectral components of the pulse are at the front edge of the pulse, and the low-frequency (red) components are at the trailing edge. In the section c-d-a, the pulse chirp is negative. The action of the nonlinear SPM effect always downshifts in frequency the front edge of the pulse and up shifts in frequency the trailing edge of the pulse. That means that the nonlinearity decreases the spectral bandwidth of positively chirped pulses (section a-b-c) and increases the spectral bandwidth of negatively chirped pulses (section c-d-a). This results in the spectral bandwidth behavior also shown in Fig. 4: The maximum spectral bandwidth is achieved in the chirp-free point in the positive section, whereas the minimum spectral bandwidth is achieved in the chirp-free point in the negative section. The condition for the pulses to be DM solitons is that the nonlinear phase shift is compensated by the dispersion-induced

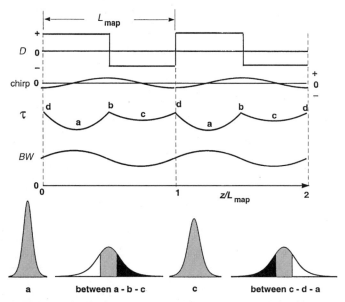

FIGURE 4 Qualitative description of dispersion-managed (DM) soliton transmission. Distance evolution of the fiber dispersion [$D(z)$], pulse chirp, pulse width [$\tau(z)$], and pulse bandwidth [$BW(z)$]. Evolution of the pulse shape in different fiber sections is shown in the bottom.

phase shift over the dispersion map length. That requires that $\int D\, BW^2\, dz > 0$ (here, BW is the pulse spectral bandwidth). Note that in the case of classical solitons, when spectral bandwidth is constant, this expression means that dispersion, D, must be positive. In the DM case, however, the pulse bandwidth is wider in the positive-D section than in the negative-D section. As a result, the integral can be positive, even when $D_{av} = \int D\,dz/L_{map}$ is zero or negative. Note that the spectral bandwidth oscillations explain also the effect of power enhancement of DM solitons.

Consider interaction of adjacent pulses in DM systems.[54] The parameter that determines the strength of the interaction is the ratio τ/T (here, τ is the pulse width and T is the spacing between adjacent pulses). As in the case of classical soliton transmission, the cross-phase modulation effect (XPM) shifts the frequencies of the interacting pulses, Δf_{XPM}, which, in turn, results in timing jitter at the output of the transmission. As it was discussed earlier, the classical soliton interaction increases very quickly with τ/T. To avoid interaction-induced penalties in classical soliton transmission systems, the pulses should not overlap significantly with each other: τ/T should be less than 0.2 to 0.3. In the DM case, the situation is different. The pulse width in the DM case oscillates with the distance $\tau(z)$; that means that the interaction also changes with distance. Also, because the pulses are highly chirped when they are significantly overlapped with each other, the sign of the interaction is essentially independent of the mutual phases of the pulses. Cross-phase modulation always shifts the leading pulse to the red spectral region, and the trailing pulse shifts to the blue spectral region. The XPM-induced frequency shifts of interacting solitons per unit distance is:

$$\frac{d\,\Delta f_{XPM}}{dz} \approx \pm 0.15\,\frac{2\pi n_2 \varepsilon}{\lambda T^2 A}\,\Phi(\tau/T) \qquad (28)$$

The minus sign in Eq. 28 corresponds to the leading pulse, and the plus sign corresponds to the trailing pulse. Numerically calculated dimensionless function, $\Phi(\tau/T)$, is shown in Fig. 5. As it follows from Eq. 28, $\Phi(\tau/T)$ describes the strength of the XPM-induced interaction of the pulses as a function of the degree of the pulse overlap. One can see that the interaction is very small when τ/T is smaller than 0.4 (i.e., when the pulses barely overlap), which is similar to the classical soliton propagation. The strength of the interaction of DM solitons also increases with τ/T, but only in the region $0 < \tau/T < 1$. In fact, the interaction reaches its maximum at $\tau/T \approx 1$ and then decreases and becomes very small again when $\tau/T \gg 1$ (i.e., when the pulses overlap nearly completely). There are two reasons for such an interesting behavior at $\tau/T \gg 1$. The XPM-induced frequency shift is proportional to the time derivative of the

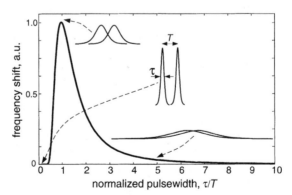

FIGURE 5 Dimensionless function, $\Phi(\tau/T)$, describing the XPM-induced frequency shift of two interacting chirped Gaussian pulses as a function of the pulse width normalized to the pulse separation.

interacting pulse's intensity, and the pulse derivative reduces with the pulse broadening. Also, when the pulses nearly completely overlap, the sign of the derivative changes across the region of overlap so that the net effect tends to be canceled out.

Based on Eq. 28 and Fig. 5, one can distinguish three main regimes of data transmission in DM systems. In all these regimes, the minimum pulse width is, of course, less than the bit slot, T. The regimes differ from each other by the maximum pulse breathing with the distance. In the first, "non-pulse-overlapped," regime, adjacent pulses barely overlap during most of the transmission, so that the pulse interaction is not a problem in this case. This is the most stable regime of transmission. In the "partially-pulse-overlapped" regime, the adjacent pulses spend a considerable portion of the transmission being partially overlapped [$\tau(z)$ being around T]. Cross-phase modulation causes the frequency and timing jitter in this case. In the third, "pulse-overlapped," regime, the adjacent pulses are almost completely overlapped with each other during most of the transmission [τ_{min} ($L_{map}/z_{d, local}$) $\gg T$]. The XPM-induced pulse-to-pulse interaction is greatly reduced in this case in comparison with the previous one. The main limiting factor for this regime of transmission is the intrachannel four-wave mixing taking place during strong overlap of adjacent pulses.[54] The intrachannel four-wave mixing leads to the amplitude fluctuations of the pulses and "ghost" pulse generation in the "zero" slots of the data stream.

7.9 WAVELENGTH-DIVISION MULTIPLEXED DISPERSION-MANAGED SOLITON TRANSMISSION

One of the advantages of DM transmission over classical soliton transmission is that the local dispersion can be very high ($|D_+|, |D_-| \gg |D_{av}|$), which efficiently suppresses the four-wave mixing from soliton-soliton collisions in WDM. Consider the timing jitter induced by collisions in the non-pulse-overlapped DM transmission. The character of the pulse collisions in DM systems is quite different from the case of a transmission line with uniform dispersion: In the former, the alternating sign of the high local dispersion causes the colliding solitons to move rapidly back and forth with respect to each other, with the net motion determined by D_{av}.[56–59] Because of this rapid breathing of the distance between the pulses, each net collision actually consists of many fast or "mini" collisions. The net collision length can be estimated as:[59]

$$L_{coll} \approx \frac{2\tau}{D_{av}\,\Delta\lambda} + \frac{(D_+ - D_{av})L_+}{D_{av}} \approx \frac{2\tau}{D_{av}\,\Delta\lambda} + \frac{\tau_{eff}}{D_{av}\,\Delta\lambda} \qquad (29)$$

Here, τ is the minimum (unchirped) pulse width. Here, we also defined the quantity $\tau_{eff} \equiv L_+ D_+ \Delta\lambda$, which plays the role of an effective pulse width. For strong dispersion management, τ_{eff} is usually much bigger than τ. Thus, L_{coll} becomes almost independent of $\Delta\lambda$ and much longer than it is for classical solitons subject to the same D_{av}. As a result, the residual frequency shift caused by complete pulse collisions tends to become negligibly small for transmission using strong maps.[58] The maximum frequency excursion during the DM soliton collision is:[59]

$$\delta f_{max} \approx \pm\frac{2n_2\varepsilon}{L_+ D_+ A D_{av}\lambda\,\Delta\lambda^2} = \pm\frac{2n_2\varepsilon}{A\,D_{av}\,\lambda\,\Delta\lambda\tau_{eff}} \qquad (30)$$

Now, we can estimate the time shift of the solitons per complete collision:

$$\delta t_{cc} \approx D_{av}\,\lambda^2/c \int \delta f dz \approx \alpha D_{av} L_{coll}\,\delta f_{max}\,\lambda^2/c \approx \pm\alpha\,\frac{2n_2\varepsilon\lambda}{cA D_{av}\,\Delta\lambda^2} \qquad (31)$$

Here, $\alpha \leq 1$ is a numerical coefficient that takes into account the particular shape of the frequency shift as a function of distance. Consider now the time shifts caused by all collisions. In a two-channel transmission, the distance between subsequent collisions is $l_{coll} = T/(D_{av}\Delta\lambda)$. The maximum number of complete collisions at the transmission distance, L, is $(L - L_{coll})/l_{coll}$ (we assume that $L > L_{coll}$), and the number of incomplete collisions at the end of transmission is L_{coll}/l_{coll}. The timing shift caused by all these collisions can be estimated as

$$\delta t_{\Sigma c} \approx \delta t_{cc} (L - L_{coll}/2)/l_{coll} = \pm\alpha \frac{2n_2\varepsilon\lambda}{cAT\,\Delta\lambda} (L - L_{coll}/2) \qquad (32)$$

Consider the problem of initial partial collisions. As it was discussed earlier for the case of classical solitons, initial partial collisions can be a serious problem by introducing large timing jitter at the output of transmission. On the other hand, for the classical case, one could avoid the half-collisions by staggering the pulse positions of the WDM channels at the transmission input. The situation is very different for the DM case. In the DM case, the collision length is usually longer than the distance between subsequent collisions (i.e., $L_{coll} > l_{coll}$). Thus, a pulse can collide *simultaneously* with several pulses of another channel. The maximum number of such simultaneous collisions is $N_{sc} \approx L_{coll}/l_{coll} = 2\tau/T + [(D_+ - D_{av})L_+ \Delta\lambda]/T$. Note that N_{sc} increases when the channel spacing, $\Delta\lambda$, increases. The fact that the collision length is greater than the distance between collisions also means that initial partial collisions are inevitable in DM systems. Moreover, depending on the data pattern in the interacting channel, each pulse can experience up to N_{sc} initial partial collisions with that channel (not just one as in the classical case). As a consequence, the residual frequency shifts can be bigger than δf_{max}. The total time shift caused by the initial partial collisions at distance $L > L_{coll}$ can be estimated as:

$$\delta\tau_{pc} \approx \beta\delta f_{max}N_{sc}(L - L_{coll}/2)D_{av}\lambda^2/c \approx \pm\beta \frac{2n_2\varepsilon\lambda}{cAT\,\Delta\lambda} (L - L_{coll}/2) \qquad (33)$$

Here, $\beta \leq 1$ is a numerical coefficient that takes into account the particular shape of the frequency shift as a function of distance for a single collision.

Equations 32 and 33 assume that the transmission distance is greater than the collision length. When $L > L_{coll}$, these equations should be replaced by:

$$\delta t_{\Sigma c,\,pc} \approx (\alpha, \beta) \, D_{av}\delta f_{max} \frac{\lambda^2}{c} \frac{L^2}{2l_{coll}} \approx \pm(\alpha, \beta) \frac{n_2\varepsilon\lambda}{cAT\,\Delta\lambda} \frac{L^2}{L_{coll}} \qquad (34)$$

Note that the signs of the timing shifts caused by initial partial collisions and by complete collisions are opposite. Thus, the maximum (worst-case) spread of the pulse arriving times caused by pulse collisions in the two-channel WDM transmission is described by:

$$\delta t_{max} = |\delta t_{pc}| + |\delta t_{\Sigma c}| \qquad (35)$$

In a WDM transmission with more than two channels, one has to add contributions to the time shift from all the channels. Note that the biggest contribution makes the nearest neighboring channels, because the time shift is inversely proportional to the channel spacing, $\Delta\lambda$. Now, we can summarize the results of Eqs. 32 through 35 as follows. When $L > L_{coll}$ (Eqs. 32–33), corresponding to very long distance transmission, δt_{max} increases linearly with the distance and almost independently of the path-average dispersion, D_{av}. When $L < L_{coll}$ (Eq. 34), which corresponds to short-distance transmission and/or very low path-average dispersion, δt_{max} increases quadratically with the distance and in proportion to D_{av}. Note also that the WDM data transmission at near zero path-averaged dispersion, $D_{av} = 0$, may not be desirable, because $L_{coll} \to \infty$ and frequency excursions $\delta f_{max} \to \infty$ when $\overline{D} \to 0$ (see Eq. 30). Thus, even though Eq. 34 predicts the time shift to be zero when D_{av} is exactly zero, the frequency shifts of the solitons can be unacceptably large and Eq. 34 may be no longer valid. There are also practical difficulties in making maps with $D_{av} < 0.1$ ps nm^{-1} km^{-1} over the wide spectral range required for dense WDM transmission.

It is interesting to compare these results with the results for the case of classical solitons (Eqs. 17–22). The time shifts per complete collisions (Eqs. 18 and 31) are about the same, the time shifts from all initial partial collisions (Eqs. 20 and 33) are also close to each other. The total maximum time shifts from all collisions are also close to each other for the case of long distance transmission. That means that, similar to the classical case, one has to control the collision-induced timing jitter when it becomes too large. As it was discussed earlier, the sliding frequency-guiding filters are very effective in suppressing the timing jitter. Because the collision length in DM systems is much longer than in classical systems, and, at the same time, it is almost independent of the channel wavelength separation, the requirement that the collision length is much greater than the filter spacing, $L_{coll} \gg L_f$, is easy to meet. As a result, the guiding filters suppress the timing jitter in DM systems even more effective than in classical soliton systems. The fact that the frequency excursions during collisions are much smaller in DM case, also makes the filters to work more effectively.

As we have discussed previously, many important features of DM solitons come from the fact that the soliton spectral bandwidth oscillates with the distance. That is why guiding filters alter the dispersion management itself and give an additional degree of freedom in the system design.[60] Note also that the position of the filters in the dispersion map can change the soliton stability in some cases.[61] It should also be noted that because of the weak dependence of the DM soliton spectral bandwidth on the soliton pulse energy, the energy fluctuations damping length provided by the guided filters is considerably longer than the frequency damping length.[62] This is the price one has to pay for many advantages of DM solitons. From the practical point of view, the most important advantage is the flexibility in system design and freedom in choosing the transmission fibers. For example, one can upgrade existing systems by providing an appropriate dispersion compensation with dispersion compensation fibers or with lumped dispersion compensators (fiber Bragg gratings, for example). The biggest advantage of DM systems is the possibility to design dispersion maps with essentially zero dispersion slope of the path-average dispersion, $dD_{av}/d\lambda$, by combining commercially available fibers with different signs of dispersion and dispersion slopes. (Note that it was a nonzero dispersion slope that limited the maximum number of channels in classical soliton long distance WDM transmission.) This was demonstrated in the experiment where almost flat average dispersion, $D_{av} = 0.3$ ps nm^{-1} km^{-1} was achieved by combining standard, dispersion-compensating, and True-Wave (Lucent nonzero dispersion-shifted) fibers.[63] By using sliding frequency-guiding filters and this dispersion map, "error-free" DM soliton transmission of twenty-seven 10-Gbit/s WDM channels was achieved over more than 9000 km without using forward error correction. It was shown that once the error-free transmission with about 10 channels is achieved, adding additional channels practically does not change performance of the system. (This is because, for each channel, only the nearest neighboring channels degrade its performance.) The maximum number of WDM channels in this experiment was limited only by the power and bandwidth of optical amplifiers used in the experiment. One can expect that the number of channels can be increased by a few times if more powerful and broader-bandwidth amplifiers are used.

7.10   CONCLUSION

We considered the basic principles of soliton transmission systems. The main idea of the "soliton philosophy" is to put under control, balance, and even to extract the maximum benefits from otherwise detrimental effects of the fiber dispersion and nonlinearity. The "soliton approach" is to make transmission systems intrinsically stable. Soliton technology is a very rapidly developing area of science and engineering, which promises a big change in the functionality and capacity of optical data transmission and networking.

7.11 REFERENCES

1. V. E. Zaharov and A. B. Shabat, "Exact Theory of Two Dimentional Self Focusing and One-Dimentional Self-Modulation of Waves in Nonlinear Media," *Zh. Eksp. Teor. Fiz.* **61**:118–134 (1971) [*Sov. Phys. JETP* **34**:62–69 (1972)].

2. A. Hasegawa and F. D. Tappert, "Transmission of Stationary Nonlinear Optical Pulses in Dispersive Dielectric Fibers. I. Anomalous Dispersion," *Applied Phys. Letters* **23**:142–144 (1973).

3. J. Satsuma and N. Yajima, "Initial Value Problem of One-Dimentional Self-Modulation of Nonlinear Waves in Dispersive Media," *Prog. Theor. Phys. Suppl.* **55**:284–306 (1980).

4. P. V. Mamyshev and S. V. Chernikov, "Ultrashort Pulse Propagation in Optical Fibers," *Optics Letters* **15**:1076–1078 (1990).

5. R. H. Stolen, in *Optical Fiber Telecommunications,* S. E. Miller and H. E. Chynoweth (eds.), Academic Press, New York, 1979, Chap. 5.

6. L. F. Mollenauer, J. P. Gordon, and P. V. Mamyshev, "Solitons in High Bit Rate, Long Distance Transmission," in *Optical Fiber Telecommunications III,* Academic Press, 1997, Chap. 12.

7. L. F. Mollenauer, J. P. Gordon, and M. N. Islam, "Soliton Propagation in Long Fibers with Periodically Compensated Loss," *IEEE J. Quantum Electron.* **QE-22**:157 (1986).

8. L. F. Mollenauer, M. J. Neubelt, S. G. Evangelides, J. P. Gordon, J. R. Simpson, and L. G. Cohen, "Experimental Study of Soliton Transmission Over More Than 10,000 km in Dispersion Shifted Fiber," *Opt. Lett.* **15**:1203 (1990).

9. L. F. Mollenauer, S. G. Evangelides, and H. A. Haus, "Long Distance Soliton Propagation Using Lumped Amplifiers and Dispersion Shifted Fiber," *J. Lightwave Technol.* **9**:194 (1991).

10. K. J. Blow and N. J. Doran, "Average Soliton Dynamics and the Operation of Soliton Systems with Lumped Amplifiers," *Photonics Tech. Lett.* **3**:369 (1991).

11. A. Hasegawa and Y. Kodama, "Guiding-Center Soliton in Optical Fibers," *Opt. Lett.* **15**:1443 (1990).

12. G. P. Gordon, "Dispersive Perturbations of Solitons of the Nonlinear Schroedinger Equation," *JOSA B* **9**:91–97 (1992).

13. A. Hasegawa and Y. Kodama, "Signal Transmission by Optical Solitons in Monomode Fiber," *Proc. IEEE* **69**:1145 (1981).

14. K. J. Blow and N. J. Doran, "Solitons in Optical Communications," *IEEE J. of Quantum Electronics,"* **QE-19**:1883 (1982).

15. P. B. Hansen, H. A. Haus, T. C. Damen, J. Shah, P. V. Mamyshev, and R. H. Stolen, "Application of Soliton Spreading in Optical Transmission," Dig. ECOC, Vol. 3, Paper WeC3.4, pp. 3.109–3.112, Oslo, Norway, September 1996.

16. K. Tajima, "Compensation of Soliton Broadening in Nonlinear Optical Fibers with Loss," *Opt. Lett.* **12**:54 (1987).

17. H. H. Kuehl, "Solitons on an Axially Nonuniform Optical Fiber," *J. Opt. Soc. Am. B* **5**:709–713 (1988).

18. E. M. Dianov, L. M. Ivanov, P. V. Mamyshev, and A. M. Prokhorov, "High-Quality Femtosecond Fundamental Soliton Compression in Optical Fibers with Varying Dispersion," Topical Meeting on Nonlinear Guided-Wave Phenomena: Physics and Applications, 1989, Technical Digest Series, vol. 2, OSA, Washington, D.C., 1989, pp. 157–160, paper FA-5.

19. P. V. Mamyshev, "Generation and Compression of Femtosecond Solitons in Optical Fibers," *Bull. Acad. Sci. USSR, Phys. Ser.,* **55**(2):374–381 (1991) [*Izv. Acad. Nauk, Ser. Phys.* **55**(2):374–381 (1991).

20. S. V. Chernikov and P. V. Mamyshev, "Femtosecond Soliton Propagation in Fibers with Slowly Decreasing Dispersion," *J. Opt. Soc. Am. B.* **8**(8):1633–1641 (1991).

21. P. V. Mamyshev, S. V. Chernikov, and E. M. Dianov, "Generation of Fundamental Soliton Trains for High-Bit-Rate Optical Fiber Communication Lines," *IEEE J. of Quantum Electron.* **27**(10):2347–2355 (1991).

22. V. A. Bogatyrev, M. M. Bubnov, E. M. Dianov, A. S. Kurkov, P. V. Mamyshev, A. M. Prokhorov, S. D. Rumyantsev, V. A. Semeonov, S. L. Semeonov, A. A. Sysoliatin, S. V. Chernikov, A. N. Gurianov, G. G. Devyatykh, S. I. Miroshnichenko, "Single-Mode Fiber with Chromatic Dispersion Varying along the Length," *IEEE J. of Lightwave Technology* **LT-9**(5):561–566 (1991).

23. V. I. Karpman and V. V. Solov'ev, "A Perturbation Approach to the Two-Soliton System," *Physica D* **3**:487–502 (1981).

24. J. P. Gordon, "Interaction Forces among Solitons in Optical Fibers," *Optics Letters* **8**:596–598 (1983).

25. J. P. Gordon and L. F. Mollenauer, "Effects of Fiber Nonlinearities and Amplifier Spacing on Ultra Long Distance Transmission," *J. Lightwave Technol.* **9**:170 (1991).

26. J. P. Gordon and H. A. Haus, "Random Walk of Coherently Amplified Solitons in Optical Fiber," *Opt. Lett.* **11**:665 (1986).

27. K. Smith and L. F. Mollenauer, "Experimental Observation of Soliton Interaction over Long Fiber Paths: Discovery of a Long-Range Interaction," *Opt. Lett.* **14**:1284 (1989).

28. E. M. Dianov, A. V. Luchnikov, A. N. Pilipetskii, and A. N. Starodumov, "Electrostriction Mechanism of Soliton Interaction in Optical Fibers," *Opt. Lett.* **15**:314 (1990).

29. E. M. Dianov, A. V. Luchnikov, A. N. Pilipetskii, and A. M. Prokorov, "Long-Range Interaction of Solitons in Ultra-Long Communication Systems," *Soviet Lightwave Communications* **1**:235 (1991).

30. E. M. Dianov, A. V. Luchnikov, A. N. Pilipetskii, and A. M. Prokhorov "Long-Range Interaction of Picosecond Solitons Through Excitation of Acoustic Waves in Optical Fibers," *Appl. Phys. B* **54**:175 (1992).

31. A. Mecozzi, J. D. Moores, H. A. Haus, and Y. Lai, "Soliton Transmission Control," *Opt. Lett.* **16**:1841 (1991).

32. Y. Kodama and A. Hasegawa, "Generation of Asymptotically Stable Optical Solitons and Suppression of the Gordon-Haus Effect," *Opt. Lett.* **17**:31 (1992).

33. M. Nakazawa, E. Yamada, H. Kubota, and K. Suzuki, "10 Gbit/s Soliton Transmission over One Million Kilometers," *Electron. Lett.* **27**:1270 (1991).

34. T. Widdowson and A. D. Ellis, "20 Gbit/s Soliton Transmission over 125 Mm," *Electron. Lett.* **30**:1866 (1994).

35. L. F. Mollenauer, J. P. Gordon, and S. G. Evangelides, "The Sliding-Frequency Guiding Filter: An Improved Form of Soliton Jitter Control," *Opt. Lett.* **17**:1575 (1992).

36. P. V. Mamyshev and L. F. Mollenauer, "Stability of Soliton Propagation with Sliding Frequency Guiding Filters," *Opt. Lett.* **19**:2083 (1994).

37. L. F. Mollenauer, P. V. Mamyshev, and M. J. Neubelt, "Measurement of Timing Jitter in Soliton Transmission at 10 Gbits/s and Achievement of 375 Gbits/s-Mm, error-free, at 12.5 and 15 Gbits/s," *Opt. Lett.* **19**:704 (1994).

38. D. LeGuen, F. Fave, R. Boittin, J. Debeau, F. Devaux, M. Henry, C. Thebault, and T. Georges, "Demonstration of Sliding-Filter-Controlled Soliton Transmission at 20 Gbit/s over 14 Mm," *Electron. Lett.* **31**:301 (1995).

39. P. V. Mamyshev and L. F. Mollenauer, "NRZ-to-Soliton Data Conversion by a Filtered Transmission Line," in *Optical Fiber Communication Conference OFC-95*, Vol. 8, 1995 OSA Technical Digest Series, OSA, Washington, D.C., 1995, Paper FB2, pp. 302–303.

40. P. V. Mamyshev and L. F. Mollenauer, "WDM Channel Energy Self-Equalization in a Soliton Transmission Line Using Guiding Filters," *Optics Letters* **21**(20):1658–1660 (1996).

41. L. F. Mollenauer, S. G. Evangelides, and J. P. Gordon, "Wavelength Division Multiplexing with Solitons in Ultra Long Distance Transmission Using Lumped Amplifiers," *J. Lightwave Technol.* **9**:362 (1991).

42. P. A. Andrekson, N. A. Olsson, J. R. Simpson, T. Tanbun-ek, R. A. Logan, P. C. Becker, and K. W. Wecht, *Electron. Lett.* **26**:1499 (1990).

43. P. V. Mamyshev, and L. F. Mollenauer, "Pseudo-Phase-Matched Four-Wave Mixing in Soliton WDM Transmission," *Opt. Lett.* **21**:396 (1996).

44. L. F. Mollenauer, P. V. Mamyshev, and M. J. Neubelt, "Demonstration of Soliton WDM Transmission at 6 and 7 × 10 GBit/s, Error-Free over Transoceanic Distances," *Electron. Lett.* **32**:471 (1996).

45. L. F. Mollenauer, P. V. Mamyshev, and M. J. Neubelt, "Demonstration of Soliton WDM Transmission at up to 8 × 10 GBit/s, Error-Free over Transoceanic Distances," OFC-96, Postdeadline paper PD-22.

46. M. Suzuki, I Morita, N, Edagawa, S. Yamamoto, H. Taga, and S. Akiba, "Reduction of Gordon-Haus Timing Jitter by Periodic Dispersion Compensation in Soliton Transmission," *Electron. Lett.* **31**:2027–2029 (1995).

47. N. J. Smith, N. J. Doran, F. M. Knox, and W. Forysiak, "Energy-Scaling Characteristics of Solitons in Strongly Dispersion-Managed Fibers," *Opt. Lett.* **21**:1981–1983 (1996).

48. I. Gabitov and S. K. Turitsyn, "Averaged Pulse Dynamics in a Cascaded Transmission System with Passive Dispersion Compensation," *Opt. Lett.* **21**:327–329 (1996).

49. N. J. Smith, W. Forysiak, and N. J. Doran, "Reduced Gordon-Haus Jitter Due to Enhanced Power Solitons in Strongly Dispersion Managed Systems," *Electron. Lett.* **32**:2085–2086 (1996).

50. V. S. Grigoryan, T. Yu, E. A. Golovchenko, C. R. Menyuk, and A. N. Pilipetskii, "Dispersion-Managed Soliton Dynamics," *Opt. Lett.* **21**:1609–1611 (1996).

51. G. Carter, J. M. Jacob, C. R. Menyuk, E. A. Golovchenko, and A. N. Pilipetskii, "Timing Jitter Reduction for a Dispersion-Managed Soliton System: Experimental Evidence," *Opt. Lett.* **22**:513–515 (1997).

52. S. K. Turitsyn, V. K. Mezentsev and E. G. Shapiro, "Dispersion-Managed Solitons and Optimization of the Dispersion Management," *Opt. Fiber Tech.* **4**:384–452 (1998).

53. J. P. Gordon and L. F. Mollenauer, "Scheme for Characterization of Dispersion-Managed Solitons," *Opt. Lett.* **24**:223–225 (1999).

54. P. V. Mamyshev and N. A. Mamysheva, "Pulse-Overlapped Dispersion-Managed Data Transmission and Intra-Channel Four-Wave Mixing," *Opt. Lett.* **24**:1454–1456 (1999).

55. D. Le Guen, S. Del Burgo, M. L. Moulinard, D. Grot, M. Henry, F. Favre, and T. Georges, "Narrow Band 1.02 Tbit/s (51 × 20 gbit/s) Soliton DWDM Transmission over 1000 km of Standard Fiber with 100 km Amplifier Spans," *OFC-99,* postdeadline paper PD-4.

56. S. Wabnitz, *Opt. Lett.* **21**:638–640 (1996).

57. E. A. Golovchenko, A. N. Pilipetskii, and C. R. Menyuk, *Opt. Lett.* **22**:1156–1158 (1997).

58. A. M. Niculae, W. Forysiak, A. G. Gloag, J. H. B. Nijhof, and N. J. Doran, "Soliton Collisions with Wavelength-Division Multiplexed Systems with Strong Dispersion Management," *Opt. Lett.* **23**:1354–1356 (1998).

59. P. V. Mamyshev and L. F. Mollenauer, "Soliton Collisions in Wavelength-Division-Multiplexed Dispersion-Managed Systems," *Opt. Lett.* **24**:448–450 (1999).

60. L. F. Mollenauer, P. V. Mamyshev, and J. P. Gordon, "Effect of Guiding Filters on the Behavior of Dispersion-Managed Solitons," *Opt. Lett.* **24**:220–222 (1999).

61. M. Matsumoto, *Opt. Lett.* **23**:1901–1903 (1998).

62. M. Matsumoto, *Electron. Lett.* **33**:1718 (1997).

63. L. F. Mollenauer, P. V. Mamyshev, J. Gripp, M. J. Neubelt, N. Mamysheva, L. Gruner-Nielsen, and T. Veng, "Demonstration of Massive WDM over Transoceanic Distances Using Dispersion Managed Solitons," *Optics Letters* **25**:704–706 (2000).

CHAPTER 8
TAPERED-FIBER COUPLERS, MUX AND DEMUX

Daniel Nolan
Corning Inc.
Corning, New York

8.1 INTRODUCTION

Fiber-optic couplers, including splitters and wavelength-division multiplexing (WDM) components, have been used extensively over the last two decades. This use continues to grow both in quantity and in the ways in which the devices are used. The uses today include, among other applications, simple splitting for signal distribution and wavelength multiplexing and demultiplexing multiple wavelength signals.

Fiber-based splitters and WDM components are among the simplest devices. Other technologies that can be used to fabricate components that exhibit similar functions include the planar waveguide and micro-optic technologies. These devices are, however, most suitable for integrated-optics in the case of planar or more complex devices in the case of micro-optic components. In this chapter, we will show the large number of optical functions that can be achieved with simple tapered fiber components. We will also describe the physics of the propagation of light through tapers in order to better understand the breadth of components that can be fabricated with this technology. The phenomenon of coupling includes an exchange of power that can depend both on wavelength and on polarization. Beyond the simple 1×2 power splitter, other devices that can be fabricated from tapered fibers include $1 \times N$ devices, wavelength multiplexing, polarization multiplexing, switches, attenuators, and filters.

Fiber-optic couplers have been fabricated since the early seventies. The fabrication technologies have included fusion tapering,[1-3] etching,[4] and polishing.[5-7] The tapered single-mode fiber-optic power splitter is perhaps the most universal of the single-mode tapered devices.[8] It has been shown that the power transferred during the tapering process involves an initial adiabatic transfer of the power in the input core to the cladding/air interface.[9] The light is then transferred to the adjacent core-cladding mode. During the up-tapering process, the input light will transfer back onto the fiber cores. In this case, it is referred to as a *cladding mode coupling device*. Light that is transferred to a higher-order mode of the core-cladding structure leads to an excess loss. This is because these higher-order modes are not bounded by the core and are readily stripped by the higher index of the fiber coating.

In the tapered fiber coupler process, two fibers are brought into close proximity after the protective plastic jacket is removed. Then, in the presence of a torch, the fibers are fused and stretched (see Fig. 1). The propagation of light through this tapered region is described using Maxwell's vector equations, but for a good approximation the scalar wave equation is valid. The scalar wave equation, written in cylindrical coordinates, is expressed as

$$[1/r \, d/dr \, r \, d/dr - \hat{v}^2 \, /r^2 + k^2 n l^2 - \beta^2 - (V/a)^2 \, f(r/a)] \, \psi = \varepsilon\mu \, d^2 \, \psi/dt^2 \tag{1}$$

In Eq. (1), $n1$ is the index value at $r = 0$, β is the propagation constant, which is to be determined, a is the core radius, $f(r/a)$ is a function describing the index distribution with radius, and V is the modal volume

$$V = 2\pi a n1 \, \frac{[\sqrt{[2\Delta]}]}{\lambda} \tag{2}$$

With

$$\Delta = \frac{[n1^2 - n2^2]}{[2n1^2]} \tag{3}$$

As light propagates in the single-mode fiber, it is not confined to the core region, but extends out into the surrounding region. As the light propagates through the tapered region, it is bounded by the shrinking, air-cladding boundary.

In the simplest case, the coupling from one cladding to the adjacent one can be described by perturbation theory.[10] In this case, the cladding air boundary is considered as the waveguide outer boundary, and the exchange of power along z is described as

$$P = \sin^2 [CZ] \tag{4}$$

$$C = 2\pi/[\lambda\alpha^2] \, \sqrt{[1 - (n1/n2)^2]} \tag{5}$$

$$[n1^2 - n2^2]/(n1^2 - n2^2)^{1.5}$$

$$K0[2(\alpha + (2\pi d/\lambda)) \, \sqrt{[n1^2 - n2^2]}]/$$

$$K1^2 \left\{ \alpha\sqrt{[n1^2 - n^2]} \right\}$$

with

$$\alpha = 2\pi n1/\lambda \tag{6}$$

It is important to point out that Eqs. (4) and (5) are only a first approximation. These equations are derived using first-order perturbation theory. Also, the scalar wave equation is not strictly

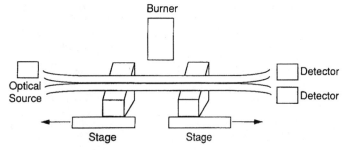

FIGURE 1 Fused biconic tapered coupler process. The fibers are stripped of their coating and fused and tapered using a heat source.

valid under the presence of large index differences, such as at a glass/air boundary. However, these equations describe a number of important effects. The sinusoidal dependence of the power coupled with wavelength, as well as the dependence of power transfer with cladding diameter and other dependencies, is well described with the model.

Equation (4) can be described by considering the light input to one core as a superposition of symmetric and antisymmetric modes.[10] These modes are eigen solutions to the composite two-core structure. The proper superposition of these two modes enables one to impose input boundary conditions for the case of a two-core structure. The symmetric and antisymmetric modes are written

$$\Psi s = \frac{[\psi 1 + \psi 2]}{\sqrt{2}} \tag{7}$$

$$\Psi a = \frac{[\psi 1 - \psi 2]}{\sqrt{2}} \tag{8}$$

Light input onto one core is described with $\Psi 1$ at $z = 0$,

$$\Psi 1 = \frac{[\psi s + \psi a]}{\sqrt{2}} \tag{9}$$

Propagation through the coupler is characterized with the superposition of Ψs and Ψa. This superposition describes the power transfer between the two guides along the direction of propagation.[10] The propagation constants of Ψs and Ψa are slightly different, and this value can be used to estimate excess loss under certain perturbations.

8.2 ACHROMATICITY

The simple sinusoidal dependence of the coupling with wavelength as just described is not always desired, and often a more achromatic dependence of the coupling is required. This can be achieved when dissimilar fibers[10] are used to fabricate the coupler. Fibers are characterized as dissimilar when the propagation constants of the guides are of different values. When dissimilar fibers (see Fig. 2) are used, Eqs. (4) and (5) can be replaced with

$$P1(z) = P1(0) + F^2 \{P2(0) - P1(0) + [(B_1 - B_2)/C] [P1(0) P2(0)]^5 \} \sin^2 (Cz/F) \tag{10}$$

where

$$F = 1./\left\{1 + \frac{B_1^2 - B_c^2}{4C^2}/[4\,C^2]\right\}^{1/2} \tag{11}$$

In most cases, the fibers are made dissimilar by changing the cladding diameter of one of the fibers. Etching or pre-tapering one of the fibers can do this. Another approach is to slightly change the cladding index of one of the fibers.[11] When dissimilar fibers are used, the total amount of power coupled is limited. As an example, an achromatic 3 dB coupler is made achromatic by operating at the sinusoidal maximum with wavelength rather than at the power of maximum power change with wavelength. Another approach to achieve achromaticity is to taper the device such that the modes expand well beyond the cladding boundaries.[12] This condition greatly weakens the wavelength dependence of the coupling. This has been achieved by encapsulating the fibers in a third matrix glass with an index very close to that of the fiber's cladding index. The difference in index between the cladding and the matrix glass is on the order of 0.001. The approach of encapsulating the fibers in a third-index material[13,14] is also

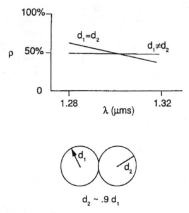

FIGURE 2 Achromatic couplers are fabricated by tapering two fibers with different propagating characteristics in the region of coupling.

useful for reasons other than achromaticity. One reason is that the packaging process is simplified. Also, a majority of couplers made for undersea applications use this method because it is a proven approach to ultra high reliability.

The wavelength dependence of the couplers just described is most often explained using mode coupling and perturbation theory. Often, numerical analysis is required in order to explain the effects that the varying taper angles have on the overall coupling. An important numerical approach is the beam propagation method.[15] In this approach, the propagation of light through a device is solved by an expansion of the evolution operator using a Taylor series and with the use of fast Fourier transforms to evaluate the appropriate derivatives. In this way, the propagation of the light can be studied as it couples to the adjacent guides or to higher order modes.

8.3   WAVELENGTH DIVISION MULTIPLEXING

Besides power splitting, tapered couplers can be used to separate wavelengths. To accomplish this separation, we utilize the wavelength dependence of Eqs. (4) and (5). By proper choice of the device length and taper ratio, two predetermined wavelengths can be put out onto two different ports. Wavelengths from 50 to 600 nms can be split using this approach. Applications include the splitting and/or combining of 1480 nm and 1550 nm light, as well as multiplexing 980 nm and 1550 nm onto an erbium fiber for signal amplification. Also important is the splitting of the 1310 to 1550 nm wavelength bands, which can be achieved using this approach.

8.4   1 × N POWER SPLITTERS

Often it is desirable to split a signal onto a number of output ports. This can be achieved by concatenating 1 × 2 power splitters. Alternatively, one can split the input simultaneously onto multiple output ports[16,17] (see Fig. 3). Typically, the output ports are of the form 2^N (i.e., 2, 4, 8, 16). The configuration of the fibers in the tapered region affects the distribution of the output power per port. A good approach to achieve uniform 1 × 8 splitting is described in Ref. 18.

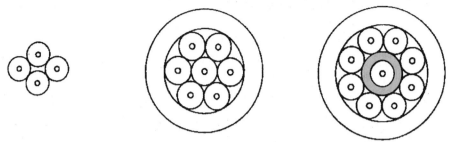

FIGURE 3   MXN couplers are fabricated by fusing and tapering fibers of the appropriate configuration. These configurations have been commercialized by Gould, BT&D, and Corning.

8.5   SWITCHES AND ATTENUATORS

In a tapered device, the power coupled over to the adjacent core can be significantly affected by bending the device at the midpoint. By encapsulating two fibers before tapering in a third index medium (see Fig. 4), the device is rigid and can be reliably bent in order to frustrate the coupling. The bending establishes a difference in the propagation constants of the two guiding media, preventing coupling or power transfer.

This approach can be used to fabricate both switches and attenuators. Switches with up to 30 dB crosstalk and attenuators with variable crosstalk up to 30 dB as well over the erbium wavelength band have been fabricated. Displacing one end of a 1-cm taper by 1 millimeter is enough to alter the crosstalk by the 30-dB value. Applications for attenuators have been increasing significantly over the last few years. An important reason is to maintain the gain in erbium-doped fiber amplifiers. This is achieved by limiting the amount of pump power into the erbium fiber. Over time, as the pump degrades, the power output of the attenuator is increased in order to compensate for the pump degradation.

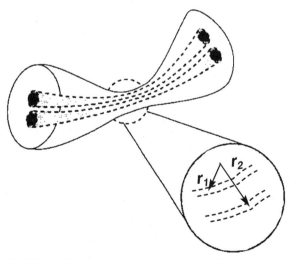

FIGURE 4   The coupling can be affected by bending the coupler at the midsection. Switched and variable attenuators are fabricated in this manner.

8.6 *MACH-ZEHNDER DEVICES*

Devices to split narrowly spaced wavelengths are very important. As previously mentioned, tapers can be designed such that wavelengths from 60 nm to 600 nm can be split in a tapered device. Dense WDM networks require splitting of wavelengths with separations on the order of nms. Fiber-based Mach-Zehnder devices enable such splitting. Monolithic fiber-based Mach-Zehnders can be fabricated using fibers with different cores (see Fig. 5),[20] (i.e., different propagation constants). Two or more tapers can be used to cause light from two different optical paths to interfere. The dissimilar cores enable light to propagate at different speeds between the tapers, causing the required constructive and destructive interference. These devices are environmentally stable due to the monolithic structure. Mach-Zehnders can also be fabricated using fibers with different lengths between the tapers. In this approach, it is the packaging that enables an environmentally stable device.

Multiple tapers can be used to fabricate devices with a wavelength spectra with higher-order Fourier components.[23] Figure 6 shows the spectrum of a three-tapered band splitter.

Mach-Zehnders and lattice filters can also be fabricated by tapering single-fiber devices.[24] In the tapered regions, the light couples to a cladding mode. The cladding mode propagates between tapers since a lower index overcladding replaces the higher index coating material. An interesting application for these devices is as gain-flattening filters for amplifiers.

8.7 *POLARIZATION DEVICES*

It is well-known that two polarization modes propagate in single-mode fiber. Most optical fiber modules allow both polarizations to propagate, but specify that the performance of the components be insensitive to the polarization states of the propagating light. However, this is often not the situation for fiber-optic sensor applications. Often, the state of polarization is important to the operation of the sensor itself. In these situations, polarization-maintaining fiber is used. Polarization components such as polarization-maintaining couplers and also single-

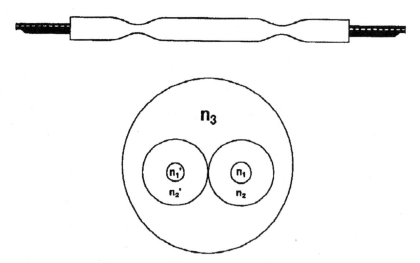

FIGURE 5 Narrow-band WDM devices can be fabricated by multiply tapering two fibers with different cores.

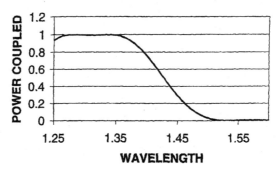

FIGURE 6 Band splitters are fabricated with three tapers.

polarization devices are used. In polarization-maintaining fiber, a difference in propagation constants of the polarization modes prevents mode coupling or exchange of energy. This is achieved by introducing stress or shape birefringence within the fiber core. A significant difference between the two polarization modes is maintained as the fiber twists in a cable or package.

In many fiber sensor systems, tapered fiber couplers are used to couple light from one core to another. Often the couplers are composed of birefringent fibers[24] (see Fig. 7). This is done in order to maintain the alignment of the polarizations to the incoming and outgoing fibers and also to maintain the polarization states within the device. The axes of the birefringent fibers are aligned before tapering, and care is taken not to excessively twist the fibers during the tapering process.

The birefringent fibers contain stress rods, elliptical core fibers, or inner claddings in order to maintain the birefringence. The stress rods in some birefringent fibers have an index higher than the silica cladding. In the tapering process, this can cause light to be trapped in these

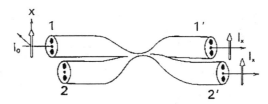

(a) Polarization-Maintaining FC

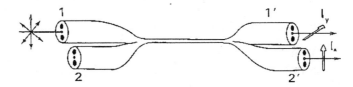

(b) Polarization-Splitting FC

FIGURE 7 Polarization-maintaining couplers and polarization splitters are fabricated using polarization-maintaining fibers.

rods, resulting in an excess loss in the device. Stress rods with an index lower than that of silica can be used in these fibers, resulting in very low-loss devices.

8.8 SUMMARY

Tapered fiber couplers are extremely useful devices. Such devices include 1×2 and $1 \times N$ power splitters, wavelength-division multiplexers and filters, and polarization-maintaining and splitting components. Removing the fiber's plastic coating and then fusing and tapering two or more fibers in the presence of heat forms these devices. The simplicity and flexibility of this fabrication process is in part responsible for the widespread use of these components. The mechanism involved in the fabrication process is reasonably understood and simple, which is in part responsible for the widespread deployment of these devices. These couplers are found in optical modules for the telecommunication industry and in assemblies for the sensing industry. They are also being deployed as standalone components for fiber-to-the-home applications.

8.9 REFERENCES

1. T. Ozeki and B. S. Kawaski, "New Star Coupler Compatible with Single Multimode Fiber Links," *Elect. Lett.* **12**:151–152 (1976).
2. B. S. Kawaski and K. O. Hill, "Low Loss Access Coupler for Multimode Optical Fiber Distribution Networks," *Applied Optics* **16**:1794–1795 (1977).
3. G. E. Rawson and M. D. Bailey, "Bitaper Star Couplers with Up to 100 Fiber Channels," *Electron Lett.* **15**:432–433 (1975).
4. S. K. Sheem and T. G. Giallorenzi, "Single-Mode Fiber Optical Power Divided; Encapsulated Etching Technique," *Opt. Lett.* **4**:31 (1979).
5. Y. Tsujimoto, H. Serizawa, K. Hatori, and M. Fukai, "Fabrication of Low Loss 3 dB Couplers with Multimode Optical Fibers," *Electron Lett.* **14**:157–158 (1978).
6. R. A. Bergh, G. Kotler, and H. J. Shaw, "Single-Mode Fiber Optic Directional Coupler," *Electron Lett.* **16**:260–261 (1980).
7. O. Parriaux, S. Gidon, and A. Kuznetsov, "Distributed Coupler on Polished Single-Mode Fiber," *Appl. Opt.* **20**:2420–2423 (1981).
8. B. S. Kawaski, K. O. Hill, and R. G. Lamont, "Biconical—Taper Single-Mode Fiber Coupler," *Opt. Lett.* **6**:327 (1981).
9. R. G. Lamont, D. C. Johnson, and K. O. Hill, *Appl Opt.* **24**:327–332 (1984).
10. A. Snyder and J. D. Love, *Optical Waveguide Theory*, Chapman and Hall, 1983.
11. W. J. Miller, C. M. Truesdale, D. L. Weidman, and D. R. Young, U.S. Patent 5,011,251 (April 1991).
12. D. L. Weidman, "Achromat Overclad Coupler," U.S. Patent 5,268,979 (December 1993).
13. C. M. Truesdale and D. A. Nolan, "Core-Clad Mode Coupling in a New Three-Index Structure," European Conference on Optical Communications, Barcelona Spain, 1986.
14. D. B. Keck, A. J. Morrow, D. A. Nolan, and D. A. Thompson, *J. of Lightwave Technology* **7**:1623–1633 (1989).
15. M. D. Feit and J. A. Fleck, "Simple Spectral Method for Solving Propagation Problems in Cylcindrical Geometry with Fast Fourier Transforms," *Optics Letters* **14**:662–664 (1989).
16. D. B. Mortimore and J. W. Arkwright, "Performance of Wavelength-Flattened 1×7 Fused Couplers," Optical Fiber Conference, TUG6 (1990).
17. D. L. Weidman, "A New Approach to Achromaticity in Fused $1 \times N$ Couplers," Optical Fiber Conference, Post Deadline papers (1994).

18. W. J. Miller, D. A. Nolan, and G. E. Williams, "Method of Making a 1 × N Coupler," U.S. Patent 5,017,206.

19. M. A. Newhouse and F. A. Annunziata, "Single-Mode Optical Switch," *Technical Digest of the National Fiber Optic Conference,* 1990.

20. D. A. Nolan and W. J. Miller, "Wavelength Tunable Mach-Zehnder Device," Optical Fiber Conference (1994).

21. B. Malo, F. Bilodeau, K. O. Hill, and J. Albert, *Electron. Lett.* **25**:1416, (1989).

22. C. Huang, H. Luo, S. Xu, and P. Chen, "Ultra Low Loss, Temperature Insensitive 16 channel 100 Ghz Dense WDMs Based on Cascaded All Fiber Unbalanced Mach-Zehnder Structure," Optical Fiber Conference, TUH2 (1999).

23. D. A. Nolan, W. J. Miller, and R. Irion, "Fiber Based Band Splitter," Optical Fiber Conference (1998).

24. D. A. Nolan, W. J. Miller, G. Berkey, and L. Bhagavatula, "Tapered Lattice Filters," Optical Fiber Conference, TUH4 (1999).

25. I. Yokohama, M. Kawachi, K. Okamoto, and J. Noda, *Electron. Lett.* **22**:929, 1986.

CHAPTER 9
FIBER BRAGG GRATINGS

Kenneth O. Hill
Communications Research Centre
Ottawa, Ontario, Canada

Nu-Wave Photonics
Ottawa, Ontario, Canada

9.1   GLOSSARY

FBG	fiber Bragg grating
FWHM	full width measured at half-maximum intensity
N_{eff}	effective refractive index for light propagating in a single mode
pps	pulses per second
β	propagation constant of optical fiber mode
Δn	magnitude of photoinduced refractive index change
κ	grating coupling coefficient
Λ	spatial period (or pitch) of spatial feature measured along optical fiber
λ	vacuum wavelength of propagating light
λ_B	Bragg wavelength
L	length of grating

9.2   INTRODUCTION

A fiber Bragg grating (FBG) is a periodic variation of the refractive index of the fiber core along the length of the fiber. The principal property of FBGs is that they reflect light in a narrow bandwidth that is centered about the Bragg wavelength, λ_B, which is given by $\lambda_B = 2N_{eff}\Lambda$, where Λ is the spatial period (or pitch) of the periodic variation and N_{eff} is the effective refractive index for light propagating in a single mode, usually the fundamental mode of a monomode optical fiber. The refractive index variations are formed by exposure of the fiber core to an intense optical interference pattern of ultraviolet light. The capability of light to induce permanent refractive index changes in the core of an optical fiber has been named photosensitivity. Photosensitivity was discovered by Hill et al. in 1978 at the Communications

Research Centre in Canada (CRC).[1,2] The discovery has led to techniques for fabricating Bragg gratings in the core of an optical fiber and a means for manufacturing a wide range of FBG-based devices that have applications in optical fiber communications and optical sensor systems.

This chapter reviews the characteristics of photosensitivity, the properties of Bragg gratings, the techniques for fabricating Bragg gratings in optical fibers, and some FBG devices. More information on FBGs can be found in the following references, which are reviews on Bragg grating technology,[3,4] the physical mechanisms underlying photosensitivity,[5] applications for fiber gratings,[6] and the use of FBGs as sensors[7].

9.3 PHOTOSENSITIVITY

When ultraviolet light radiates an optical fiber, the refractive index of the fiber is changed permanently; the effect is termed *photosensitivity*. The change in refractive index is permanent in the sense that it will last for several years (lifetimes of 25 years are predicted) if the optical waveguide after exposure is annealed appropriately; that is, by heating for a few hours at a temperature of 50°C above its maximum anticipated operating temperature.[8] Initially, photosensitivity was thought to be a phenomenon that was associated only with germanium-doped-core optical fibers. Subsequently, photosensitivity has been observed in a wide variety of different fibers, many of which do not contain germanium as dopant. Nevertheless, optical fiber with a germanium-doped core remains the most important material for the fabrication of Bragg grating–based devices.

The magnitude of the photoinduced refractive index change (Δn) obtained depends on several different factors: the irradiation conditions (wavelength, intensity, and total dosage of irradiating light), the composition of glassy material forming the fiber core, and any processing of the fiber prior and subsequent to irradiation. A wide variety of different continuous-wave and pulsed-laser light sources, with wavelengths ranging from the visible to the vacuum ultraviolet, have been used to photoinduce refractive index changes in optical fibers. In practice, the most commonly used light sources are KrF and ArF excimer lasers that generate, respectively, 248- and 193-nm light pulses (pulse width ~10 ns) at pulse repetition rates of 50 to 100 pps. Typically, the fiber core is exposed to laser light for a few minutes at pulse levels ranging from 100 to 1000 mJ cm^{-2} pulse^{-1}. Under these conditions, Δn is positive in germanium-doped monomode fiber with a magnitude ranging between 10^{-5} and 10^{-3}.

The refractive index change can be enhanced (photosensitization) by processing the fiber prior to irradiation using such techniques as *hydrogen loading*[9] or *flame brushing*.[10] In the case of hydrogen loading, a piece of fiber is put in a high-pressure vessel containing hydrogen gas at room temperature; pressures of 100 to 1000 atmospheres (atm; 101 kPa/atm) are applied. After a few days, hydrogen in molecular form has diffused into the silica fiber; at equilibrium the fiber becomes saturated (i.e., loaded) with hydrogen gas. The fiber is then taken out of the high-pressure vessel and irradiated before the hydrogen has had sufficient time to diffuse out. Photoinduced refractive index changes up to 100 times greater are obtained by hydrogen loading a Ge-doped-core optical fiber. In flame brushing, the section of fiber that is to be irradiated is mounted on a jig and a hydrogen-fueled flame is passed back and forth (i.e., brushed) along the length of the fiber. The brushing takes about 10 minutes, and upon irradiation, an increase in the photoinduced refractive index change by about a factor of 10 can be obtained.

Irradiation at intensity levels higher than 1000 mJ/cm^2 marks the onset of a different nonlinear photosensitive process that enables a single irradiating excimer light pulse to photoinduce a large index change in a small localized region near the core/cladding boundary of the fiber. In this case, the refractive index changes are sufficiently large to be observable with a phase contrast microscope and have the appearance of physically damaging the fiber. This phenomenon has been used for the writing of gratings using a single-excimer light pulse.

Another property of the photoinduced refractive index change is *anisotropy*. This characteristic is most easily observed by irradiating the fiber from the side with ultraviolet light that is polarized perpendicular to the fiber axis. The anisotropy in the photoinduced refractive index change results in the fiber becoming birefringent for light propagating through the fiber. The effect is useful for fabricating polarization mode-converting devices or rocking filters.[11]

The physical processes underlying photosensitivity have not been fully resolved. In the case of germanium-doped glasses, photosensitivity is associated with GeO color center defects that have strong absorption in the ultraviolet (~242 nm) wavelength region. Irradiation with ultraviolet light bleaches the color center absorption band and increases absorption at shorter wavelengths, thereby changing the ultraviolet absorption spectrum of the glass. Consequently, as a result of the Kramers-Kronig causality relationship,[12] the refractive index of the glass also changes; the resultant refractive index change can be sensed at wavelengths that are far removed from the ultraviolet region extending to wavelengths in the visible and infrared. The physical processes underlying photosensitivity are, however, probably much more complex than this simple model. There is evidence that ultraviolet light irradiation of Ge-doped optical fiber results in structural rearrangement of the glass matrix leading to densification, thereby providing another mechanism for contributing to the increase in the fiber core refractive index. Furthermore, a physical model for photosensitivity must also account for the small anisotropy in the photoinduced refractive index change and the role that hydrogen loading plays in enhancing the magnitude of the photoinduced refractive change. Although the physical processes underlying photosensitivity are not completely known, the phenomenon of glass-fiber photosensitivity has the practical result of providing a means, using ultraviolet light, for photoinducing permanent changes in the refractive index at wavelengths that are far removed from the wavelength of the irradiating ultraviolet light.

9.4 *PROPERTIES OF BRAGG GRATINGS*

Bragg gratings have a periodic index structure in the core of the optical fiber. Light propagating in the Bragg grating is backscattered slightly by Fresnel reflection from each successive index perturbation. Normally, the amount of backscattered light is very small except when the light has a wavelength in the region of the Bragg wavelength, λ_B, given by

$$\lambda_B = 2N_{eff}\Lambda$$

where N_{eff} is the modal index and Λ is the grating period. At the Bragg wavelength, each back reflection from successive index perturbations is in phase with the next one. The back reflections add up coherently and a large reflected light signal is obtained. The reflectivity of a strong grating can approach 100 percent at the Bragg wavelength, whereas light at wavelengths longer or shorter than the Bragg wavelength pass through the Bragg grating with negligible loss. It is this wavelength-dependent behavior of Bragg gratings that makes them so useful in optical communications applications. Furthermore, the optical pitch ($N_{eff}\Lambda$) of a Bragg grating contained in a strand of fiber is changed by applying longitudinal stress to the fiber strand. This effect provides a simple means for sensing strain optically by monitoring the concomitant change in the Bragg resonant wavelength.

Bragg gratings can be described theoretically by using coupled-mode equations.[4, 6, 13] Here, we summarize the relevant formulas for tightly bound monomode light propagating through a uniform grating. The grating is assumed to have a sinusoidal perturbation of constant amplitude, Δn. The reflectivity of the grating is determined by three parameters: (1) the coupling coefficient, κ, (2) the mode propagation constant, $\beta = 2\pi N_{eff}/\lambda$, and (3) the grating length, L. The coupling coefficient, κ, which depends only on the operating wavelength of the light and the amplitude of the index perturbation, Δn, is given by $\kappa = (\pi/\lambda)\Delta n$. The most interesting case is when the wavelength of the light corresponds to the Bragg wavelength. The grating reflectivity, R, of the grating is then given by the simple expression, $R = \tanh^2(\kappa L)$, where κ is the coupling coefficient at the Bragg wavelength and L is the length of the grating. Thus, the product κL can be used as a measure of grating strength. For $\kappa L = 1, 2, 3$, the grating reflectivity is, respectively, 58, 93, and 99 percent. A grating with a κL greater than one is termed a strong grating, whereas a weak grating has κL less than one. Figure 1 shows the typical reflection spectra for weak and strong gratings.

The other important property of the grating is its bandwidth, which is a measure of the wavelength range over which the grating reflects light. The bandwidth of a fiber grating that is most easily measured is its full width at half-maximum, $\Delta\lambda_{\text{FWHM}}$, of the central reflection peak, which is defined as the wavelength interval between the 3-dB points. That is the separation in the wavelength between the points on either side of the Bragg wavelength where the reflectivity has decreased to 50 percent of its maximum value. However, a much easier quantity to calculate is the bandwidth, $\Delta\lambda_0 = \lambda_0 - \lambda_B$, where λ_0 is the wavelength where the first zero in the reflection spectra occurs. This bandwidth can be found by calculating the difference in the propagation constants, $\Delta\beta_0 = \beta_0 - \beta_B$, where $\beta_0 = 2\pi N_{\text{eff}}/\lambda_0$ is the propagation constant at wavelength λ_0 for which the reflectivity is first zero, and $\beta_B = 2\pi N_{\text{eff}}/\lambda_B$ is the propagation constant at the Bragg wavelength for which the reflectivity is maximum.

In the case of weak gratings ($\kappa L < 1$), $\Delta\beta_0 = \beta_0 - \beta_B = \pi/L$, from which it can be determined that $\Delta\lambda_{\text{FWHM}} \sim \Delta\lambda_0 = \lambda_B^2/2N_{\text{eff}}L$; the bandwidth of a weak grating is inversely proportional to the grating length, L. Thus, long, weak gratings can have very narrow bandwidths. The first

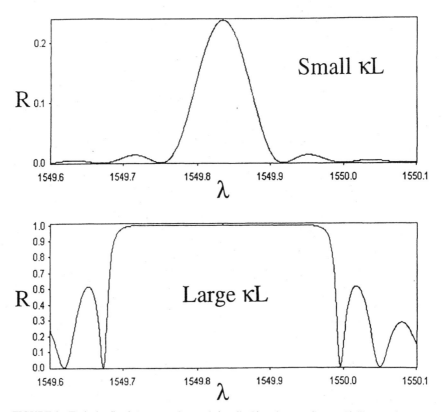

FIGURE 1 Typical reflection spectra for weak (small κL) and strong (large κL) fiber gratings.

Bragg grating written in fibers[1,2] was more than 1 m long and had a bandwidth less than 100 MHz, which is an astonishingly narrow bandwidth for a reflector of visible light. On the other hand, in the case of a strong grating ($\kappa L > 1$), $\Delta\beta_0 = \beta_0 - \beta_B = 4\kappa$ and $\Delta\lambda_{FWHM} \sim 2\Delta\lambda_0 = 4\lambda_B^2\kappa/\pi N_{eff}$. For strong gratings, the bandwidth is directly proportional to the coupling coefficient, κ, and is independent of the grating length.

9.5 FABRICATION OF FIBER GRATINGS

Writing a fiber grating optically in the core of an optical fiber requires irradiating the core with a periodic interference pattern. Historically, this was first achieved by interfering light that propagated in a forward direction along an optical fiber with light that was reflected from the fiber end and propagated in a backward direction.[1] This method for forming fiber gratings is known as the *internal writing technique,* and the gratings were referred to as *Hill gratings.* The Bragg gratings, formed by internal writing, suffer from the limitation that the wavelength of the reflected light is close to the wavelength at which they were written (i.e., at a wavelength in the blue-green spectral region).

A second method for fabricating fiber gratings is the *transverse holographic technique,*[14] which is shown schematically in Fig. 2. The light from an ultraviolet source is split into two beams that are brought together so that they intersect at an angle, θ. As Fig. 2 shows, the intersecting light beams form an interference pattern that is focused using cylindrical lenses (not shown) on the core of the optical fiber. Unlike the internal writing technique, the fiber core is irradiated from the side, thus giving rise to its name *transverse holographic technique.* The technique works because the fiber cladding is transparent to the ultraviolet light, whereas the core absorbs the light strongly. Since the period, Λ, of the grating depends on the angle, θ, between the two interfering coherent beams through the relationship $\Lambda = \lambda_{UV}/2 \sin(\theta/2)$, Bragg gratings can be made that reflect light at much longer wavelengths than the ultraviolet light that is used in the fabrication of the grating. Most important, FBGs can be made that function in the spectral regions that are of interest for fiber-optic communication and optical sensing.

A third technique for FBG fabrication is the *phase mask technique,*[15] which is illustrated in Fig. 3. The phase mask is made from a flat slab of silica glass, which is transparent to ultraviolet light. On one of the flat surfaces, a one-dimensional periodic surface relief structure is etched using photolithographic techniques. The shape of the periodic pattern approximates a square wave in profile. The optical fiber is placed almost in contact with and at right angles to the corrugations of the phase mask, as shown in Fig. 3. Ultraviolet light, which is incident normal to the

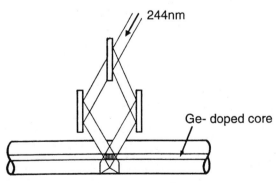

FIGURE 2 Schematic diagram illustrating the writing of an FBG using the transverse holographic technique.

phase mask, passes through and is diffracted by the periodic corrugations of the phase mask. Normally, most of the diffracted light is contained in the 0, +1, and −1 diffracted orders. However, the phase mask is designed to suppress the diffraction into the zero order by controlling the depth of the corrugations in the phase mask. In practice, the amount of light in the zero order can be reduced to less than 5 percent with approximately 80 percent of the total light intensity divided equally in the ±1 orders. The two ±1 diffracted-order beams interfere to produce a periodic pattern that photoimprints a corresponding grating in the optical fiber. If the period of the phase mask grating is Λ_{mask}, the period of the photoimprinted index grating is $\Lambda_{mask}/2$. Note that this period is independent of the wavelength of ultraviolet light that irradiates the phase mask.

The phase mask technique has the advantage of greatly simplifying the manufacturing process for Bragg gratings, while yielding high-performance gratings. In comparison with the holographic technique, the phase mask technique offers easier alignment of the fiber for photoimprinting, reduced stability requirements on the photoimprinting apparatus, and

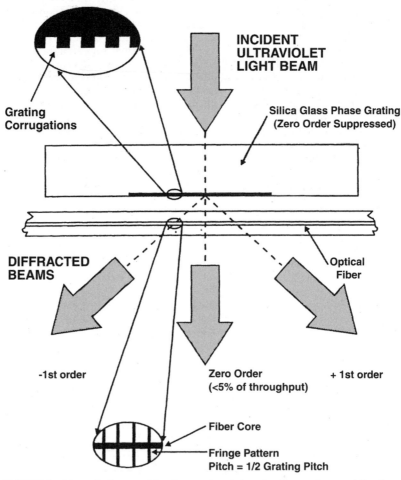

FIGURE 3 Schematic diagram of the phase mask technique for the manufacture of fiber Bragg gratings.

lower coherence requirements on the ultraviolet laser beam, thereby permitting the use of a cheaper ultraviolet excimer laser source. Furthermore, there is the possibility of manufacturing several gratings at once in a single exposure by irradiating parallel fibers through the phase mask. The capability to manufacture high-performance gratings at a low per-unit grating cost is critical for the economic viability of using gratings in some applications. A drawback of the phase mask technique is that a separate phase mask is required for each different Bragg wavelength. However, some wavelength tuning is possible by applying tension to the fiber during the photoimprinting process; the Bragg wavelength of the relaxed fiber will shift by ~2 nm.

The phase mask technique not only yields high-performance devices, but is also very flexible in that it can be used to fabricate gratings with controlled spectral response characteristics. For instance, the typical spectral response of a finite-length grating with a uniform index modulation along the fiber length has secondary maxima on both sides of the main reflection peak. In applications like wavelength-division multiplexing, this type of response is not desirable. However, if the profile of the index modulation, Δn, along the fiber length is given a bell-like functional shape, these secondary maxima can be suppressed.[16] The procedure is called *apodization.* Apodized fiber gratings have been fabricated using the phase mask technique, and suppressions of the sidelobes of 30 to 40 dB have been achieved,[17,18]

Figure 4 shows the spectral response of two Bragg gratings with the same full width at half-maximum (FWHM). One grating exhibits large sidebands, whereas the other has much-reduced sidebands. The one with the reduced sidebands is a little longer and has a coupling coefficient, κ, apodized as a second-degree cosine ($\cos^2$) along its length. Apodization has one disadvantage: It decreases the effective length of the Bragg grating. Therefore, to obtain fiber gratings having the same FWHM, the apodized fiber grating has a longer length than the equivalent-bandwidth unapodized fiber grating.

The phase mask technique has been extended to the fabrication of chirped or aperiodic fiber gratings. *Chirping* means varying the grating period along the length of the grating in order to broaden its spectral response. Aperiodic or chirped gratings are desirable for making dispersion compensators[19] or filters having broad spectral responses. The first chirped fiber gratings were made using a double-exposure technique.[20] In the first exposure, an opaque mask is positioned between the fiber and the ultraviolet beam blocking the light from irradiating the fiber. The mask is then moved slowly out of the beam at a constant velocity to increase continuously the length of the fiber that is exposed to the ultraviolet light. A continuous change in the photoinduced refractive index is produced that varies linearly along the fiber length with the largest index change occurring in the section of fiber that is exposed to ultraviolet light for the longest duration. In a second exposure, a fiber grating is photoimprinted in the fiber by using the standard phase mask technique. Because the optical pitch of a fiber grating depends on both the refractive index and the mechanical pitch (i.e., optical pitch = $N_{eff}\Lambda$), the pitch of the photoimprinted grating is effectively chirped, even though its mechanical period is constant. Following this demonstration, a variety of other methods have been developed to manufacture gratings that are chirped permanently[21,22] or that have an adjustable chirp.[23,24]

The phase mask technique can also be used to fabricate tilted or blazed gratings. Usually, the corrugations of the phase mask are oriented normal to the fiber axis, as shown in Fig. 3. However, if the corrugations of the phase mask are oriented at an angle to the axis of the fiber, the photoimprinted grating is tilted or blazed. Such fiber gratings couple light out from the bound modes of the fiber to either the cladding modes or the radiation modes. Tilted gratings have applications in fabricating fiber taps.[25] If the grating is simultaneously blazed and chirped, it can be used to fabricate an optical spectrum analyzer.[26]

Another approach to grating fabrication is the *point-by-point technique,*[27] also developed at CRC. In this method, each index perturbation of the grating is written point by point. For gratings with many index perturbations, the method is not very efficient. However, it has been used to fabricate micro-Bragg gratings in optical fibers,[28] but it is most useful for making coarse gratings with pitches of the order of 100 μm that are required for LP_{01} to LP_{11} mode

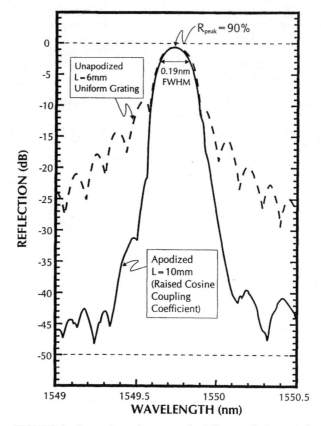

FIGURE 4 Comparison of an unapodized fiber grating's spectral response with that of an apodized fiber grating having the same bandwidth (FWHM).

converters[27] and polarization mode converters.[11] The interest in coarse period gratings has increased lately because of their use in long-period fiber-grating band-rejection filters[29] and fiber-amplifier gain equalizers.[30]

9.6 THE APPLICATION OF FIBER GRATINGS

Hill and Meltz[6] provide an extensive review of the many potential applications of fiber gratings in lightwave communication systems and in optical sensor systems. Our purpose here is to note that a common problem in using FBGs is that a transmission device is usually desired, whereas FBGs function as reflection devices. Thus, means are required to convert the reflection spectral response into a transmission response. This can be achieved using a Sagnac loop,[31] a Michleson (or Mach-Zehnder) interferometer,[32] or an optical circulator. Figure 5 shows an example of how this is achieved for the case of a multichannel dispersion compensator using chirped or aperiodic fiber gratings.

In Fig. 5a, the dispersion compensator is implemented using a Michelson interferometer. Each wavelength channel ($\lambda_1, \lambda_2, \lambda_3$) requires a pair of identically matched FBGs, one in each

a)

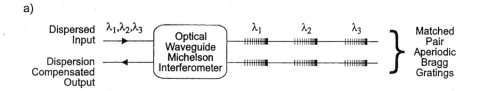

b)

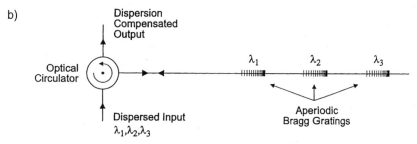

FIGURE 5 Schematic diagram of a multichannel dispersion compensator that is formed by using (*a*) a Michelson interferometer and (*b*) an optical circulator.

arm of the interferometer. Since it is difficult to fabricate identical Bragg gratings (i.e., having the same resonant wavelength and chirp), this configuration for the dispersion compensator has not yet been demonstrated. However, a wavelength-selective device that requires matched grating pairs has been demonstrated.[33,34] An additional disadvantage of the Michelson interferometer configuration being an interferometric device is that it would require temperature compensation. The advantage of using a Michelson interferometer is that it can be implemented in all-fiber or planar-integrated optics versions.

Figure 5*b* shows the dispersion compensator implemented using an optical circulator. In operation, light that enters through the input port is routed by the circulator to the port with the Bragg gratings. All of the light that is reflected by the FBGs is routed to the output channel. This configuration requires only one chirped FBG per wavelength channel and is the preferred method for implementing dispersion compensators using FBGs. The only disadvantage of this configuration is that the optical circulator is a bulk optic device (or microoptic device) that is relatively expensive compared with the all-fiber Michelson interferometer.

9.7 *REFERENCES*

1. K. O. Hill, Y. Fujii, D. C. Johnson, et al., "Photosensitivity in Optical Fiber Waveguides: Application to Reflection Filter Fabrication," *Applied Physics Letters* **32**(10):647–649 (1978).

2. B. S. Kawasaki, K. O. Hill, D. C. Johnson, et al., "Narrow-Band Bragg Reflectors in Optical Fibers," *Optics Letters* **3**(8):66–68 (1978).

3. K. O. Hill, B. Malo, F. Bilodeau, et al., "Photosensitivity in Optical Fibers," *Annual Review of Material Science* **23**:125–157 (1993).

4. I. Bennion, J. A. R. Williams, L. Zhang, et al., "Tutorial Review, UV-Written In-Fibre Bragg Gratings," *Optical and Quantum Electronics* **28**:93–135 (1996).

5. B. Poumellec, P. Niay, M. Douay, et al., "The UV-Induced Refractive Index Grating in Ge:SiO$_2$ Preforms: Additional CW Experiments and the Macroscopic Origin of the Change in Index," *Journal of Physics D, Applied Physics* **29**:1842–1856 (1996).

6. Kenneth O. Hill and Gerald Meltz, "Fiber Bragg Grating Technology Fundamentals and Overview," *Journal of Lightwave Technology* **15**(8):1263–1276 (1997).

7. A. D. Kersey, M. A. Davis, H. J. Patrick, et al., "Fiber Grating Sensors," *Journal of Lightwave Technology* **15**(8):1442–1463 (1997).

8. T. Erdogan, V. Mizrahi, P. J. Lemaire, et al., "Decay of Ultraviolet-Induced Fiber Bragg Gratings," *Journal of Applied Physics* **76**(1):73–80 (1994).

9. P. J. Lemaire, R. M. Atkins, V. Mizrahi, et al., "High Pressure H$_2$ Loading as a Technique for Achieving Ultrahigh UV Photosensitivity and Thermal Sensitivity in GeO$_2$ Doped Optical Fibres," *Electronics Letters* **29**(13):1191–1193 (1993).

10. F. Bilodeau, B. Malo, J. Albert, et al., "Photosensitization of Optical Fiber and Silica-on-Silicon/Silica Waveguides," *Optics Letters* **18**(12):953–955 (1993).

11. K. O. Hill, F. Bilodeau, B. Malo, et al., "Birefringent Photosensitivity in Monomode Optical Fibre: Application to the External Writing of Rocking Filters," *Electronic Letters* **27**(17):1548–1550 (1991).

12. Alan Miller, "Fundamental Optical Properties of Solids," in *Handbook of Optics,* edited by Michael Bass, McGraw-Hill, New York, 1995, vol. 1, pp. 9–15.

13. D. K. W. Lam and B. K. Garside, "Characterization of Single-Mode Optical Fiber Filters," *Applied Optics* **20**(3):440–445 (1981).

14. G. Meltz, W. W. Morey, and W. H. Glenn, "Formation of Bragg Gratings in Optical Fibers by a Transverse Holographic Method," *Optics Letters* **14**(15):823–825 (1989).

15. K. O. Hill, B. Malo, F. Bilodeau, et al., "Bragg Gratings Fabricated in Monomode Photosensitive Optical Fiber by UV Exposure Through a Phase Mask," *Applied Physics Letters* **62**(10):1035–1037 (1993).

16. M. Matsuhara and K. O. Hill, "Optical-Waveguide Band-Rejection Filters: Design," *Applied Optics* **13**(12):2886–2888 (1974).

17. B. Malo, S. Thériault, D. C. Johnson, et al., "Apodised In-Fibre Bragg Grating Reflectors Photoimprinted Using a Phase Mask," *Electronics Letters* **31**(3):223–224 (1995).

18. J. Albert, K. O. Hill, B. Malo, et al., "Apodisation of the Spectral Response of Fibre Bragg Gratings Using a Phase Mask with Variable Diffraction Efficiency," *Electronics Letters* **31**(3):222–223 (1995).

19. K. O. Hill, "Aperiodic Distributed-Parameter Waveguides for Integrated Optics," *Applied Optics* **13**(8):1853–1856 (1974).

20. K. O. Hill, F. Bilodeau, B. Malo, et al., "Chirped In-Fibre Bragg Grating for Compensation of Optical-Fiber Dispersion," *Optics Letters* **19**(17):1314–1316 (1994).

21. K. Sugden, I. Bennion, A. Molony, et al., "Chirped Gratings Produced in Photosensitive Optical Fibres by Fibre Deformation during Exposure," *Electronics Letters* **30**(5):440–442 (1994).

22. K. C. Byron and H. N. Rourke, "Fabrication of Chirped Fibre Gratings by Novel Stretch and Write Techniques," *Electronics Letters* **31**(1):60–61 (1995).

23. D. Garthe, R. E. Epworth, W. S. Lee, et al., "Adjustable Dispersion Equaliser for 10 and 20 Gbit/s over Distances up to 160 km," *Electronics Letters* **30**(25):2159–2160 (1994).

24. M. M. Ohn, A. T. Alavie, R. Maaskant, et al., "Dispersion Variable Fibre Bragg Grating Using a Piezoelectric Stack," *Electronics Letters* **32**(21):2000–2001 (1996).

25. G. Meltz, W. W. Morey, and W. H. Glenn, "In-Fiber Bragg Grating Tap," presented at the Conference on Optical Fiber Communications, OFC'90, San Francisco, CA, 1990 (unpublished).

26. J. L. Wagener, T. A. Strasser, J. R. Pedrazzani, et al., "Fiber Grating Optical Spectrum Analyzer Tap," presented at the IOOC-ECOC'97, Edinburgh, UK, 1997 (unpublished).

27. K. O. Hill, B. Malo, K. A. Vineberg, et al., "Efficient Mode Conversion in Telecommunication Fibre Using Externally Written Gratings," *Electronics Letters* **26**(16):1270–1272 (1990).

28. B. Malo, K. O. Hill, F. Bilodeau, et al., "Point-by-Point Fabrication of Micro-Bragg Gratings in Photosensitive Fibre Using Single Excimer Pulse Refractive Index Modification Techniques," *Electronic Letters* **29**(18):1668–1669 (1993).

29. A. M. Vengsarkar, P. J. Lemaire, J. B. Judkins, et al., "Long-Period Fiber Gratings as Band-Rejection Filters," presented at the Optical Fiber Communication conference, OFC'95, San Diego, CA, 1995 (unpublished).

30. A. M. Vengsarkar, J. R. Pedrazzani, J. B. Judkins, et al., "Long-Period Fiber-Grating-Based gain equalizers," Optics Letters **21**(5):336–338 (1996).

31. K. O. Hill, D. C. Johnson, F. Bilodeau, et al., "Narrow-Bandwidth Optical Waveguide Transmission Filters: A New Design Concept and Applications to Optical Fibre Communications," *Electronics Letters* **23**(9):465–466 (1987).

32. D. C. Johnson, K. O. Hill, F. Bilodeau, et al., "New Design Concept for a Narrowband Wavelength-Selective Optical Tap and Combiner," *Electronics Letters* **23**(13):668–669 (1987).

33. F. Bilodeau, K. O. Hill, B. Malo, et al., "High-Return-Loss Narrowband All-Fiber Bandpass Bragg Transmission Filter," *IEEE Photonics Technology Letters* **6**(1):80–82 (1994).

34. F. Bilodeau, D. C. Johnson, S. Thériault, et al., "An All-Fiber Dense-Wavelength-Division Multiplexer/Demultiplexer Using Photoimprinted Bragg Gratings," *IEEE Photonics Technology Letters* **7**(4):388–390 (1995).

CHAPTER 10
MICRO-OPTICS-BASED COMPONENTS FOR NETWORKING

Joseph C. Palais
Department of Electrical Engineering
College of Engineering and Applied Sciences
Arizona State University
Tempe, Arizona

10.1 INTRODUCTION

The optical portion of many fiber networks requires a number of functional devices, some of which can be fabricated using small optical components (so-called *micro-optic* components). Micro-optic components are made up of parts that have linear dimensions on the order of a few millimeters. The completed functional device may occupy a space a few centimeters on a side. Components to be described in this section have the common feature that the fiber transmission link is opened and small (micro-optic) devices are inserted into the gap between the fiber ends to produce a functional component. Network components constructed entirely of fibers or constructed in integrated-optic form are described elsewhere in this Handbook.

The following sections describe, in order: a generalized component, specific useful network functions, microoptic subcomponents required to make up the final component, and complete components.

10.2 GENERALIZED COMPONENTS

A generalized fiber-optic component is drawn in Fig. 1. As indicated, input fibers are on the left and output fibers are on the right. Although some components have only a single input port and a single output port, many applications require more than one input and/or output ports. In fact, the number of ports in some devices can exceed 100. The *coupling loss* between any two ports is given, in decibels, by

$$L = -10 \log(P_{\text{out}}/P_{\text{in}}) \tag{1}$$

With respect to Fig. 1, P_{in} refers to the input power at any of the ports on the left, and P_{out} refers to the output power at any of the ports on the right. Because we are only considering

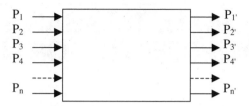

FIGURE 1   The generalized component.

passive components in this section, P_{out} will be less than P_{in}, and the loss will be a positive number.

Insertion loss refers to the coupling loss between any two ports where coupling is desired, and *isolation* (or *directionality*) refers to the coupling loss between any two ports where coupling is unwanted. *Excess loss* is the fraction of input power that does not emerge from any of the desired output ports, as expressed in decibels. It is the sum of all the useful power out divided by the input power.

10.3   NETWORK FUNCTIONS

Functions useful for many fiber-optic communications applications are described in the following paragraphs.

Attenuators

Attenuators reduce the amount of power flowing through the fiber system. Both fixed and variable attenuators are available. The applications include testing of receiver sensitivities (varying the attenuation changes the amount of power incident on the receiver) and protecting a receiver from saturating due to excess incident power. Attenuation from a few tenths of a dB to more than 50 dB are sometimes required.

Power Splitters and Directional Couplers

These devices distribute input power from a single fiber to two or more output fibers. The component design controls the fraction of power delivered to each of the output ports. Applications include power distribution in local area networks (LANs) and in subscriber networks. The most common splitters and couplers have a single input and equal distribution of power among each of two outputs, as shown schematically in Fig. 2a. For an ideal three-port splitter (one with no excess loss), half the input power emerges from each of the two output ports. The insertion loss, as calculated from Eq. 1 with a ratio of powers of 0.5, yields a 3 dB loss to each of the two output ports. Any excess loss is added to the 3 dB.

A splitter with more than two output ports can be constructed by connecting several three-port couplers in a tree pattern as indicated schematically in Fig. 2b. Adding more splitters in the same manner allows coupling from one input port to 8, 16, 32 (and more) output ports.

Adding a fourth port, as in Fig. 3, creates a *directional coupler.* The arrows in the figure show the allowed directions of wave travel through the coupler. An input beam is split between two output ports and is isolated from the fourth. By proper component design, any

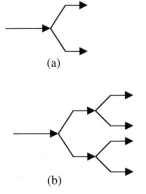

(a)

(b)

FIGURE 2   Power splitters: (*a*) 1:2 split and (*b*) 1:4 split.

desired power splitting ratio can be obtained. One application of the directional coupler is to the distribution network of a local area network, where simultaneous transmission and reception are required. Figure 4 illustrates this usage at one LAN terminal.

Isolators

An isolator is a one-way transmission line. It permits the flow of optical power in just one direction (the forward direction). Applications include protection of a transmitting laser diode from back reflections. Such reflections increase the noise in the system by disrupting the diode's operation. Isolators also improve the stability of fiber amplifiers by minimizing the possibility of feedback, which causes unwanted oscillations in such devices.

Circulators

In a circulator, power into the first port emerges from the second, while power into the second port emerges from a third. This behavior repeats at each successive input port until power into the last port emerges from the first. Practical circulators are typically three- or four-port devices.

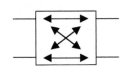

FIGURE 3   Four-port directional coupler.

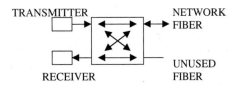

FIGURE 4   LAN terminal, illustrating application of the directional coupler.

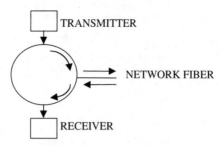

FIGURE 5   An optical circulator separates transmitted and received messages at a terminal.

Using a circulator, efficient two-way (*full-duplex*) transmission along a single fiber at a single wavelength is possible. The circulator separates the transmitting and receiving beams of light at each terminal, as illustrated in Fig. 5.

Multiplexers/Demultiplexers/Duplexers

The multiplexer and demultiplexer are heavily utilized in fiber-optic wavelength-division multiplexing (WDM) systems. The *multiplexer* combines beams of light from the different transmitters (each at a slightly shifted wavelength) onto the single transmission fiber. The *demultiplexer* separates the individual wavelengths transmitted and guides the separate channels to the appropriate optical receivers. These functions are illustrated in Fig. 6. Requirements for multiplexers/demultiplexers include combining and separating independent channels less than a nanometer apart, and accommodating numerous (in some cases over 100) channels. A frequency spacing between adjacent channels of 100 GHz corresponds to a

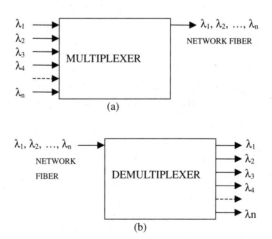

FIGURE 6   (*a*) A multiplexer combines different wavelength channels onto a single fiber for transmission. (*b*) A demultiplexer separates several incoming channels at different wavelengths and directs them to separate receivers.

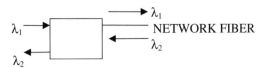

FIGURE 7 A duplexer allows two-way transmission along a single network fiber.

wavelength spacing of 0.8 nm for wavelengths near 1.55 μm. Insertion losses can be as low as a few tenths of a dB and isolations of 40 dB or more.

The *duplexer* allows for simultaneous two-way transmission along a single fiber. The wavelengths are different for the transmitting and receiving light beam. The duplexer separates the beams as indicated in Fig. 7, where λ_1 is the transmitting wavelength and λ_2 is the receiving wavelength.

Mechanical Switches

Operationally, an optical switch acts just like an electrical switch. Mechanical movement of some part (as implied schematically in Fig. 8) causes power entering one port to be directed to one of two or more output ports. Such devices are useful in testing of fiber components and systems and in other applications, such as bypassing inoperative nodes in a local area network. Insertion losses less than 0.10 dB and isolations greater than 50 dB are reasonable requirements.

10.4 SUBCOMPONENTS

Micro-optic subcomponents that form part of the design of many complete microoptic components are described in this section.

Prisms

Because of the dispersion in glass prisms, they can operate as multiplexers, demultiplexers, and duplexers. The dispersive property is illustrated in Fig. 9.

Right-angle glass prisms also act as excellent reflectors, as shown in Fig. 10, owing to perfect reflection (total internal reflection) at the glass-to-air interface. The critical angle for the glass-to-air interface is about 41°, and the incident ray is beyond that at 45°.

The beam-splitting cube, drawn in Fig. 11, consists of two right-angle prisms cemented together with a thin reflective layer between them. This beam splitter has the advantage over a flat reflective plate in that no angular displacement occurs between the input and output beam directions. This simplifies the alignment of the splitter with the input and output fibers.

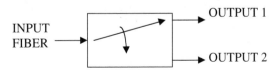

FIGURE 8 Mechanical optical switch.

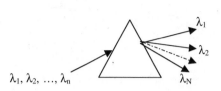

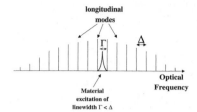

FIGURE 9 A dispersive prism spatially separates different wavelengths. This represents demultiplexing. Reversing the directions of the arrows illustrates combining of different wavelengths. This is multiplexing.

FIGURE 10 Totally reflecting prism.

Gratings

Ruled reflection gratings are also used in multiplexers and demultiplexers. As illustrated in Fig. 12, the dispersion characteristics of the grating perform the wavelength separation function required of a demultiplexer. The grating has much greater dispersive power than a prism, permitting increased wavelength spatial separation. The relationship between the incident and reflected beams, for an incident collimated light beam, is given by the diffraction equation

$$\sin \theta_i + \sin \theta_r = m\lambda/d \tag{2}$$

where θ_i and θ_r are the incident and reflected beam angles, d is the separation between adjacent reflecting surfaces, and m is the *order* of the diffraction. Typically, gratings are blazed so as to maximize the power into the first-order beams. As deduced from this equation for $m = 1$, the diffracted peak occurs at a different angle for different wavelengths. This feature produces the demultiplexing function needed in WDM systems. Reversing the arrows in Fig. 12 illustrates the multiplexing capability of the grating.

Filters

Dielectric-layered filters, consisting of very thin layers of various dielectrics deposited onto a glass substrate, are used to construct multiplexers, demultiplexers, and duplexers. Filters have unique reflectance/transmittance characteristics. They can be designed to reflect at certain wavelengths and transmit at others, thus spatially separating (or combining) different wavelengths as required for WDM applications.

Beam Splitters

A beam-splitting plate, shown in Fig. 13, is a partially silvered glass plate. The thickness of the silvered layer determines the fraction of light transmitted and reflected. In this way, the input beam can be divided in two parts of any desired ratio.

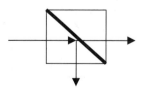

FIGURE 11 Beam-splitting cube.

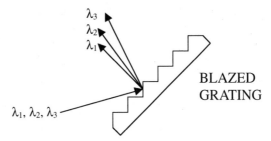

FIGURE 12   Blazed reflection grating operated as a demultiplexer.

Faraday Rotators

The Faraday rotator produces a nonreciprocal rotation of the plane of polarization. The amount of rotation is given by

$$\theta = VHL \tag{3}$$

where θ is the rotation angle, V is the *Verdet constant* (a measure of the strength of the Faraday effect), H is the applied magnetic field, and L is the length of the rotator. A commonly used rotator material is YIG (yttrium-iron garnet), which has a high value of V.

Figure 14 illustrates the nonreciprocal rotation of the state of polarization (SOP) of the wave. The rotation of a beam traveling from left-to-right is 45°, while the rotation for a beam traveling from right-to-left is an additional 45°.

The Faraday rotator is used in the isolator and the circulator.

Polarizers

Polarizers based upon dichroic absorbers and polarization prisms using birefringent materials are common. The polarizing beam splitter, illustrated in Fig. 15, is useful in microoptics applications such as the optical circulator. The polarizing splitter separates two orthogonally polarized beams.

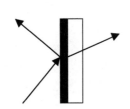

FIGURE 13   Beam-splitting plate.

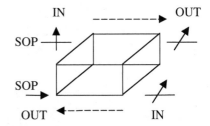

FIGURE 14   Faraday rotator. The dashed arrows indicate the direction of beam travel. The solid arrows represent the wave polarization in the plane perpendicular to the direction of wave travel.

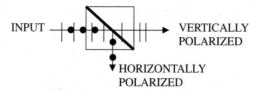

FIGURE 15 Polarizing beam splitter.

GRIN-Rod Lens

The subcomponents discussed in the last few paragraphs perform the operations indicated in their descriptions. The problem is that they cannot be directly inserted into a fiber transmission line. To insert one of the subcomponents into the fiber link requires that the fiber be opened to produce a gap. The subcomponent would then fit into the gap. Because the light emerging from a fiber diverges, with a gap present the receiving fiber does not capture much of the transmitted light. This situation is illustrated in Fig. 16. The emitted diverging light must be collimated, the required subcomponent (e.g., beam splitter, grating, etc.) inserted, and the light refocused. A commonly used device for performing this function is the *graded-index rod lens* (GRIN-rod lens). Its use is illustrated in Fig. 17. The diverging light emitted by the transmitting fiber is collimated by the first GRIN-rod lens. The collimated beam is refocused onto the receiving fiber by the second GRIN-rod lens. The collimation is sufficient such that a gap of 20 mm introduces less than 0.5 dB excess loss.[1] This allows for the insertion of beam-modifying devices of the types described in the preceding paragraphs (e.g., prisms, gratings, and beam splitters) in the gap with minimum added loss.

10.5 COMPONENTS

The subcomponents introduced in the last section are combined into useful fiber devices in the manner described in this section.

Attenuators

The simplest attenuator is produced by a gap introduced between two fibers, as in Fig. 18. As the gap length increases, so does the loss. Loss is also introduced by a lateral displacement. A variable attenuator is produced by allowing the gap (or the lateral offset) to be changeable. A disc whose absorption differs over different parts may also be placed between the fibers. The attenuation is varied by rotating the disk.

In another attenuator design, a small thin flat reflector is inserted at variable amounts into the gap to produce the desired amount of loss.[2]

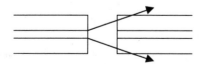

FIGURE 16 Diverging wave emitted from an open fiber couples poorly to the receiving fiber.

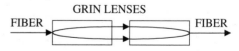

FIGURE 17 Collimating light between two fibers using GRIN-rod lenses.

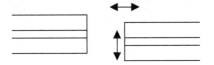

FIGURE 18   Gap attenuator showing relative displacement of the fibers to vary the insertion loss.

Power Splitters and Directional Couplers

A power splitter[3] can be constructed as illustrated in Fig. 19. A beam-splitting cube (or a beam-splitting plate) is placed in the gap between two GRIN-rod lenses to connect Ports 1 and 2. A third combination of lens and fiber collects the reflected light at Port 3. The division of power between the two output fibers is determined by the reflective properties of the splitter itself. Any desired ratio of outputs can be obtained.

If a fourth port is added (Port 4 in Fig. 19), the device is a four-port directional coupler.

Isolators and Circulators

The isolator combines the Faraday rotator and two polarizers[4] as indicated in Fig. 20. The input and output fibers can be coupled to the isolator using GRIN lenses. The vertically polarized beam at the input is rotated by 45° and passed through the output polarizer. Any reflected light is rotated an additional 45°, emerging cross-polarized with respect to the polarizer on the left. In this state, the reflected light will not pass back into the transmitting fiber. Similarly, a light beam traveling from right-to-left will be cross-polarized at the input polarizer and will not travel further in that direction. The polarizers can be polarizing beam splitters, dichroic polarizers, or polarizing fibers.

A circulator also requires a Faraday rotator and polarizers (polarizing beam splitters or polarizing fiber). Additional components include reflecting prisms, reciprocal 45° rotators, and fiber coupling devices such as GRIN-rod lenses.[5]

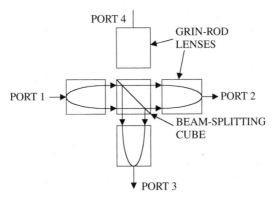

FIGURE 19   Three-power splitter and (with Port 4 added) four-port directional coupler.

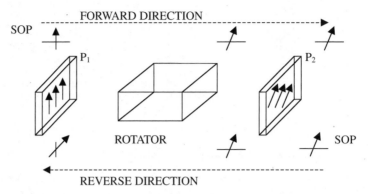

FIGURE 20   Optical isolator. P_1 and P_2 are polarizers.

Multiplexers/Demultiplexers/Duplexers

The multiplexer, demultiplexer, and duplexer are fundamentally the same device. The application determines which of the three descriptions is most appropriate. One embodiment is illustrated in Fig. 21 for a two-channel device. As a demultiplexer, the GRIN lens collimates the diverging beam from the network fiber and guides it onto the diffraction grating. The grating redirects the beam according to its wavelength. The GRIN lens then focuses the various wavelengths onto the output fibers for reception. As a multiplexer, the operation is just reversed with the *receiver fibers* replaced by transmitter fibers and all arrows reversed. As a duplexer, one of the two *receiver fibers* becomes a transmitter fiber.

Other configurations also use the diffraction grating, including one incorporating a concave reflector for properly collimating and focusing the beams between input and output fibers.[6] Microoptic grating-based devices can accommodate more than 100 WDM channels, with wavelength spacing on the order of 0.4 nm.

A filter-based multiplexer/demultiplexer appears in Fig. 22. The reflective coating transmits wavelength λ_1 and reflects wavelength λ_2. The device is illustrated as a demultiplexer. Again, by reversing the directions of the arrows, the device becomes a multiplexer. Filter-based multiplexers/demultiplexers can be extended to several channels in the microoptical form, essentially by cascading several devices of the type just described.

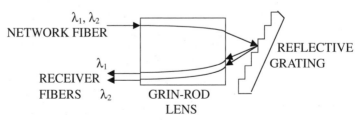

FIGURE 21   Two-channel demultiplexer. Only the beam's central rays are drawn. To operate as a multiplexer the arrows are reversed. To operate as a duplexer, the arrows for just one of the two wavelengths are reversed.

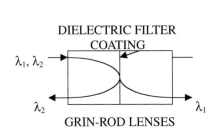

FIGURE 22   Filter-based multiplexer/demultiplexer.

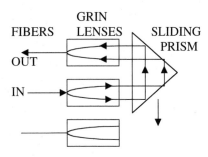

FIGURE 23   Moveable reflecting prism switch.

Mechanical Switches[7]

Switching the light beam from one fiber to another one is basically easy. Simply move the transmitting fiber to mechanically align it with the desired receiving fiber. The problem is that even very small misalignments between the two fiber cores introduce unacceptable transmission losses. Several construction strategies have been utilized. Some incorporate a moving fiber, and others incorporate a moveable reflector.[8] In a moveable fiber switch, the fiber can be positioned either manually or by using an electromagnetic force. The switching action in Fig. 23 occurs when the totally reflecting prism moves to align the beam with one or the other of the two output fibers.

10.6 REFERENCES

1. R. W. Gilsdorf and J. C. Palais, "Single-Mode Fiber Coupling Efficiency with Graded-Index Rod Lenses," *Appl. Opt.* **33:**3440–3445 (1994).

2. C. Marxer, P. Griss, and N. F. de Rooij, "A Variable Optical Attenuator Based on Silicon Micromechanics," *IEEE Photon. Technol. Lett.* **11:**233–235 (1999).

3. C.-L. Chen, *Elements of Optoelectronics and Fiber Optics,* Irwin, Chicago, 1996.

4. R. Ramaswami and K. N. Sivarajan, *Optical Networks: A Practical Perspective,* Morgan Kaufmann, San Francisco, 1998.

5. N. Kashima, *Passive Optical Components for Optical Fiber Transmission,* Artech House, Boston, 1995.

6. J. P. Laude and J. M. Lerner, "Wavelength Division Multiplexing/Demultiplexing (WDM) Using Diffraction Gratings," *SPIE-Application, Theory and Fabrication of Periodic Structures,* **503:**22–28 (1984).

7. J. C. Palais, *Fiber Optic Communications,* 4th ed., Prentice-Hall, Upper Saddle River, New Jersey, 1998.

8. W. J. Tomlinson, "Applications of GRIN-Rod Lenses in Optical Fiber Communications Systems," *Appl. Opt.* **19:**1123–1138 (1980).

CHAPTER 11
SEMICONDUCTOR OPTICAL AMPLIFIERS AND WAVELENGTH CONVERSION

Ulf Österberg

Thayer School of Engineering,
Dartmouth College, Hanover, New Hampshire

11.1 GLOSSARY

B	photodetector bandwidth
d	active layer thickness
e	electronic charge
F	noise figure
G	amplifier gain
G_s	single-pass gain
g	gain per unit length
g_0	small-signal gain
$g(N)$	material gain coefficient
h	Planck's constant
I	light intensity
I_{sat}	saturated light intensity
J	current density
L	laser amplifier length
N	carrier density
N_0	transparency carrier density
N_{ph}	photon density
$N_{ph, sat}$	saturated photon density
n	refractive index

P_{signal} signal power

P_{shot} shot noise power

P_{thermal} thermal noise power

$P_{\text{signal-sp}}$ noise power from signal-spontaneous beating

$P_{\text{sp-sp}}$ noise power from spontaneous-spontaneous beating

R reflectivity

$R(N)$ recombination rate

SNR signal-to-noise ratio

V volume of active layer

v_g group velocity of light

w laser amplifier stripe width

α absorption coefficient

β line width enhancement factor

Γ optical confinement factor

γ population inversion factor

λ_c converted wavelength

λ_s signal wavelength

ν frequency

σ_g differential gain

τ_s carrier lifetime

Φ phase shift

ω_a angular frequency, anti-Stokes light

ω_{pump} angular frequency, pump light

ω_s angular frequency, Stokes light

11.2   WHY OPTICAL AMPLIFICATION?

Despite the inherently very low losses in optical glass fibers, it is necessary to amplify the light to maintain high signal-to-noise ratios (SNRs) for low bit error rate (BER) detection in communications systems and for sensor applications. Furthermore, as the bandwidth required is getting larger, it is also necessary to perform all-optical amplification using devices that are independent of bit rates and encoding schemes. In today's telecommunication systems (Chap. 10 in Ref. 1; Chaps. 1 and 2 in this book), optical amplifiers are typically utilized in the following three ways (Fig. 1):

- As power boosters immediately following the laser source
- To provide optical regeneration or in-line amplification for long-distance communications
- As preamplifiers at the receiver end

With the recent introduction of more complex local area networks, it has also become necessary to amplify the signal over short distances to allow it to be distributed to many users.

To characterize any optical amplifier, it is important to consider the following criteria:[2,3]

- Gain—depending on input power, can vary between 5 and 50 dB
- Bandwidth—varies between 1 and 10 THz or 30 and 100 nm

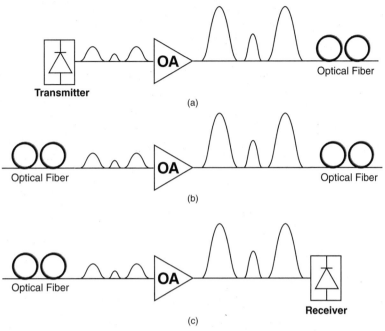

FIGURE 1 Applications of optical amplifiers.

- Pumping efficiency—varies between 1 and 10 dB/mW
- Gain saturation—on the order of 10 dBm for semiconductor amplifiers
- Noise—contributes a power penalty of at least 3 dB
- Polarization sensitivity
- Operating wavelength—most commonly 1.3 or 1.55 μm
- Crosstalk—occurs primarily in wavelength-division multiplexing (WDM) systems where many channels need to be amplified simultaneously
- Coupling loss—should be less than 0.3 dB

Naturally, these numbers are just estimates that can and will vary greatly depending on the exact operating wavelength used and the particular type of optical amplifier under consideration.

There are two main types of optical amplifiers: semiconductor and fiber. They each have strengths and weaknesses, and it becomes the task of the engineer to decide which type of amplifier to use with a specific application. This chapter is primarily about semiconductor amplifiers, but we will start with a short description of the salient features of fiber-optic amplifiers.

Fiber-optic Amplifiers

There are three types of fiber-optic amplifiers (Chap. 5): rare-earth-doped fiber amplifiers, Raman amplifiers, and Brillouin amplifiers (Chap. 38 in Ref. 1; Ref. 4; Chap. 5 in this book). Rare-earth-doped amplifiers are obtained by doping the fiberglass core with rare earth ions—neodymium for amplification around 1.06 μm, praseodymium for amplification around 1.3

µm, and erbium for amplification around 1.55 µm. An obvious advantage for optical fiber amplifiers is that they can be readily spliced to the main optical fiber, minimizing any coupling losses; furthermore, all types of fiber amplifiers are polarization insensitive. The Brillouin and Raman amplifiers rely on third-order nonlinearities in the glass to provide nonparametric interactions between photons and optical and acoustical phonons due to lattice or molecular vibrations within the core of the fiber. Brillouin amplification occurs when the pump and signal beams propagate in opposite directions. The gain is large but the bandwidth is very narrow (<100 MHz). Brillouin amplification is mostly used for receiver preamplification and for selective wavelength amplification for moderate-bit-rate communications systems (≤100 Mbit/s). Raman amplification is similar to Brillouin amplification. The most significant differences between Raman and Brillouin amplification are: (1) for Raman amplification, the pump and signal beam copropagate; (2) the Raman bandwidth is very large (>5 THz); and (3) the Stokes shift is orders of magnitude larger. Due to its broad bandwidth, Raman amplification can be used for very high-bit-rate communication systems. Unlike erbium-doped fiber amplifiers, Raman amplifiers can be used for any wavelength region, being limited only by the available pump sources.[5]

Semiconductor Amplifiers

Semiconductor laser amplifiers (SLAs) (Chap. 13 in Ref. 6) are most commonly divided into two types: (1) Fabry-Perot (FP) amplifiers and (2) traveling wave (TW) amplifiers. Both of these amplifier types are based on semiconductor lasers. The FP amplifier has facet reflectivities $R \approx 0.3$ (slightly less than the values for diode lasers) and the TW amplifier has $R \approx 10^{-3}$–10^{-4} (values as small as 10^{-5} have been reported[7]).

TW amplifiers whose bandwidth is >30 nm (bandwidth exceeding 150 nm has been obtained with the use of multiple quantum well structures[8]) are suitable for wavelength-division multiplexing (WDM) applications (Chap. 13). FP amplifiers have a very narrow bandwidth, typically 5 to 10 GHz (~0.1 nm at 1.5 µm). Due to the nonlinear properties of the semiconductor gain medium in conjunction with the feedback mechanism from the facet reflections, FP amplifiers are used for optical signal processing applications. In Fig. 2, the gain spectrum is shown for an SLA with two different facet reflectivities.

In Fig. 3 is a schematic of a typical amplifier design of length L (200 to 500 µm), thickness d (1 to 3 µm), and active region width w (10 µm). Amplification occurs when excited electrons in the active region of the semiconductor are stimulated to return to the ground level and excess energy is released as additional identical photons.

The connection between excited electrons (defined as number of electrons per cubic centimeter and referred to as the carrier density N) and the number of output photons N_{ph} is given by the rate equation

$$\frac{dN}{dt} = \frac{J}{ed} - R(N) - v_g \cdot g(N) \cdot N_{ph} \tag{1}$$

where J is the injection current density, v_g is the group velocity of light traveling through the amplifier, and $R(N)$ is the recombination rate (for a simple analytic analysis it is approximated to be linearly proportional to the carrier density, $R(N) \approx N/\tau_s$). For large injection currents this approximation breaks down and higher-order Auger recombination terms have to be incorporated.[9] $g(N)$ is the material gain coefficient, which depends on the light intensity and the specific band structure of the semiconductor used,

$$g(N) = \frac{\Gamma \cdot \sigma_g}{V} \cdot (N - N_0) \tag{2}$$

where Γ is the optical confinement factor, σ_g is the differential gain, $V = L \cdot d \cdot w$ is the volume of the active region, and N_0 is the carrier density needed for transparency—that is, no absorp-

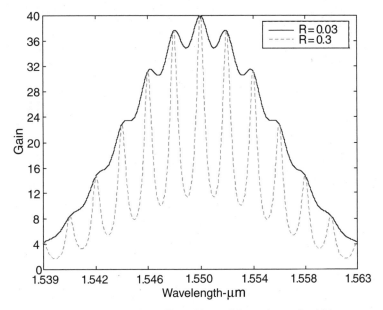

FIGURE 2   Gain spectrum for an SLA with two different facet reflectivities.

tion. The dependence on the band structure moves the peak wavelength of the gain peak toward shorter wavelengths as the carrier density is increased with increasing injection current.

The gain coefficient $g(N)$ is the parameter we wish to solve for in Eq. (1). We do that by setting the time derivative equal to 0 (steady state),

$$g(N) = \frac{g_0}{1 + \dfrac{N_{ph}}{N_{ph,\,sat}}} = \frac{g_0}{1 + \dfrac{I}{I_{sat}}} \tag{3}$$

where

$$g_0 = \frac{\Gamma \cdot \sigma_g}{V} \left(\frac{J}{ed} \cdot \tau_s - N_0 \right) \tag{4}$$

$$I_{sat} = \frac{h\nu \cdot L \cdot d \cdot w}{\Gamma^2 \cdot \sigma_g \cdot \tau_s} \tag{5}$$

g_0 is referred to as the small-signal gain. From Eq. (5) we notice that for semiconductor materials with small differential gain coefficients and short recombination times we will obtain a large saturation intensity. For typical semiconductor materials the gain saturation is comparatively large.

The net gain per unit length for the amplifier is given by

$$g = \Gamma \cdot g(N) - \alpha \tag{6}$$

where α is the total loss coefficient per unit length. If we assume that g does not vary with distance along the direction of propagation within the active gain medium, we obtain through integration the single-pass gain G_s

$$G_s = \frac{I(L)}{I(0)} = e^{g \cdot L} = e(\Gamma \cdot g_0/1 + I/I_{sat} - \alpha)L \tag{7}$$

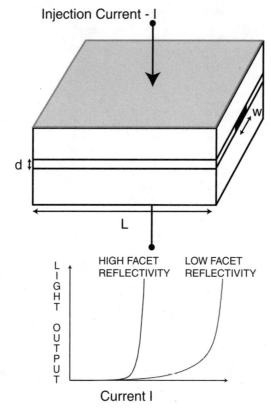

FIGURE 3 Schematic of an SLA and its light output versus injection current characteristics for two different facet reflectivities.

Notice that as the input intensity is increased above the saturation intensity the gain G_s starts to decrease rapidly. The reason for this is that there are not enough excited carriers to amplify all the incoming photons.

The phase shift for a single-pass amplifier is obtained by integrating the net gain g over the entire length L of the amplifier,[10]

$$\Phi = \frac{2\pi \cdot n \cdot L}{\lambda} + \frac{g_0 \cdot L \cdot \beta}{2}\left[\frac{I}{I + I_{\text{sat}}}\right] \tag{8}$$

where β is the line width enhancement factor and n is the refractive index. The second term, through gain saturation, will impose a frequency chirp on the amplified signal. The sign and linearity of this chirp in an SLA is such that the light pulse can be temporally compressed if it propagates in the anomalous dispersion regime of an optical fiber ($\lambda > 1.3$ μm).

From a systems point of view, noise is a very important design parameter. The noise for an amplifier is usually expressed using the noise figure F.[4,10]

$$F = \frac{SNR_{\text{in}}}{SNR_{\text{out}}} = \frac{SNR_b}{SNR_s} \tag{9}$$

where subscripts b and s refer to beat-noise-limited regime and shot-noise-limited regime, respectively.

The noise figure F is obtained by first calculating the SNR for an amplifier with gain G and for which the output light is detected with an ideal photodetector (bandwidth B) only limited by shot noise (Chaps. 17 and 18 in Ref. 1) and then calculating the SNR for a "real" amplifier for which the contribution from spontaneous emission is added as well as thermal noise for the photodetector. The SNR for the ideal case is

$$SNR_{in} = \frac{P_{signal}}{2 \cdot h\nu \cdot B} \tag{10}$$

and for the more realistic case it is

$$SNR_{out} = \frac{P_{signal}}{P_{shot} + P_{thermal} + P_{sp\text{-}sp} + P_{signal\text{-}sp}} \approx \frac{P_{signal}}{4 \cdot B \cdot h\nu \cdot \gamma} \cdot \frac{G}{G-1} \tag{11}$$

where γ is the population inversion factor,[4] and sp-sp and signal-sp are beating noise between either the spontaneously emitted light and itself or the spontaneously emitted light and the signal.

For large gain, $P_{signal\text{-}sp}$ dominates and

$$F = 2\gamma \cdot \frac{G}{G-1} \approx 2 \cdot \gamma \tag{12}$$

For an ideal amplifier, $\gamma = 1 \Rightarrow F = 3\text{dB}$; for most SLAs, $F > 5\text{dB}$.

11.3   *WHY OPTICAL WAVELENGTH CONVERSION?*

Wavelength-division multiplexed networks (Chap. 13) are already a reality, and as these networks continue to grow in size and complexity it will become necessary to use the same wavelengths in many different local parts of the network. To solve the wavelength contention problems at the connecting nodes, it is necessary to be able to perform wavelength conversion.

An optical wavelength converter should have the following characteristics:[11,12]

- Transparency to bit rates and coding schemes
- Fast setup time of output wavelength
- Conversion to both shorter and longer wavelengths
- Moderate input power levels
- Possibility for no wavelength conversion
- Polarization independence
- Small chirp
- High extinction ratio (power ratio between bit 0 and bit 1)
- Large SNR
- Simple implementation

Options for Altering Optical Wavelengths

There are two different techniques that have primarily been used for wavelength conversion. One is optoelectronic conversion (Chap. 13 in Ref. 6), in which the signal has to be converted from optical to electrical format before being transmitted at a new optical wavelength. This technique is presently good up to bit rates of 10 Gbit/s.[11] The main drawbacks of this method are power consumption and complexity. The second method is all-optical, and it can further be divided into two different approaches—nonlinear optical parametric processes (Chap. 38 in Ref. 6; Chaps. 3 and 17 in this book) and cross-modulation using an SLA.

The most common nonlinear optical method is four-photon mixing (FPM). FPM occurs naturally in the optical fiber due to the real part of the third-order nonlinear polarization. When the signal beam is mixed with a pump beam, two new wavelengths are generated at frequencies ω_s and ω_a according to the phase-matching condition

$$\omega_s - \omega_{\text{pump}} = \omega_{\text{pump}} - \omega_a \qquad (13)$$

where subscripts s and a stand for Stokes and anti-Stokes, respectively. Since the conversion efficiency is proportional to the square of the third-order nonlinear susceptibility, this is not a very efficient process. Furthermore, the FPM process is polarization sensitive and generates additional (satellite) wavelengths, which reduces the conversion efficiency to the desired wavelength and contributes to channel crosstalk. One major advantage is that no fiber splicing is necessary.

A similar nonlinear optical process that has also been used for wavelength conversion is difference-frequency generation (DFG). This process is due to the real part of the second-order nonlinear polarization and therefore cannot occur in the optical glass fiber. For DFG to be used, it is necessary to couple the light into an external waveguide, for example $LiNbO_3$. DFG does not generate any satellite wavelengths; however, it suffers from low conversion efficiency, polarization sensitivity, and coupling losses between fiber and external waveguide.

Semiconductor Optical Wavelength Converters

To date, the most promising method for wavelength conversion has been cross-modulation in an SLA in which either the gain or the phase can be modulated (XGM and XPM, respectively). A basic XGM converter is shown in Fig. 4a. The idea behind XGM is to mix the input signal with a cw beam at the new desired wavelength in the SLA. Due to gain saturation, the cw beam will be intensity modulated so that after the SLA it carries the same information as

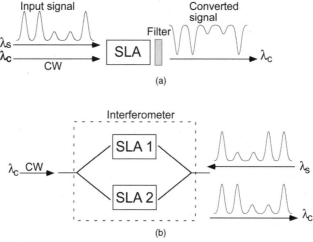

FIGURE 4 Use of an SLA for wavelength conversion. (*a*) Cross-gain modulation. (*b*) Cross-phase modulation.

the input signal. A filter is placed after the SLA to terminate the original wavelength λ_s. The signal and the cw beam can be either co- or counterpropagating. A counterpropagation approach has the advantage of not requiring the filter as well as making it possible for no wavelength conversion to take place. A typical XGM SLA converter is polarization independent but suffers from an inverted output signal and low extinction ratio.

Using an SLA in XPM mode for wavelength conversion makes it possible to generate a noninverted output signal with improved extinction ratio. The XPM relies on the fact that the refractive index in the active region of an SLA depends on the carrier density N, Eq. (1). Therefore, when an intensity-modulated signal propagates through the active region of an SLA it depletes the carrier density, thereby modulating the refractive index, which results in phase modulation of a CW beam propagating through the SLA simultaneously.

When the SLA is incorporated into an interferometer setup, the phase modulation can be transformed into an intensity modulated signal (Fig. 4b).

To improve the extinction ratio further, different setups using ring laser cavities[13] and non-linear optical loop mirrors[14] have been proposed.

11.4 REFERENCES

1. M. Bass, E. W. Van Stryland, D. R. Williams, and W. L. Wolfe (eds.), *Handbook of Optics,* 2d ed., vol. II, McGraw-Hill, New York, 1995.

2. M. J. Mahony, "Semiconductor Laser Optical Amplifiers for use in Future Fiber Systems," *IEEE J. Light. Tech.* **6**(4):531 (1988).

3. Max Ming-Kang Lin, *Principles and Applications of Optical Communications,* Irwin, 1996, chap. 17.

4. Govind P. Agrawal, *Fiber-Optic Communication Systems,* 2d ed., Wiley, New York, 1997, chap. 8.

5. M. J. Guy, S. V. Chernikov, and J. R. Taylor, "Lossless Transmission of 2 ps Pulses over 45 km of Standard Fibre at 1.3 μm using Distributed Raman Amplification," *Elect. Lett.* **34**(8):793 (1998).

6. M. Bass, E. W. Van Stryland, D. R. Williams, and W. L. Wolfe (eds.), *Handbook of Optics,* 2d ed., vol. I, McGraw-Hill, New York, 1995. [Chap 13 is P. L. Derry, L. Figueroa, and C.-S. Hong, "Semiconductor Lasers.")

7. T. Saitoh, T. Mukai, and O. Mikami, "Theoretical Analysis and Fabrication of Anti-Reflection Coatings on Laser Diode Facets," *IEEE J. Light. Tech.* **3**(2):288 (1985).

8. M. Tabuchi, "External Grating Tunable MQW Laser with Wide Tuning of 240 nm," *Elect. Lett.* **26**:742 (1990).

9. J. W. Wang, H. Olesen, and K. E. Stubkjaer, "Recombination, Gain, and Bandwidth Characteristics of 1.3 μm Semiconductor Laser Amplifiers," *Proc. IOOC-ECOC* 157 (1985).

10. I. Andonovic, "Optical Amplifiers," in O. D. D. Soares, ed., *Trends in Optical Fibre Metrology and Standards,* Kluwer, Dordrecht, 1995, sec. 5.1.

11. T. Durhuus, B. Mikkelsen, C. Joergensen, S. L. Danielsen, and K. E. Stubkjaer, "All-Optical Wavelength Conversion by Semiconductor Optical Amplifiers," *IEEE J. Light. Tech.* **14**(6):942 (1996).

12. M. S. Borella, J. P. Jue, D. Banerjee, B. Ramamurthy, and B. Mukherjee, "Optical Components for WDM Lightwave Networks," *Proc. IEEE* **85**(8):1274 (1997).

13. Y. Hibino, H. Terui, A. Sugita, and Y. Ohmori, "Silica-Based Optical Waveguide Ring Laser Integrated with Semiconductor Laser Amplifier on Si Substrate," *Elect. Lett.* **28**(20):1932 (1992).

14. M. Eiselt, W. Pieper, and H. G. Weber, "Decision Gate for All-Optical Retiming using a Semiconductor Laser Amplifier in a Loop Mirror Configuration," *Elect. Lett.* **29**:107 (1993).

CHAPTER 12

OPTICAL TIME-DIVISION MULTIPLEXED COMMUNICATION NETWORKS

Peter J. Delfyett
*School of Optics/The Center for Research
and Education in Optics and Lasers (CREOL)
University of Central Florida
Orlando, Florida*

12.1 GLOSSARY

Definitions

Bandwidth	A measure of the frequency spread of a signal or system—that is, its information-carrying capacity.
Chirping	The time dependence of the instantaneous frequency of a signal.
Commutator/decommutator	A device that assists in the sampling, multiplexing, and demultiplexing of time domain signals.
Homogeneous broadening	A physical mechanism that broadens the line width of a laser transition. The amount of broadening is exactly the same for all excited states.
Kerr effect	The dependence of a material's index of refraction on the square of an applied electric field.
Mode partition noise	Noise associated with mode competition in a multimode laser.
Multiplexing/demultiplexing	The process of combining and separating several independent signals that share a common communication channel.
Passband	The range of frequencies allowed to pass in a linear system.
Photon lifetime	The time associated with the decay in light intensity within an optical resonator.
Picosecond	One trillionth of a second.
p-n junction	The region that joins two materials of opposite doping. This occurs

	when n-type and p-type materials are joined to form a continuous crystal.
Pockel's effect	The dependence of a material's index of refraction on an applied electric field.
Quantum confined Stark effect (QCSE)	Optical absorption induced by an applied electric field across a semiconductor quantum well.
Quantum well	A thin semiconductor layer sandwiched between material with a larger band gap. The relevant dimension of the layer is on the order of 10 nm.
Sampling	The process of acquiring discrete values of a continuous signal.
Spatial hole burning	The resultant nonuniform spatial distribution of optical gain in a material owing to standing waves in an optical resonator.
Spontaneous emission	An energy decay mechanism to reduce the energy of excited states by the emission of light.
Stimulated emission	An energy decay mechanism that is induced by the presence of light in matter to reduce the energy of excited states by the emission of light.
Terabit	1 trillion bits.

Abbreviations

ADC	analog-to-digital converter
APD	avalanche photodetector
CEPT	European Conference of Postal and Telecommunications Administrations
CMI	code mark inversion
DBR	distributed Bragg reflector
DFB	distributed feedback
DS	digital signal
EDFA	erbium-doped fiber amplifier
FDM	frequency-division multiplexing
FP	Fabry-Perot
LED	light-emitting diode
NRZ	non-return-to-zero
OC-N	optical carrier (Nth level)
OOK	on-off keying
PAM	pulse amplitude modulation
PCM	pulse code modulation
PLL	phase-locked loop
PLM	pulse length modulation
PPM	pulse position modulation
RZ	return-to-zero
SDH	synchronous digital hierarchy
SLALOM	semiconductor laser amplifier loop optical mirror
SONET	synchronous optical network
SPE	synchronous payload envelope

STS	synchronous transmission signal
TDM	time-division multiplexing
TDMA	time-division multiple access
TOAD	terahertz optical asymmetric demultiplexer
UNI	unbalanced nonlinear interferometer
WDM	wavelength-division multiplexing
VCO	voltage-controlled oscillator

Symbols

B	number of bits respresenting N levels in an analog-to-digital converter
f_S	sampling frequency
N	number of levels in an analog-to-digital converter
n	index of refraction; integer
p_T	periodic sampling pulse train
$R_{1,2}$	mirror reflectivities
T	period
W (Hz)	bandwidth of a signal in hertz
$x_S(t)$	sampled version of a continuous function of time
$x(t)$	continuous analog signal
$X(\omega)$	frequency spectrum of the signal $x(t)$
δ	delta function
Λ	grating period
λ	wavelength
τ_D or τ_P	photon decay time or photon lifetime
τ_{RT}	round-trip propagation time of an optical cavity
ϕ	phase shift
ω	angular frequency (radians per second)

12.2 *INTRODUCTION*

Information and data services, such as voice, data, video, and the Internet, are integral parts of our everyday personal and business lives. By the year 2001, total information traffic on the phone lines will exceed 1 Tbit/s, with the Internet accounting for at least 50 percent of the total. More importantly, the amount of traffic is expected to grow to 100 Tbit/s by the year 2008. Clearly, there is a tremendous demand for the sharing, transfer, and use of information and related services. However, as the demand continues to increase, it should be noted that technology must evolve to meet this demand. This chapter discusses the current status of optical time-division multiplexed communication networks. This chapter is generally organized to initially provide the reader with a brief review of digital signals and sampling to show how and why time-division multiplexing (TDM) becomes a natural way of transmitting information. Following this introduction, time-division multiplexing and time-division multiple access (TDMA) are discussed in terms of their specific applications, for example voice communication/circuit-switched networks and data communication/packet-switched networks for TDM and TDMA, respectively. These two sections provide the reader with a firm understanding of the overall system perspective as to how these networks are constructed and expected to perform. To provide an understanding of the current state of the art, a review of selected high-speed optical and optoelectronic device technology is given. Before a final summary and outlook toward future directions, a specific ultra-high-speed optical time-division optical link is discussed to coalesce the concepts with the discussed device technology.

Fundamental Concepts

Multiplexing is a technique used to combine the information of multiple communication sites or users over a common communication medium and to send that information over a communication channel where the *bandwidth,* or information-carrying capacity, is shared between each user. The reason for sharing the information channel is to reduce the cost and complexity of establishing a communication network for many users. In the case where the shared medium is time, a communication link is created by combining information from several independent sources and transmitting that information from each source simultaneously without the portions of information from each source interfering with each other. This is done by temporally interleaving small portions, or *bits,* of each source of information so that each user sends data for a very short period of time over the communication channel. The user waits until all other users transmit their data before being able to transmit another bit of information. At a switch or receiver end, the user for which the data was intended picks out, or *demultiplexes,* the data that is intended for that user, while the rest of the information on the communication channel continues to its intended destination.

Sampling

An important concept in time-division multiplexing is being able to have a simple and effective method for converting real-world information into a form that is suitable for transmission by light over an optical fiber or by a direct line-of-sight connection in free space. As networks evolve, the standard for information transmission is primarily becoming digital in nature—information is transmitted by sending a coded message using two symbols (e.g., a 1 or a 0) physically corresponding to light being either present or not on a detector at the receiving location. This process of transforming real signals into a form that is suitable for reliable transmission requires one to sample the analog signal to be sent and digitize and convert the analog signal to a stream of 1s and 0s. This process is usually performed by a sample-and-hold circuit, followed by an analog-to-digital converter (ADC). In this section the concepts of signal sampling and digitization are reviewed with the motivation to convey the idea of the robustness of digital communications. It should be noted, however, that pure analog time-division multiplexed systems can still be realized, as will be shown later, and it is necessary to review this prior to examining digital TDM networks.

Sampling Theorem

The key feature of time-division multiplexing is that it relies on the fact that an analog bandwidth-limited signal may be exactly specified by taking samples of the signal, if the samples are taken sufficiently frequently. Time multiplexing is achieved by interleaving the samples of the individual signals. It should be noted that since the samples are pulses, the system is said to be *pulse modulated.* An understanding of the fundamental principle of time-division multiplexing, called the *sampling theorem,* is needed to see that any signal, including a signal continuously varying in time, can be exactly represented by a sequence of samples or pulses. The theorem states that a real valued bandwidth-limited signal that has no spectral components above a frequency of W Hz is determined uniquely by its value at uniform intervals spaced no greater than $1/(2W)$ s apart. This means that an analog signal can be completely reconstructed from a set of discrete samples uniformly spaced in time. The signal samples $x_S(t)$ are usually obtained by multiplying the signal $x(t)$ by a train of narrow pulses $p_T(t)$, with a time period $T = 1/f_s \leq \frac{1}{2}W$. The process of sampling can be mathematically represented as

$$x_S(t) = x(t) \cdot p_T(t)$$

$$= x(t) \cdot \sum_{n=-\infty}^{+\infty} \delta(t - nT)$$

$$= \sum_{n=-\infty}^{+\infty} x(nT)\delta(t - nT) \tag{1}$$

where it is assumed that the sampling pulses are ideal impulses and n is an integer. Defining the Fourier transform and its inverse as

$$X(\omega) = \int_{-\infty}^{+\infty} x(t) \exp(-j\omega t)dt \tag{2}$$

and

$$x(t) = \frac{1}{2\pi} \int_{-\infty}^{\infty} X(\omega) \exp(+j\omega t)d\omega \tag{3}$$

one can show that the spectrum $X_S(\omega)$ of the signal $x_S(t)$ is given by

$$X_S(\omega) = \frac{1}{T} \sum P\left(\frac{2\pi n}{T}\right) \cdot X\left(\omega - \frac{2\pi n}{T}\right)$$

$$= \frac{1}{T} P(\omega) \cdot \sum X\left(\omega - \frac{2\pi n}{T}\right) \tag{4}$$

In the case of the sampling pulses p being perfect delta functions, and given that the Fourier transform of $\delta(t)$ is 1, the signal spectrum is given by

$$X_S = \sum X\left(\omega - \frac{2\pi n}{T}\right) \tag{5}$$

This is represented pictorially in Fig. 1a–c. In Fig. 1a and b is an analog signal and its sampled version, where the sample interval is ~8 times the nominal sample rate of $1/(2W)$. From Fig. 1c it is clear that the spectrum of the signal is repeated in frequency every $2\pi/T$ Hz if the sample rate T is $1/(2W)$. By employing (passing the signal through) an ideal rectangular low-pass filter—that is, a uniform (constant) passband with a sharp cutoff, centered at direct current (DC) with a bandwidth of $2\pi/T$ the signal can be completely recovered. This filter characteristic implies an impulse response of

$$h(t) = 2W\frac{\sin(2\pi Wt)}{(2\pi Wt)} \tag{6}$$

The reconstructed signal can now be given as

$$x(t) = 2W\sum_{n=-\infty}^{+\infty} x(nT) \cdot \frac{\sin[2\pi W(t - nT)]}{2\pi W(t - nT)}$$

$$= x(t)/T, \ T = \frac{1}{2W} \tag{7}$$

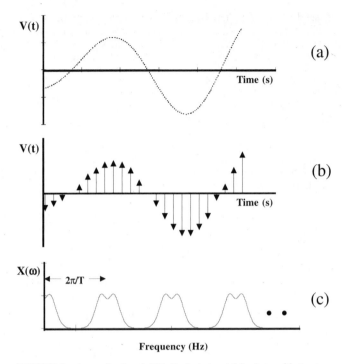

FIGURE 1 An analog bandwidth-limited signal (*a*), along with its sampled counterpart (*b*) sampled at a rate of ~8 times the Nyquist rate. (*c*) Frequency spectrum of a band-limited signal that has been sampled at a rate of $T = \frac{1}{2}W$, where W is the bandwidth of the signal.

This reconstruction is shown in Fig. 2. It should be noted that the oscillating nature of the impulse response $h(t)$ interferes destructively with other sample responses for times away from the centroid of each reconstructed sample. The sampling theorem now suggests three possibilities. (1) It is possible to interleave multiple sampled signals from several independent sources in time and transmit the total composite signal (time-division multiplexing). (2) Any parameter of the sampling train can be varied, such as its pulse length, pulse amplitude, or pulse position in direct accordance with the sampled values of the signal—that is, pulse length modulation (PLM), pulse amplitude modulation (PAM), and pulse position modulation (PPM). (3) The samples can be quantized and coded in binary or m-ary level format and transmitted as a digital signal, leading to pulse code modulation (PCM). Figure 3 shows an example of a sinusoidal signal and its representation in PAM, PPM, and PLM.

Interleaving

The sampling principle can be exploited in time-division multiplexing by considering the ideal case of a single point-to-point link connecting N users to N other users over a single communication channel, which is shown schematically in Fig. 4. At the transmitter end, a number of users with bandwidth-limited signals, each possessing a similar bandwidth, are connected to the contact points of a rotary switch called a *commutator*. For example, each user may be transmitting band-limited voice signals, each limited to 3.3 kHz. As the rotary arm of the

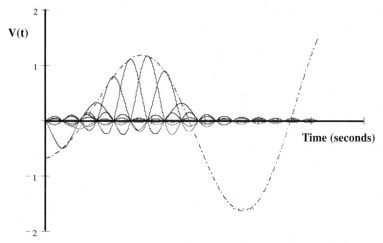

FIGURE 2   Temporal reconstruction of the sampled signal after passing the samples through a rectangular filter.

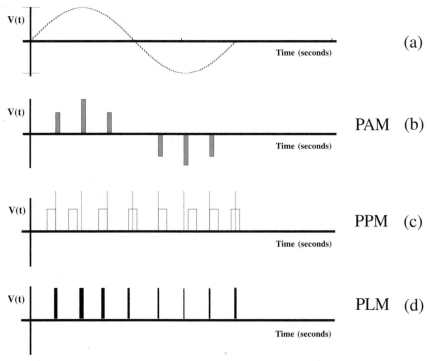

FIGURE 3   Schematic representation of three different possible methods of transmitting discrete samples of a continuous analog signal. (*a*) Analog sinusoidal. (*b*) Pulse amplitude modulation. (*c*) Pulse position modulation. (*d*) Pulse length modulation.

switch swings around, it samples each signal sequentially. The rotary switch at the receiving end is in synchrony with the switch at the sending end. The two switches make contact simultaneously at a similar number of contacts. With each revolution of the switch, one sample is taken of each input signal and presented to the correspondingly numbered contact of the switch at the receiving end. The train of samples at terminal 1 in the receiver passes through a low-pass filter and at the filter output the original signal $m(t)$ appears reconstructed. Of course, if f_M is the highest-frequency spectral component present in any of the input signals, the switches must make at least two f_M revolutions per second.

When the signals need to be multiplexed vary rapidly in time, electronic switching systems are employed, as opposed to the simple mechanical switches depicted in Fig. 4. The sampling and switching mechanism at the transmitter is called the *commutator;* while the sampling and switching mechanism at the receiver is called the *decommutator.* The commutator samples and combines samples, while the decommutator separates or *demultiplexes* samples belonging to individual signals so that these signals may be reconstructed.

The interleaving of the samples that allow multiplexing is shown in Fig. 5. For illustrative purposes, only two analog signals are considered. Both signals are repetitively sampled at a sample rate $T;$ however, the instants at which the samples of each signal are taken are different. The input signal to receiver 1 in Fig. 4 is the train of samples from transmitter 1 and the input signal to receiver 2 is the train of samples from transmitter 2. The relative timing of the sampled signals of transmitter 1 has been drawn to be exactly between the samples of transmitter 2 for clarity; however, in practice, these samples would be separated by a smaller timing interval to accommodate additional temporally multiplexed signals.

In this particular case, it is seen that the train of pulses corresponding to the samples of each signal is modulated in amplitude in direct proportion to the signal. This is referred to as pulse amplitude modulation (PAM). Multiplexing of several PAM signals is possible because the various signals are kept distinct and are separately recoverable by virtue of the fact that they are

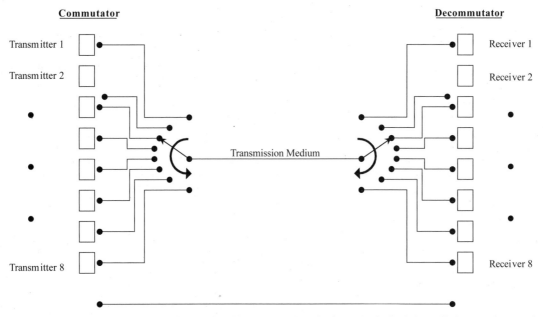

FIGURE 4   Illustration of a time multiplexer/demultiplexer based on simple mechanical switches called *commutators* and *decommutators.*

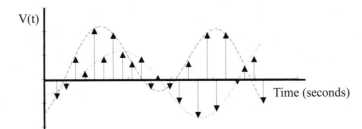

FIGURE 5 Two band-limited analog signals and their respective samples occurring at a rate of approximately 6 times the highest frequency, or 3 times the Nyquist rate.

sampled at different times; thus this is an example of a time-division multiplexed system. This is in contrast to systems that can keep the signals separable by virtue of their frequency (or optical wavelength) translation to different portions of the frequency (wavelength) domain. These systems are referred to as *frequency-division multiplexed* (FDM) or *wavelength-division multiplexed* (WDM).

In today's fiber-optic systems, formally, the sampled signals are transmitted on an optical carrier frequency, or *wavelength,* while older conventional electrical communication links transmit the multiplexed signals directly over a wire pair. It should be noted that the process of transmitting information on optical carriers is analogous to radio transmission, where the data is transmitted on carrier frequencies in the radio frequency range (kilohertz to gigahertz).

Demultiplexing—Synchronization of Transmitter and Receiver

In any type of time-division multiplexing system, it is required that the sampling at the transmitter end and the demultiplexing at the receiver end be *in step* (synchronized with each other). As an example, consider the diagram of the commutator in Fig. 4. When the transmitting multiplexer is set in a position that samples and transmits information from user 1, the receiving demultiplexer must be in a position to pick out, or *demultiplex,* and receive information that is directed for receiver 1. To accomplish this timing synchronization, the receiver has a local clock signal that controls the timing of the commutator as it switches from one time slot to the next. The clock signal may be a narrowband sinusoidal signal from which an appropriate clocking signal, with sufficiently fast rising edges of the appropriate signal strength, can be derived. The repetition rate of the clock in a simple configuration would then be equal to the sampling rate of an individual channel times the number of channels being multiplexed, thereby assigning one time slot per clock cycle.

At the receiver end, the clock signal is required to keep the decommutator synchronized to the commutator, that is, to keep both running at the same rate. As well, there must be additional timing information to provide agreement as to the relative positions or phase of the commutator-decommutator pair, which assures that information from transmitter 1 is guaranteed to be received at the desired destination of receiver 1. The time interval from the beginning of the time slot allocated to a particular channel until the next recurrence of that particular time slot is commonly referred to as a *frame.* As a result, timing information is required at both the bit (time slot) and frame levels. A common arrangement in time-division multiplexed systems is to allow for one or more time slots per frame to provide timing information, depending on the temporal duration of the transmitted frame. It should be noted that there are a variety of methods for providing timing information, such as directly using a portion of the allocated bandwidth, as just mentioned, or alternatively, recovering a clock signal by deriving timing information directly from the transmitted data.

Digital Signals—Pulse Code Modulation

In most applications that employ optical time-division multiplexing, signals are usually sent in a pulse-code-modulated format, as opposed to sending optical samples that are directly proportional to the analog signal amplitude (e.g., PAM, PPM, and PLM). The key feature of sending the information in the form of a digital code is that the analog form of the signal can be corrupted with noise that generally cannot be separated from the signal. The pulse code modulation format provides a mechanism by which the digitization and *quantization,* or coding, of the signal produces a signal that can be recovered from the noise introduced by the communication link.

The limitation of a simple analog communication system is that once noise is introduced onto the signal, it is impossible to remove. When quantization is employed, a new signal is created that is an approximation of the original signal. The main benefit of employing a quantization technique is that, in large part, the noise can be removed from the signal. The main characteristic of a general quantizer is it has an input-output characteristic that is in the form of a staircase, as shown in Fig. 6. It is observed that while the input signal $V_{in}(t)$ varies smoothly, the output $V_o(t)$ is held constant at a fixed level until the signal varies by an amount of V_{max}/N, where N is the number of levels by which the output signal changes its output level. The output quantized signal represents the sampled waveform, assuming that the quantizer is linearly related to the input. The transition between one level and the next occurs at the instant when the signal is midway between two adjacent quantized levels. As a result, the quantized signal is an approximation of the original signal. The quality of the approximation may be improved by reducing the step size or increasing the number of quantized levels. With sufficiently small step size or number of quantized levels, the distinction between the original signal and the quantized signal becomes insignificant. Now, consider that the signal is transmitted and subsequently received, with the addition of noise on the received signal. If this signal is presented to the input of another identical quantizer, and if the peak value of the noise signal is less than half the step size of the quantizer, the output of the second quantizer is identical to the original transmitted quantized signal, without the noise that was added by the transmission channel! It should be noted that this example is presented only to illustrate the concept of noise removal via quantization techniques. In reality, there is always a finite probability—no matter how small—that the noise signal will have a value that is larger than half the step size, resulting in a detected error. While this example shows the benefits of quantization and digital transmission, the system trade-off is that additional bandwidth is required to transmit the coded signal.

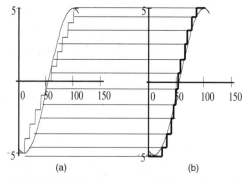

FIGURE 6 The input-output "staircase" transfer function of a digital quantizer. (*a*) Staircase function and sinusoid. (*b*) The resultant quantized function superimposed on the original sinusoid, showing a slight deviation of the quantized signal from the original sinusoid.

It should be noted that the resultant quantized signal shown in Fig. 6 possesses a slight distortion that results from the quantization process. This slight distortion generates a signal-to-noise ratio (SNR) that is not uniform for all values of received signals. This nonuniform SNR tends to increase the error in the transmitted signal due to quantization. One method of reducing this quantization error is to predistort the signal such that small-amplitude signals are received with the same SNR as large-amplitude signals. This process of predistorting the signal is called *compressing,* and is achieved in devices called *companders.* Obviously, on the receiver end, a similar process to invert the predistortion process is required, and is accomplished in an *expander.*

Pulse Code Modulation

A signal that is to be quantized prior to transmission has been sampled as well. The quantization is used to reduce the effects of noise, and the sampling allows us to time-division multiplex a number of users. The combined signal-processing techniques of sampling and quantizing generate a waveform composed of pulses whose amplitudes are limited to a discrete number of levels. Instead of these quantized sample values being transmitted directly, each quantized level can be represented as a binary code, and the code can be sent instead of the actual value of the signal. The benefit is immediately recognized when considering the electronic circuitry and signal processing required at the receiver end. In the case of binary code transmission, the receiver only has to determine whether one of two signals was received (e.g., a 1 or a 0), as compared to a receiver system, which would need to discern the difference between the N distinct levels used to quantize the original signal. The process of converting the sampled values of the signal into a binary coded signal is generally referred to as *encoding.* Generally, the signal-processing operations of sampling and encoding are usually performed simultaneously, and as such, the entire process is referred to as analog-to-digital (A-to-D) conversion.

Analog-to-Digital Conversion

The sampled signal, as shown in Fig. 5, represents the actual values of the analog signal at the sampling instants. In a practical communication system or in a realistic measurement setup, the received or measured values can never be absolutely correct because of the noise introduced by the transmission channel or small inaccuracies impressed on the received data owing to the detection or measurement process. It turns out that it is sufficient to transmit and receive only the quantized values of the signal samples. The quantized values of sampled signals, represented to the nearest digit, may be represented in a binary form or in any coded form using only 1s and 0s. For example, sampled values between 2.5 and 3.4 would be represented by the quantized value of 3, and could be represented as 11, using two bits (in base 2 arithmetic). This method of representing a sampled analog signal, as noted earlier, is known as pulse code modulation. An error is introduced on the signal by this quantization process. The magnitude of this error is given by

$$\varepsilon = \frac{0.4}{N} \tag{8}$$

where N is the number of levels determined by $N = 2^B$, and B is the B-bit binary code—for example, $B = 8$ for eight-bit words representing 256 levels. Thus one can minimize the error by increasing the number of levels, which is achieved by reducing the step size in the quantization process. It is interesting to note that using only four bits (16 levels), a maximum error of 2.5 percent is achieved, while increasing the number of bits to eight (256 levels) gives a maximum error of 0.15 percent.

Optical Representation of Binary Digits and Line Coding

The binary digits can be represented and transmitted on an optical beam and passed through an optical fiber or transmitted in free space. The optical beam is modulated to form pulses to represent the sampled and digitized information. A family of four such representations is shown in Fig. 7. There are two particular forms of data transmission that are quite common in optical communications owing to the fact that their modulation formats occur naturally in both direct and externally modulated optical sources. These two formats are referred to as *non-return-to-zero* (NRZ) and *return-to-zero* (RZ). In addition to NRZ and RZ data formats, pulse-code-modulated data signals are transmitted in other codes that are designed to optimize the link performance, owing to channel constraints. Some important data transmission formats for optical time-division multiplexed networks are code mark inversion (CMI) and Manchester coding or bi-phase coding. In CMI, the coded data has no transitions for logical 1 levels. Instead, the logic level alternates between a high and low level. For logical 0, on the other hand, there is always a transition from low to high at the middle of the bit interval. This transition for every logical 0 bit ensures proper timing recovery. For Manchester coding, logic 1 is represented by a return-to-zero pulse with a 50 percent duty cycle over the bit period (a half-cycle square wave), and logic 0 is represented by a similar return-to-zero waveform of opposite phase, hence the name *bi-phase*. The salient feature of both bi-phase and CMI coding is that their power spectra have significant energy at the bit rate, owing to the guarantee of a significant number of transitions from logic 1 to 0. This should be compared to the power spectra of RZ and NRZ data, which are shown in Fig. 8. The NRZ spectrum has no energy at the bit rate, while the RZ power spectrum does have energy at the bit rate—but the RZ spectrum is also broad, having twice the width of the NRZ spectrum. The received data power spectrum is important for TDM transmission links, where a clock or synchronization signal is required at the receiver end to demultiplex the data. It is useful to be able to recover a clock or synchronization signal derived from the transmitted data, instead of using a portion of the channel bandwidth to send a clock signal. Therefore, choosing a transmission format with a large power spectral component at the transmitted bit rate provides an easy method of recovering a clock signal.

Consider for example the return-to-zero (RZ) format just discussed. If the transmitted bits are random independent 1s and 0s with equal probability, the transmitted waveform can be

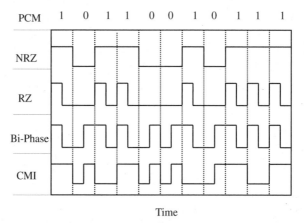

FIGURE 7 Line-coded representations of the pulse-code-modulated logic signal 10110010111. NRZ: non-return-to-zero format; RZ: return-to-zero format; bi-phase, also commonly referred to as Manchester coding; CMI: code mark inversion format.

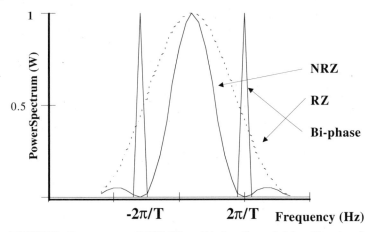

FIGURE 8   Power spectra of NRZ, RZ, and bi-phase line coded data. Note the relative power at the bit rate.

considered to be the sum of a periodic clock sequence with half of the amplitude and a random sequence with zero mean as shown in Fig. 9. The Fourier transform of the clock component has a peak at the bit frequency, and the Fourier transform of the random component is 0 at the bit frequency. Therefore, if there is a narrow-bandpass filter at the receiver with the received signal as the input, the clock component will pass through and the random part will be rejected. The output is thus a pure sinusoid oscillating at the clock frequency or bit rate. This concept of line filtering for clock recovery is schematically represented in Fig. 10.

Generally, pulse-code-modulated signals are transmitted in several different formats to fit within the constraints determined by the transmission channel (bandwidth and so on). It is clear from Fig. 8 that the power spectrum of return-to-zero PCM data has a spectral spread

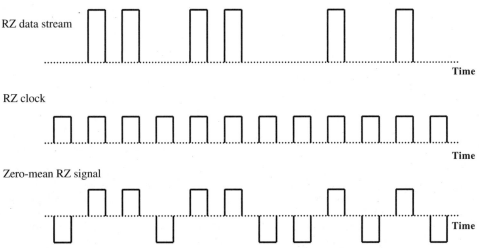

FIGURE 9   Illustration showing a random RZ data stream, along with its RZ clock component and its zero-mean counterpart. Note that the zero-mean signal results from the difference between the RZ data and the clock component.

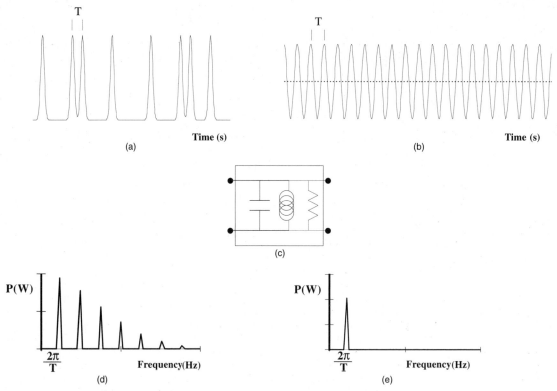

FIGURE 10   Principle of clock recovery using line filtering. (*a*) Input RZ data stream. (*b*) Filtered time-domain clock signal. (*c*) Schematic of an electrical tank circuit for realizing a bandpass filter. (*d*) Power spectrum of a periodic RZ sequence. (*e*) Power spectrum of the filtered signal.

that is approximately twice that of non-return-to-zero PCM data. Both formats have a large amount of power in the DC and low-frequency components of their power spectra. In contrast, the bi-phase code has very low power in the DC and low-frequency portion of the power spectrum, and as a result is a very useful format for efficient timing recovery.

Timing Recovery

Time-division multiplexing and time-division multiple-access networks inherently require timing signals to assist in demultiplexing individual signals from their multiplexed counterparts. One possible method is to utilize a portion of the communication bandwidth to transmit a timing signal. Technically, this is feasible; however (1) this approach requires hardware dedicated to timing functions distributed at each network node that performs multiplexing and demultiplexing functions, and (2) network planners want to optimize the channel bandwidth without resorting to dedicating a portion of the channel bandwidth to timing functions. The desired approach is to derive a timing signal directly from the transmitted data. This allows the production of the required timing signals for multiplexing and demultiplexing without the need to use valuable channel bandwidth.

As suggested by Fig. 10, a simple method for recovering a timing signal from transmitted return-to-zero data is to use a bandpass filter to pass a portion of the power spectrum of the transmitted data. The filtered output from the tank circuit is a pure sinusoid that provides the timing information. An important parameter to consider in line filtering is the quality factor, designated as the filter Q. Generally, the Q factor is defined as

$$Q = \frac{\omega_o}{\Delta\omega}$$

where ω_o is the resonant frequency and $\Delta\omega$ is the bandwidth of the filter. It should also be noted that Q is a measure of the amount of energy stored in the bandpass filter, such that the output from the filter decays exponentially at a rate directly proportional to Q. In addition, for bandpass filters based on passive electrical circuits, the output peak signal is directly proportional to Q. These two important physical features of passive line filtering imply that the filter output will provide a large and stable timing signal if the Q factor is large. However, since Q is inversely proportional to the filter bandwidth, a large Q typically implies a small filter bandwidth. As a result, if the transmitter bit rate and the resonant frequency of the tank circuit do not coincide, the clock output could be zero. In addition, the clock output is very sensitive to the frequency offset between the transmitter and resonant frequency. Therefore, line filtering can provide a large and stable clock signal for large filter Q, but the same filter will not perform well when the bit rate of the received signal has a large frequency variation. In TDM bit timing recovery, the ability to recover the clock of an input signal over a wide frequency range is called *frequency acquisition* or *locking range,* and the ability to tolerate timing jitter and a long interval of zero transitions is called *frequency tracking* or *hold over time.* Therefore, the tradeoff exists between the locking range (low Q) and hold over time (large Q) in line filtering.

A second general scheme to realize timing recovery and overcome the drawbacks of line filtering using passive linear components is the use of a phase-locked loop (PLL) in conjunction with a voltage-controlled oscillator (VCO) (see Fig. 11a). In this case, two signals are fed into

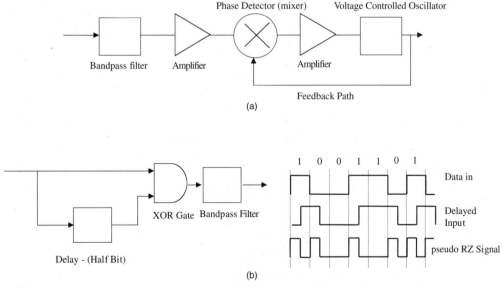

(a)

(b)

FIGURE 11 (*a*) Schematic diagram of a phase-locked loop using a mixer as a phase detector and a voltage-controlled oscillator to provide the clock signal that can track phase wander in the data stream. (*b*) Data format conversion between input NRZ data and RZ output data using an electronic logic gate. The subsequent RZ output is then suitable for use in a clock recovery device.

the mixer. One signal is derived from the data (e.g., from a line filtered signal possessing energy at the bit rate), while the second signal is a sinusoid generated from the VCO. The mixer is used as a phase detector and produces a DC voltage that is applied to the VCO to adjust its frequency of operation. The overall function of the PLL is to adjust its own voltage to track the frequency and phase of the input data signal. Owing to the active components in the PLL, this approach for timing recovery can realize a broad locking range, low insertion loss, and good phase-tracking capabilities. It should be noted that while the concepts for timing recovery described in this section were illustrated using techniques that are not directly applicable to ultra-high-speed optical networking, the underlying principles will still hold for high-speed all-optical techniques. These approaches are discussed in more detail later in the chapter.

While both these techniques require the input data to be in the return-to-zero format, many data transmission links use non-return-to-zero line coding owing to its bandwidth efficiency. Unfortunately, in the NRZ format there is no component in the power spectrum at the bit rate. As a result, some preprocessing of the input data signal is required before clock recovery can be performed. A simple method for achieving this is illustrated in Fig. 11b. The general concept is to present the data signal with a delayed version of the data at the input ports of a logic gate that performs the exclusive OR operation. The temporal delay, in this case, should be equal to half a bit. The output of the XOR gate is a pseudo-RZ data stream that can then be line filtered for clock recovery.

12.3 TIME-DIVISION MULTIPLEXING AND TIME-DIVISION MULTIPLE ACCESS

Overview

In today's evolving telecommunication (voice and real-time video) and data (e.g., Internet) networks, the general mode of transmitting information can be adapted to make maximum use of a network's bandwidth. In addition, the general characteristics of the user application may also require a specific mode of transmission format. For example, in classic circuit-switched voice communications, real-time network access is desired since voice communications are severely hampered in links that have large timing delays or *latency*. In contrast, data networks are not hampered if the communications link has small delays in the transmission of information. In this case, packet-switched data is sent in bursts, and the user does not require continuous, real-time access to the network. These two different ways of achieving access to the bandwidth are generally referred to as *time-division multiplexing* (TDM), typically used in circuit-switched voice communication networks, and *time-division multiple access* (TDMA), which is used in packet-switched data networks.

In communication links such as TDM and TDMA, since the transmission medium bandwidth is shared in the time domain, the transmitting node is required to know when (at what time) it can transmit, and the duration (or for how long) it can transmit the data. These two aspects of time multiplexing immediately imply constraints on the bit or frame synchronization and bit period or packet rate for TDM and TDMA, respectively. We will now review both TDM and TDMA access, emphasizing these two aspects.

Time-Domain Multiple Access

In time-domain multiple access (TDMA), communication nodes send their data to the shared medium during an assigned time slot. A key characteristic of TDMA is that it first stores lower-bit-rate signals in a buffer prior to transmission. As seen in Fig. 12, when a node is assigned a time slot and allowed to transmit, it transmits all the bits stored in the buffer at a high transmission rate. To relax the synchronization requirement, data bursts or time slots are

TDMA separated by a guard time. With this guard time, transmissions within different time slots may have different bit clocks. This key feature allows the simplification of the timing recovery process and removes the need for frequency justification.

Owing to the fact that there is no need for bit timing and synchronization between the multiple users, TDMA can be directly performed in the optical transmission domain. The user obtains access to the transmission medium by having an optical transmitter transmitting a burst of optical data in a pulse-code-modulation format within a time slot. It should be noted that in optical networking scenarios, optical TDMA (OTDMA) is preferred over optical TDM (OTDM), owing to the ease of implementation of OTDMA. However, it must be stressed that the OTDMA approach has a lower bandwidth efficiency because some of the time slots are required to realize required timing guard bands.

The TDMA frame in Fig. 12 consists of a reference burst and a specific number of time slots. The reference burst is used for timing and establishing a synchronization reference, in addition to carrying information regarding the signaling (the communication process that sets up the communication call and monitors the communication link). The rest of the frame, which contains additional guard bands and time slots, carries the data. The reference burst primarily contains three main components: (1) a preamble, (2) a start code, and (3) control data. The preamble is a periodic bit stream that provides bit timing synchronization. Depending on the technology employed, the temporal duration or number of bits required to establish synchronization is on the order of a few bit periods. Once bit timing is achieved, the content in the remaining reference burst can be read. Following the preamble is a unique start code indicating the end of the preamble and the start of the information portion of the reference burst. When the word is recognized, control data can be interpreted correctly. In general, control data carries information such as station timing, call setup status, and signal information.

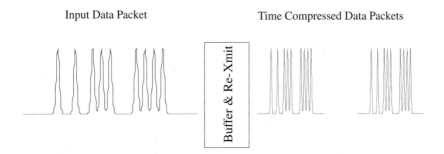

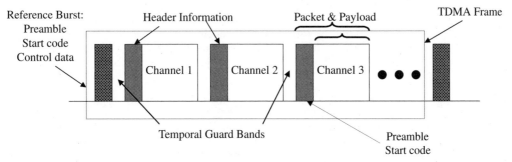

FIGURE 12 Representation illustrating the concepts of time-division multiple access, showing time-compressed data packets and the detailed layout of a TDMA packet, including header overhead and payload.

The reference burst in a TDMA frame is the overhead and occupies only a small portion of the frame. The remaining portion of the frame is divided into time slots separated by guard time bands. As in the reference burst, each time slot consists of a preamble, a unique start code, and the information payload. Owing to the different propagation delays between stations, the guard time between time slots is necessary to avoid overlap between two consecutive time slots. It should be noted that in TDMA networks, the transmitted data from the nodes must wait for time slots to become available. This occurs through an assigning process termed the *call setup*. Once a node obtains a time slot, it can use the same time slot in every frame for the duration of the communication session. In this case, the access is deterministic and as a result TDMA is generally used for constant-bit-rate transmission. While the node waits for the assigning of a time slot, the user stores its information into a buffer. Once the time slot is assigned, the bits are read out at a higher bit rate, and as a result the transmitted data bits have been compressed in time during the high-speed readout and transmission. When the input bits are stored and read out at a later time, a compression delay is introduced that is generally equal to the frame size. In real-time applications, it is critical to reduce the compression delay, and as a result the frame size should be as small as possible. However, since each frame has associated overhead in the preamble burst, the bandwidth or access efficiency is reduced. As a result, there is a trade-off between the network access efficiency and the compression delay.

Optical Domain TDMA

Even though there is an inherent trade-off between network access efficiency and compression delay, OTDMA is very attractive owing to the lack of any global, or network-wide, synchronization needs. As a result, the independent receiver nodes can have independent bit clocks. In an optical implementation, OTDMA bit rates are usually high, and this clock independence makes this approach attractive. One embodiment of an optical domain TDMA network is schematically illustrated in Fig. 13. To synchronize access, master frame timing needs to be distributed to all nodes. To achieve this, one of the nodes in the network, called the *master node,* generates a reference burst every T seconds, where T is the duration of a frame. Having the receiving nodes detect the reference burst means that the frame timing can be known at all receiving nodes; if the number of slots per frame is also known, the slot timing is obtained.

To allow the data to be received over a specific time slot in TDMA, a gate signal turns on during the slot interval, which is generated from the derived slot timing. As shown in Fig. 13, data in this slot interval can pass through the gate, be detected, and then be stored in the decompression buffer. The received slot timing derived is also sent to the local transmitter to determine its slot timing for transmission. The optical TDMA signal is first photodetected and then detected during a given slot interval. Data in all other time slots is suppressed by the gating operation. To preserve the received signal waveform, the bandwidth of the gating device is required to be much larger than the instantaneous bit rate. As a result, the bandwidth of the gate can limit the total TDMA throughput. To solve this problem, the gating function can be performed in the optical domain, whereby an electrooptical gate is used for a larger transmission bandwidth.

Time-Division Multiplexing

Historically, time-division multiplexing was first used in conventional digital telephony, where multiple lower-bit-rate digital data streams are interleaved in the time domain to form a higher-rate digital signal. These lower-bit-rate signals are referred to as *tributary signals.* Like TDMA, TDM is a time-domain multiple access approach, and each of its frames consists of a specific number of time slots. In contrast to the case with TDMA, data carried by different

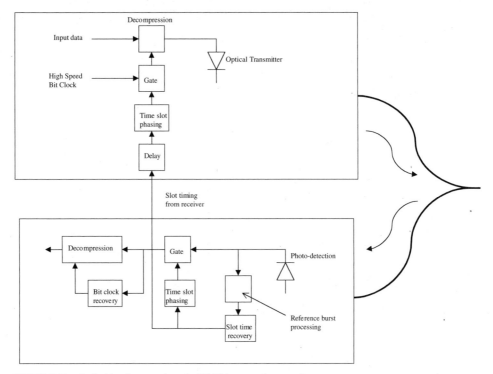

FIGURE 13 Optical implementation of a TDMA transmitter receiver.

slots is first synchronized in bit timing and then interleaved by a higher bit clock. This process of bit timing synchronization, called *frequency justification,* is necessary when upstream signals have different bit clock frequencies. Owing to the fact that all the tributary signals that feed into the overall network are synchronized at the bit level, no temporal guard band is required between different time slots, as is needed in the TDMA approach. In addition, a preamble signal at the beginning of each time slot is not required. As a result, if bit-level temporal synchronization is achievable, TDM is a better choice than TDMA, since the access and network bandwidth efficiency is higher (i.e., there are no wasted time slots used for preamble and guard band signals).

In TDM, lower-bit-rate signals are bit or byte interleaved into a higher-bit-rate signal. Accordingly, the multiplexed output consists of time slots, each of which carries one bit or byte for one input signal. To demultiplex time slots or to recognize which slots belong to which original inputs at the receiver end, time slots are grouped into frames that have additional overhead bits for frame and slot synchronization. As shown in Fig. 14, the number of time slots in a frame is equal to the total number of input signals, and when one input gets access to one slot, it continues to use the same slot in each frame for transmission. To multiplex a number of independent signals in TDM, the input signals must have the same bit clock. If there is any frequency mismatch between the bit rate of the independent signals, a premultiplexing signal processing step is required that adjusts the input bit rate of the signals to a common or master clock. This premultiplexing signal-processing step is referred to as *frequency justification* and can generally be achieved by adding additional bits to the frame, or by slip control, which may drop a byte and retransmit that byte in the next assigned time slot. These preprocessing steps of temporally aligning a number of independent signals to a common clock form one of the key challenges in high-bit-rate optical TDM systems, and for some applications this is a major drawback.

Owing to the fact that time-division multiplexing requires bit-timing synchronization, its implementation is more involved and complex. In order to synchronize the bit rates of the input signals, timing is generally performed at low bit rates directly on the input electrical signals. In order to facilitate the timing synchronization of the lower-bit-rate electrical signals that will ultimately be transmitted optically, an electronic synchronization standard has been developed that is referred to as the *synchronous optical network* (SONET) or the *synchronous digital hierarchy* (SDH). The key concept behind this synchronization process is the use of a floating payload, which eases the requirements of frequency justification, bit stuffing, and slip control.

Frame and Hierarchy

Like TDMA, TDM has a frame structure for data transmission and is composed of time slots that carry information, or data, from the lower-bit-rate or *tributary* signal. Since there is no temporal guard band or preamble signal for TDM time slots, the amount of data within a TDM time slot is generally one byte. While there is less overhead in TDM, this approach nonetheless does require the transmission of bits that assist in synchronization for the identification of frame boundaries and frequency justification, signaling for the setup and maintenance of the circuit connection, and maintenance bits for error correction and bit error rate monitoring.

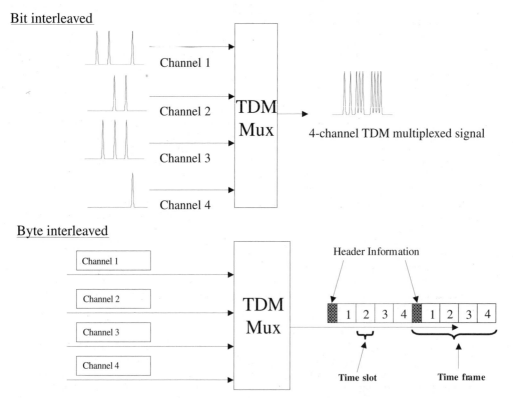

FIGURE 14 Representation illustrating the concepts of time-division multiplexing, showing schemes based on bit and byte interleaving.

In conventional TDM networks, two primary digital multiplexed systems are the 24- and 30-channel pulse-code-modulation formats for voice signals. In the 24-channel PCM-TDM format, 24 8-bit voice channels are time multiplexed to give 192 bits per frame, where each frame has a duration of 125 μs. One additional bit is inserted to provide frame synchronization, resulting in a total of 193 bits per frame. With a sampling rate of 8 kHz for standard voice communications, the overall clock rate is 1.544 Mbit/s; this is referred to as a *T1* signal or frame. Signaling information is usually transmitted over the eighth bit of the code word. A simplified block diagram of a 24-channel PCM coder/decoder is shown in Fig. 15.

A counterpart to the T1 frame of the 24-channel PCM-TDM is the 30-channel system, most generally deployed in Europe and referred to as the CEPT1 30-channel system. In this system, the frame size is also 125 μs, but each frame consists of 32 slots, with two slots (0 and 16) used for framing and signaling while the remaining 30 slots are used to carry 30 64kbit/s channels. From this design, the resulting bit rate of CEPT1 is 2.048 Mbit/s.

In TDM systems and telephony, the network is configured hierarchically—that is, higher-rate signals are multiplexed into continually higher-rate signals. In the AT&T digital hierarchy, the 24-channel PCM-TDM signals or T1 carriers are used as the basic system, and higher-order channel banks, referred to as T2, T3, and T4, are obtained by combining the lower-order channel banks. The multiplexing hierarchy is illustrated for both 24- and 30-channel systems in Fig. 16.

SONET and Frequency Justification

The synchronous optical network (SONET) is a TDM standard for transmission over optical fibers in the terrestrial United States. An international standard operating with the same

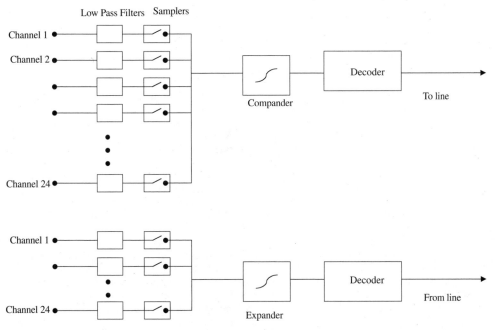

FIGURE 15 Schematic showing a 24-channel TDM transmitter/receiver. Included are compander/expander modules that compensate for quantization error and increase the system signal-to-noise ratio.

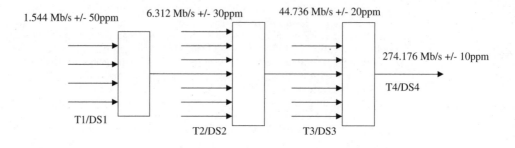

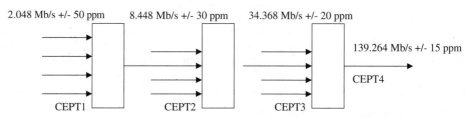

FIGURE 16 Schematic representation of the telephonic AT&T digital hierarchy and its European counterpart, CEPT.

underlying principles is called the *synchronous digital hierarchy* (SDH). These transmission standards were designed to simplify the process of frequency justification so that multiplexing and demultiplexing can be done at high speeds. To achieve this goal, SONET introduces the concept of a *floating payload,* where the information part of the packet floats with respect to the header information and the overall frame and the location of the payloads are identified by a process called *pointer processing.* A SONET frame has a two-dimensional frame structure to assist in examining its logical structure (see Fig. 17*a*). The sequence of data on the transmission line is obtained by traversing the table row by row, moving from left to right. The frame consists of 90 columns by nine rows. Since SONET transmission is to be compatible with voice communications, the frame duration is 125 µs, to be consistent with carrying at least one 8-bit digital sample of a voice channel. Therefore the basic bit rate of a SONET channel is $90 \times 9 \times 64$ kbit/s or 51.84 Mbit/s. This basic SONET signal is called synchronous transmission signal (STS)-1. STS-1 is the lowest rate in SONET, with all other SONET signals being multiples of this basic rate. It should be noted that the international version of SONET (SDH) has a two-dimensional frame structure of nine rows and 270 columns, existing for 125 µs, making the nominal SDH rate 3 times higher than that for SONET, or 155.52 Mbit/s. In this case STS-3 for SONET operates at the same rate as STS-1 (synchronous transport module) for SDH. When SONET signals are used to modulate a laser diode, the signals are then referred to as optical carrier (OC)-*N* signals.

In the SONET framing structure, the first four columns contain overhead information, and the remaining 86 columns contain the information payload. The fourth column and the remaining 86 columns make up a structure called the *synchronous payload envelope* (SPE). The salient feature of the SONET transmission is that the SPE can float with respect to the SONET frame—that is, the first byte of the SPE can be located anywhere within the 9×87 area. As one reads the SONET frame from left to right and top to bottom, the location of the overhead information is repeated in the same place in each frame. If these framing bytes continue to be present at the appropriate time, there is an extremely high probability that the signal is the framing signal and that the alignment of all other bytes is known. To identify the specific position of each payload, pointer processing becomes the critical aspect of SONET

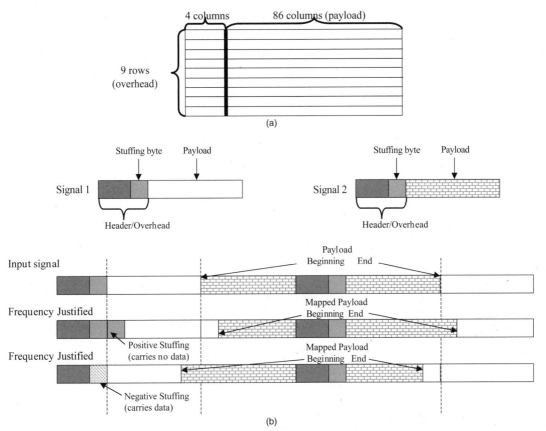

FIGURE 17   (*a*) The two-dimensional data structure of a TDM frame for SONET transmission. (*b*) The process of frequency justification, showing both positive and negative bit stuffing, to accommodate signals that are not at the same bit clock at a multiplexer.

transmission. In the classic T1 hierarchy, lower-speed signals generally arrive at the multiplexer at an arbitrary position with respect to their frame boundaries. The input data is then buffered to allow all the incoming signals to be aligned with the frame of the high-speed multiplex signal. These buffers were also necessary to allow for slight differences in clocks in the transmission lines that feed the multiplexer. The payload pointer eliminates the need for these buffers by providing a specific set of overhead bytes whose value can be used to determine the offset of the payload from the frame boundary.

The floating SPE concept and the use of pointer processing were developed to facilitate simpler implementation of frequency justification. In contrast to T carriers, where a tributary input at a multiplexer is frequency justified with respect to the frame of its next higher hierarchy, SONET performs frequency justification at the lowest STS-1 level. For example, when *N* STS-1 signals are multiplexed, the overhead of the input signals is removed, and the payloads of each input signal are mapped to the synchronous payload envelope (SPE) of the internal STS-1 signal of the multiplexer. Since each input signal is now synchronized and frequency justified after mapping to the internal STS-1 signal and its local clock, all *N* STS-1 signals can now be byte interleaved, resulting in a nominal outgoing bit rate of *N* times STS-1 for an STS-N signal. When *M* STS-N signals are multiplexed, each STS-N signal is first demultiplexed into *N* STS-1 signals, each of which is then frequency justified by the STS-1 clock of the multiplexer. Byte interleaving can then be done for the $M \times N$ STS-1 signals. It should be noted that for T1 carriers, multiplexing occurs with four T1 signals to create a T2 signal, with seven T2 signals to create a T3 signal, and so on. This hierarchical multiplexing complicates the bit clock relationship at the different T-carrier levels.

To illustrate the process of frequency justification, consider the interleaving of a TDM packet with slightly different bit clocks as compared to the local bit clock of the multiplexer, as illustrated in Fig. 17b. In order to handle the possibility of each signal having a slightly different bit rate, the frame structure must possess extra space, or stuffing bits, to accommodate this difference. If the two signals, signal 1 and signal 2, have the same bit clock and as a result are frequency justified, only the payloads are copied to the outgoing frame. If the bit clocks are different, both payloads cannot fit within the outgoing frame, owing to bit conservation. In the case where the input bit clock of signal 1 has a higher bit rate than that of signal 2, the stuffing space from the header of signal 2 must be used to carry payload data from signal 1. Since the payloads of each signal possess the same number of bits, there is a one-byte shift in the mapping, that is, the start of the payload of signal 2 is advanced by one byte and floats with respect to the header. If, on the other hand, signal 1 has a lower bit rate than signal 2, an extra *dummy* byte is inserted into the payload of signal 2, and the mapping is delayed for one byte. Given these two extremes, it is clear that the payload floats with respect to the header within the TDM frame and can advance or be delayed to accommodate the timing difference between the signals.

12.4   *INTRODUCTION TO DEVICE TECHNOLOGY*

Thus far, a general description of the concepts of digital communications and the salient features of TDM and TDMA has been presented. Next we address specific device technology that is employed in OTDM networks (e.g., sources, modulators, receivers, clock recovery oscillators, demultiplexers, and so on) to provide an understanding of how and why specific device technology may be employed in a system to optimize network performance, minimize cost, or provide maximum flexibility in supporting a wide variety of user applications.

Optical Time-Division Multiplexing—Serial vs. Parallel

Optical time-division multiplexing can generally be achieved by two main methods. The first method is referred to as *parallel multiplexing;* the second method is classified as *serial multiplexing*. These two approaches are schematically illustrated in Fig. 18. The advantage of the parallel type of multiplexer is that it employs simple, linear passive optical components, not including the intensity modulator, and that the transmission speed is not limited by the modulator or any other high-speed switching element. The drawback is that the relative temporal delays between each channel must be accurately controlled and stabilized, which increases the complexity of this approach. Alternatively, the serial approach to multiplexing is simple to configure. In this approach a high-speed optical clock pulse train and modulation signal pulses are combined and introduced into an all-optical switch to create a modulated channel on the high-bit-rate clock signal. Cascading this process allows all the channels to be independently modulated, with the requirement that the relative delay between each channel must be appropriately adjusted.

Device Technology—Transmitters

For advanced lightwave systems and networks, it is the semiconductor laser that dominates as the primary optical source that is used to generate the light that is modulated and transmitted as information. The reason for the dominance of these devices is that they are very small, typically a few hundred micrometers on a side; that they achieve excellent efficiency in converting electrons to photons; and that their cost is low. In addition, semiconductor diode lasers

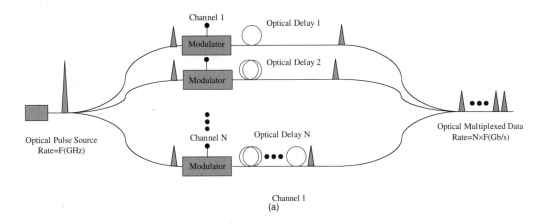

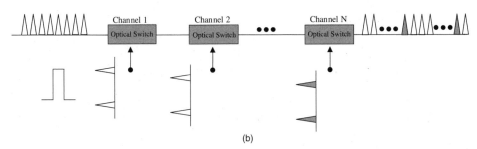

FIGURE 18 Schematic of optical time-division multiplexing for interleaving high-speed RZ optical pulses. (*a*) Parallel implementation. (*b*) Serial implementation.

can generate optical signals at wavelengths of 1.3 and 1.55 μm. These wavelengths are important because they correspond to the spectral regions where optical signals experience minimal dispersion (spreading of the optical data bits) and minimal loss.

These devices initially evolved from simple light-emitting diodes (LEDs) composed of a simple *p-n* junction, to Fabry-Perot (FP) semiconductor lasers, to distributed feedback (DFB) lasers and distributed Bragg reflector (DBR) lasers, and finally to mode-locked semiconductor diode lasers and optical fiber lasers. A simple description of each of these devices is given in the following text, along with advantages and disadvantages that influence how these optical transmitters are deployed in current optical systems and networks.

Fabry-Perot Semiconductor Lasers

Generally, the light-emitting diode is the simplest of all forms of all semiconductor light sources. These devices are quite popular for displays and indicator lights. Their use, however, is limited for communication and signal processing owing to the low modulation speeds and resulting low bandwidths achievable with these devices. In addition, owing to the fact that LEDs emit with a relatively broad optical spectrum, typically 10 to 30 nm, effects such as chromatic dispersion in the optical fiber tend to temporally broaden the optical bits and add additional constraints to the data transmission rates achievable with these devices. As a result, LEDs have a limited use in telecommunications, even though the device structure is quite simple and the cost is very low. Given this, it is the simple Fabry-Perot semiconductor laser that will be initially considered as a potential source for OTDM systems and networks.

The Fabry-Perot semiconductor laser diode is made up of a semiconductor *p-n* junction that is heavily doped and fabricated from a direct-gap semiconductor material. The injected current is sufficiently large to provide optical gain. The optical feedback is provided by mirrors, which are usually obtained by cleaving the semiconductor material along its crystal planes. The large refractive index difference between the crystal and the surrounding air causes the cleaved surfaces to act as reflectors. As a result, the semiconductor crystal acts both as the gain medium and as an optical resonator or *cavity* (see Fig. 19). Provided that the gain coefficient is sufficiently large, the feedback transforms the device into an optical oscillator or laser diode. It should be noted that the laser diode is very similar to the light-emitting diode. Both devices have a source of pumping energy that is a small electric current injected into the *p-n* junction. To contrast the devices, the light emitted from the LED is generated from spontaneous emission, whereas the light produced from an FP laser diode is generated from stimulated emission.

To contrast semiconductor lasers with conventional gas laser sources, the spectral width of the output light is quite broad for semiconductor lasers owing to the fact that transitions between electrons and holes occur between two energy bands rather than two well-defined discrete energy levels. In addition, the energy and momentum relaxation processes in both conduction and valence band are very fast, typically ranging from 50 fs to 1 ps, and the gain medium tends to behave as a homogeneously broadened gain medium. Nonetheless, effects such as spatial hole burning allow the simultaneous oscillation of many longitudinal modes. This effect is compounded in semiconductor diode lasers because the cavity lengths are short and, as a result, have only a few longitudinal modes. This allows the fields of different longitudinal modes, which are distributed along the resonator axis, to overlap less, thereby allowing partial spatial hole burning to occur. Considering that the physical dimensions of the semiconductor diode laser are quite small, the short length of the diode forces the longitudinal mode spacing $c/2nL$ to be quite large. Here c is the speed of light, L is the length of the diode chip, and n is the refractive index. Nevertheless, many of these modes can generally fit within the broad gain bandwidth allowed in a semiconductor diode laser. As an example, consider an FP laser diode operating at 1.3 µm, fabricated from the InGaAsP material system. If $n = 3.5$ and $L = 400$ µm, the modes are spaced by 107 GHz, which corresponds to a wavelength spacing of 0.6 nm. In this device, the gain bandwidth can be 1.2 THz, corresponding to a wavelength spread of 7 nm, and as many as 11 modes can oscillate. Given that the mode spacing can be modified by cleaving the device so that only one axial mode exists within the gain bandwidth, the resulting device length would be approximately 36 µm, which is difficult to

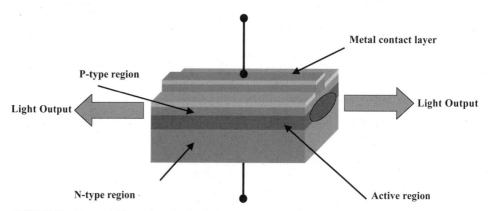

FIGURE 19 Schematic illustration of a simple Fabry-Perot semiconductor diode laser.

achieve. It should be noted that if the bias current is increased to well above threshold, the device can tend to oscillate on a single longitudinal mode. For telecommunications, it is very desirable to directly modulate the laser, thus avoiding the cost of an external modulator. However, in the case of direct modulation, the output emission spectrum will be multimode, and as a result, effects of dispersion will broaden the optical data bits and force the data rate to be reduced to avoid intersymbol interference. Given this effect, Fabry-Perot lasers tend to have a limited use in longer optical links.

Distributed Feedback Lasers

As indicated, the effects of dispersion and the broad spectral emission from semiconductor LEDs and semiconductor Fabry-Perot laser diodes tend to reduce the overall optical data transmission rate. Thus, methods have been developed to design novel semiconductor laser structures that will only operate on a single longitudinal mode. This will permit these devices to be directly modulated and allow for longer transmission paths since the overall spectral width is narrowed and the effect of dispersion is minimized.

There are several methods of achieving single-longitudinal-mode operation from semiconductor diode lasers. A standard semiconductor injection laser may be operated on a single transverse mode by reducing the waveguide's transverse dimensions, such as the width and height, while single-frequency operation may be obtained by reducing the length L of the diode chip so that the frequency spacing between adjacent longitudinal modes exceeds the spectral width of the gain medium. Other methods of single-mode operation include the use of a device known as a coupled-cleaved-cavity (C^3) laser, which is achieved by cleaving or etching a groove parallel to the end faces of the normal diode chip but placed between the end facets, thus creating two cavities. The standing-wave criteria must be satisfied by the boundary conditions at the surfaces of both cavities, and are generally only satisfied by a single frequency. In practice, however, the usefulness of this approach is limited by thermal drift, which results in both a wandering of the emission and abrupt, discrete changes in the spectral emission.

The preferred method of achieving single-frequency operation from semiconductor diode lasers is to incorporate frequency-selective reflectors at both ends of the diode chip, or alternately to fabricate the grating directly adjacent to the active layer. These two approaches result in devices referred to as *distributed Bragg reflector* (DBR) lasers and *distributed feedback* (DFB) lasers, respectively. In practice, it is easier to fabricate a single grating structure above the active layer as opposed to two separate gratings at each end. As a result, the DFB laser has become the laser of choice for telecommunications applications. These devices operate with spectral widths on the order of a few megahertz and have modulation bandwidths over 10 GHz. Clearly, the high modulation bandwidth and low spectral width make these devices well suited for direct modulation or on-off-keyed (OOK) optical networks. It should be noted that the narrow line width of a few megahertz is for the device operating in a continuous-wave mode, while modulating the device will necessarily broaden the spectral width.

In DFB lasers, Bragg reflection gratings are employed along the longitudinal direction of the laser cavity and are used to suppress the lasing of additional longitudinal modes. As shown in Fig. 20a, a periodic structure similar to a corrugated washboard is fabricated over the active layer, where the periodic spacing is denoted as Λ. Owing to this periodic structure, both forward- and backward-traveling waves must interfere constructively with each other. In order to achieve this constructive interference between the forward and backward waves, the round-trip phase change over one period should be $2\pi m$, where m is an integer and is called the order of the Bragg diffraction. With $m = 1$, the first-order Bragg wavelength λ_B is

$$2\pi = 2\Lambda(2\pi n/\lambda_B) \tag{9}$$

or

$$\lambda_B = 2\Lambda n \tag{10}$$

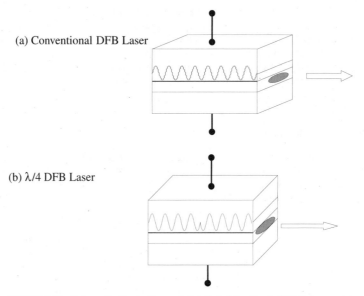

FIGURE 20 Schematic illustrations of distributed feedback (DFB) lasers. (*a*) Conventional DFB. (*b*) Quarter-wave DFB, showing the discontinuity of the Bragg grating structure to achieve single-wavelength operation.

where n is the refractive index of the semiconductor. Therefore, the period of the periodic structure determines the wavelength for the single-mode output. In reality, a periodic DFB structure generates two main modes symmetrically placed on either side of the Bragg wavelength λ_B. In order to suppress this dual-frequency emission and generate only one mode at the Bragg wavelength, a phase shift of $\lambda/4$ can be used to remove the symmetry. As shown in Fig. 20*b*, the periodic structure has a phase discontinuity of $\pi/2$ at the middle, which gives an equivalent $\lambda/4$ phase shift. Owing to the ability of the $\lambda/4$ DFB structure to generate a single-frequency, narrow spectral line width, these are the preferred devices for telecommunications at present.

Mode-locked Lasers

Mode-locking is a technique for obtaining very short bursts of light from lasers, and can be easily achieved employing both semiconductor and fiber gain media. As a result of mode-locking, the light that is produced is automatically in a pulsed form that produces return-to-zero (RZ) data if passed through an external modulator being electrically driven with non-return-to-zero data. More importantly, the temporal duration of the optical bits produced by mode-locking is much shorter than the period of the driving signal! In contrast, consider a DFB laser whose light is externally modulated. In this case, the temporal duration of the optical bits will be equal to the temporal duration of the electrical pulses driving the external modulator. As a result, the maximum possible data transmission rate achievable from the DFB will be limited to the speed of the electronic driving signal. With mode-locking, however, a low-frequency electrical drive signal can be used to generate ultrashort optical bits. By following the light production with external modulation and optical bit interleaving, one can realize the ultimate in OTDM transmission rates. To show the difference between a mode-locked pulse train and its drive, Fig. 21 plots a sinusoid and a mode-locked pulse train consisting of five locked optical modes.

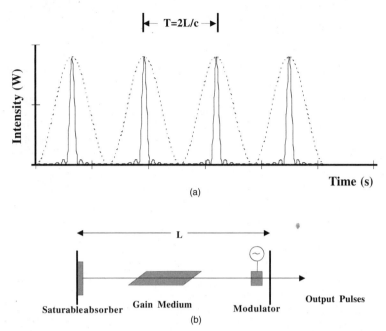

FIGURE 21 Optical intensity distribution of five coherent, phase-locked modes of a laser (*a*), and a schematic diagram of an external-cavity mode-locked laser (*b*). Superimposed on the optical pulse train is a typical sinusoid that could be used to mode-lock the laser, showing that much shorter optical pulses can be obtained from a low-frequency signal.

To understand the process of mode-locking, it should be recalled that a laser can oscillate on many longitudinal modes that are equally spaced by the longitudinal mode spacing $c/(2nL)$. Normally these modes oscillate independently; however, techniques can be employed to couple and lock their relative phases together. The modes can then be regarded as the components of a Fourier-series expansion of a periodic function of time of period $T = (2nL)/c$ that represents a periodic train of optical pulses. Consider for example a laser with multiple longitudinal modes separated by $c/2nL$. The output intensity of a perfectly mode-locked laser as a function of time t and axial position z with M locked longitudinal modes, each with equal intensity, is given by

$$I(t,z) = M^2 |A|^2 \frac{\sin c^2 [M(t-z/c)/T]}{\sin c^2 [(t-z/c)T]} \tag{11}$$

where T is the periodicity of the optical pulses and sin $c(x)$ is sin $(x)/x$. In practice, there are several methods of generating optical pulse trains by mode-locking. These generally fall into two categories: (1) active mode-locking and (2) passive mode-locking. In both cases, to lock the longitudinal modes in phase, the gain of the laser is increased above its threshold for a short duration by opening and closing a shutter that is placed within the optical cavity. This allows a pulse of light to form. Allowing the light to propagate around the cavity and continually reopening and closing the shutter at a rate inversely proportional to the round-trip time forms a stable, well-defined optical pulse. If the shutter is realized by using an external modulator, the technique is referred to as *active mode-locking,* whereas if the shutter is realized by a device or material that is activated by the light intensity itself, the process is called *passive mode-locking.* Both techniques can be used simultaneously; this is referred to as *hybrid mode-locking* (see Fig. 21*b*).

From the preceding equation, it is observed that the pulse duration is determined by the number of modes *M*, which in practice is generally limited by the gain bandwidth of the medium. Since the gain bandwidth of semiconductor and optical fiber lasers can be very broad, the resultant pulse width can be very short. In addition, since the modes are added coherently, the peak intensity is *M* times the average power, making these optical pulses sufficiently intense to induce nonlinear optical effects. Generally, high optical power in optical communication is useful for large signal-to-noise ratios in the detection process; however, other effects, such as nonlinear optical effects, can be detrimental. While nonlinear optical effects are typically avoided in data transmission, the peak intensity may exploit novel forms of optical propagation, such as optical soliton propagation. In addition, ultrafast all-optical switching and demultiplexing only become possible with such high-peak-intensity pulses. As a result, mode-locked semiconductor and fiber lasers may ultimately become the preferred laser transmitters for telecommunications.

Direct and Indirect Modulation

To transmit information in OTDM networks, the light output of the laser source must be modulated in intensity. Depending on whether the output light is modulated by directly modulating the current source to the laser or whether the light is modulated externally (after it has been generated), the process of modulation can be classified as either (1) direct or (2) indirect or external (see Fig. 22*a* and *b*). With direct modulation, the light is directly modulated inside the light source, while external modulation uses a separate external modulator placed after the laser source.

Direct modulation is used in many optical communication systems owing to its simple and cost-effective implementation. However, due to the physics of laser action and the finite response of populating the lasing levels owing to current injection, the light output under direct modulation cannot respond to the input electrical signal instantaneously. Instead, there are turn-on delays and oscillations that occur when the modulating signal, which is used as the pumping current, has large and fast changes. As a result, direct modulation has several undesirable effects, such as frequency chirping and line width broadening. In frequency chirping, the spectrum of the output generated light is time varying; that is, the wavelength and spectrum change over time. This is because as the laser is turned on and off, the gain is changed from a very low value to a high value. Since the index of refraction of the laser diode is closely related to the optical gain of the device, as the gain changes, so does its index. It is this time-varying refractive index that leads to frequency chirping, sometimes referred to as *phase modulation.* In addition, in Fabry-Perot lasers, if the device is turned on and off, the temporal behavior of the spectrum will vary from being multimode to nearly single mode within an optical bit, leading to line width broadening. The line width broadening results from measuring the time-integrated optical spectrum. In this case, since the instantaneous frequency or spectral width of the laser source varies rapidly over time, a measurement of the optical spectrum over a time interval that is long compared to the instantaneous frequency changes results in a broadened spectral width of the source as compared to a continuous wave measurement.

External Modulation

External modulation provides an alternative approach to achieving light modulation with the added benefit of avoiding the undesirable frequency chirping effects in DFB lasers and mode partition noise in FP lasers associated with direct modulation. A typical external modulator consists of an optical waveguide in which the incident light propagates through and the refrac-

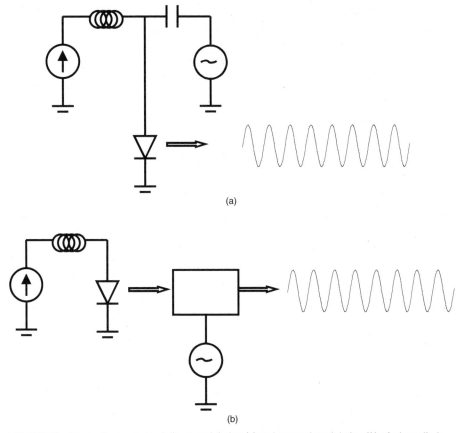

FIGURE 22   Illustrative example of direct modulation (*a*) and external modulation (*b*) of a laser diode.

tive index or absorption of the medium is modulated by a signal that represents the data to be transmitted. Depending on the specific device, three basic types of external modulators can be used: (1) electrooptic, (2) acoustooptic, and (3) electroabsorption (EA). Generally, acoustooptic modulators respond slowly—on the order of several nanoseconds—and as a result are not used for external modulators in telecommunications applications. Electroabsorption modulators rely on the fact that the band edge of a semiconductor can be frequency shifted to realize an intensity modulation for a well-defined wavelength that is close to the band edge of the modulator. Linear frequency responses up to 50 GHz are possible; however, the fact that the wavelength of the laser and the modulator must be accurately matched makes this approach more difficult to implement with individual devices. It should be noted, however, that EA modulators and semiconductor lasers can be integrated in the same devices, helping to remove restrictions on matching the transmitter's and modulator's wavelengths.

The typical desirable properties of an external modulator from a communications perspective are a large modulation bandwidth, a large depth of modulation, a small insertion loss (loss of the signal light passing through the device), and a low electrical drive power. In addition, for some types of communication TDM links, a high degree of linearity between the drive signal and modulated light signal is required (typical for analog links), and an independence of input polarization (polarization diversity) is desired. Finally, the low costs and small sizes of these devices make them extremely useful for cost-effective and wide-area deployment.

Electrooptic Modulators

An electrooptic modulator can be a simple optical channel or waveguide propagated by the light to be modulated. The material that is chosen to realize the electrooptic modulator must possess an optical birefringence that can be controlled or adjusted by an external electrical field that is applied along or transverse to the direction of propagation of the light to be modulated. This birefringence means that the index of refraction is different for light that propagates in different directions in the crystal. If the input light has a well-defined polarization state, this light can be made to see, or experience, different refractive indexes for different input polarization states. By adjusting the applied voltage to the electrooptic modulator, the polarization can be made to rotate or the speed of the light can be slightly varied. This modification of the input light property can be used to realize a change in the output light intensity by the use of a crossed polarizer or by interference of the modulated light with an exact copy of the unmodulated light. This can easily be achieved by using a waveguide interferometer, such as a Mach-Zehnder interferometer. If the refractive index is directly proportional to the applied electric field, the effect is referred to as *Pockel's effect.* In contrast, if the refractive index responds to the square of the applied electric field, the effect is referred to as the *Kerr effect.* This second effect has an interesting implication for all optical switching and modulation, since the intensity of a light beam is proportional to the square of the electric field and can therefore be used as a driving signal to modulate a second light beam.

Generally, for high-speed telecommunications applications, device designers employ the use of the electrooptic effect as a phase modulator in conjunction with an integrated Mach-Zehnder interferometer or an integrated directional coupler. Phase modulation (or delay/retardation modulation) does not affect the intensity of the input light beam. However, if a phase modulator is incorporated in one branch of an interferometer, the resultant output light from the interferometer will be intensity modulated. Consider an integrated Mach-Zehnder interferometer in Fig. 23. If the waveguide divides the input optical power equally, the transmitted intensity is related to the output intensity by the well-known interferometer equation $I_o = I_i \cos^2(\phi/2)$, where ϕ is the phase difference between the two light beams and the transmittance function is defined as $I_o/I_i = \cos^2(\phi/2)$.

Owing to the presence of the phase modulator in one of the interferometer arms, and with the phase being controlled by the applied voltage in accordance with a linear relation for

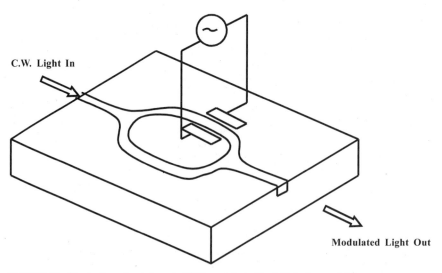

C.W. Light In

Modulated Light Out

FIGURE 23 Illustration of an integrated lithium niobate Mach-Zehnder modulator.

Pockel's effect, $\phi = \phi_o - \pi V/V\pi$. In this equation, ϕ_o is determined by the optical path difference between the two beams and $V\pi$ is the voltage required to achieve a π phase shift between the two beams. The transmittance of the device therefore becomes a function of the applied voltage V,

$$T(V) = \cos^2(\phi/2 - \pi V/2V\pi) \tag{12}$$

This function is plotted in Fig. 24 for an arbitrary value of ϕ_o. The device can be used as a linear intensity modulator by adjusting the optical path difference so that $\phi_o = \pi/2$ and conducting operation in the linear region near $T = 0.5$. In contrast, the optical phase difference may be adjusted so that ϕ_o is a multiple of 2π. In this case, $T(0) = 1$ and $T(V\pi) = 0$, so that the modulator switches the light on and off as V is switched between 0 and $V\pi$, providing digital modulation of the light intensity, or on-off keying (OOK). Commercially available integrated devices operate at speeds of up to 40 GHz and are quite suitable for OTDM applications such as modulation and demultiplexing.

Electroabsorption Modulators

Electroabsorption modulators are intensity modulators that rely on the quantum confined Stark effect. In this device, thin layers of semiconductor material are grown on a semiconductor substrate to generate a multiplicity of semiconductor quantum wells, or multiple quantum wells (MQW). For telecommunication applications, the semiconductor material family that is generally used is InGaAsP/InP. The number of quantum wells can vary, but is typically on the order of 10, with an overall device length of a few hundred micrometers. Owing to the dimensions of the thin layers, typically 100 Å or less, the electrons and holes bind to form excitons. These excitons have sharp and well-defined optical absorption peaks that occur near the band gap of the semiconductor material. When an electric field or bias voltage is applied in a direction perpendicular to the quantum well layers, the relative position of the exciton absorption peak can be made to shift to longer wavelengths. As a result, an optical field that passes through these wells can be preferentially absorbed, if the polarization of the light field is parallel to the quantum well layers. Therefore, the input light can be modulated by modu-

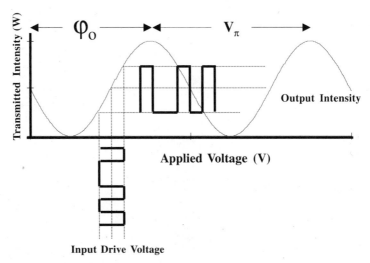

FIGURE 24 Input-output relations of an external modulator based on Pockel's effect. Superimposed on the transfer function is a modulated drive signal and the resultant output intensity from the modulator.

lating the bias voltage across the MQWs. These devices can theoretically possess modulation speeds as high as 50 GHz, with contrasts approaching 50 dB. A typical device schematic and absorption curve is shown in Fig. 25a and b.

Optical Clock Recovery

In time-division-multiplexed and multiple-access networks, it is necessary to regenerate a timing signal to be used for demultiplexing. A general discussion of clock extraction has already been given; in this section, an extension to those concepts is outlined for clock recovery in the optical domain. As in the conventional approaches to clock recovery, optical clock extraction has three general approaches: (1) the optical tank circuit, (2) high-speed phase-locked loops, and (3) injection locking of pulsed optical oscillators. The optical tank circuit can be easily real-

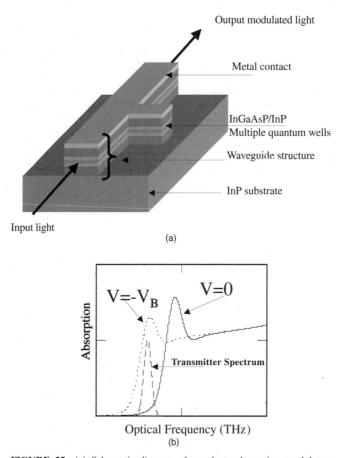

(a)

(b)

FIGURE 25 (*a*) Schematic diagram of an electroabsorption modulator. Light propagation occurs along the fabricated waveguide structure, in the plane of the semiconductor multiple quantum wells. (*b*) Typical absorption spectrum of a multiple quantum well stack under reverse bias and zero bias. Superimposed is a spectrum of a laser transmitter, showing how the shift in the absorption edge can either allow passage or attenuate the transmitted light.

ized by using a simple Fabry-Perot cavity. For clock extraction, the length L of the cavity must be related to the optical transmission bit rate. For example, if the input optical bit rate is 10 Gbit/s, the effective length of the optical tank cavity is 15 mm. The concept of the optical tank circuit is intuitively pleasing because it has many of the same features as electrical tank circuits—that is, a cavity Q and its associated decay time. In the case of a simple Fabry-Perot cavity as the optical tank circuit, the optical decay time or photon lifetime is given by

$$\tau_D = \frac{\tau_{RT}}{1 - R_1 R_2} \tag{13}$$

where τ_{RT} is the round-trip time given as $2L/c$, and R_1 and R_2 are the reflection coefficients of the cavity mirrors. One major difference between the optical tank circuit and its electrical counterpart is that the output of the optical tank circuit never exceeds the input optical intensity (see Fig. 26a).

A second technique that builds on the concept of the optical tank is optical injection seeding or injection locking. In this technique, the optical data bits are injected into a nonlinear device such as a passively mode-locked semiconductor laser diode (see Fig. 26b). The key difference between this approach and the optical tank circuit approach is that the injection-locking technique has internal gain to compensate for the finite photon lifetime, or decay, of the empty cavity. In addition to the gain, the cavity also contains a nonlinear element (e.g., a saturable absorber to initiate and sustain pulsed operation). Another important characteristic of the injection-locking technique using passively mode-locked laser diodes is that clock extraction can be prescaled—that is, a clock signal can be obtained at bit rates exactly equal to the input data bit rate or at harmonics or subharmonics of the input bit rate. In this case of generating a prescaled clock signal at a subharmonic of the input data stream, the resultant signal can be used directly for demultiplexing without any addition signal processing.

The operation of the injection seeded optical clock is as follows: The passively mode-locked laser produces optical pulses at its natural rate, which is proportional to the longitudinal mode spacing of the device cavity $c/(2L)$. Optical data bits from the transmitter are

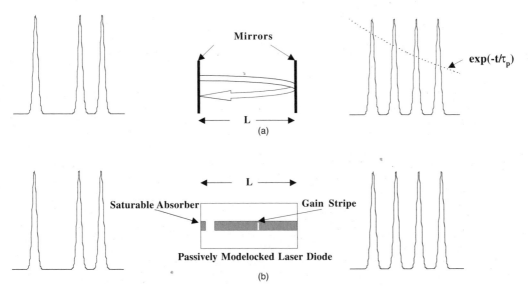

FIGURE 26   All-optical clock recovery based on optical injection of (*a*) an optical tank circuit (Fabry-Perot cavity) and (*b*) a mode-locked semiconductor diode laser.

injected into the mode-locked laser, where the data transmission rate is generally a harmonic of the clock rate. This criterion immediately provides the prescaling required for demultiplexing. The injected optical bits serve as a seeding mechanism to allow the clock to build up pulses from the injected optical bits. As the injected optical bits and the internal clock pulse compete for gain, the continuous injection of optical bits forces the internal clock pulse to evolve and shift in time to produce pulses that are synchronized with the input data. It should be noted that it is not necessary for the input optical bit rate to be equal to or greater than the nominal pulse rate of the clock—for example, the input data rate can be lower than the nominal bit rate of the clock. This is analogous to the transmitter sending data with primarily 0s, with logic 1 pulses occurring infrequently. The physical operating mechanism can also be understood by examining the operation in the frequency domain. From a frequency domain perspective, since the injected optical data bits are injected at a well-defined bit rate, the optical spectrum has a series of discrete line spectra centered around the laser emission wavelength and separated in frequency by the bit rate. Since the optical clock emits a periodic train of optical pulses, its optical spectrum is also a series of discrete line spectra separated by the clock repetition frequency. If the line spectra of the injected data bits fall within optical gain bandwidth of the optical clock, the injected line spectra will serve as seeding signals to force the optical clock to emit with line spectra similar to the injected signals. Since the injected data bits are repetitively pulsed, the discrete line spectra have the proper phase relation to force the clock to emit synchronously with the injected data.

It should be noted that the all optical clock recovery techniques discussed inherently rely on the fact that the transmitted optical data is in the return-to-zero (RZ) format. However, in present-day optical communication systems, non-return-to-zero (NRZ) is the line code that is primarily used. As shown in the preceding text, in the electrical domain there is a method to convert electrical NRZ signals to RZ signals by preprocessing using an exclusive OR logic function. In the optical domain, optical logic is possible but difficult to implement, so in theory a similar approach could be employed but would generally not be practical. Fortunately, by employing a simple optical interferometer, one can create a device that converts an optical NRZ signal to a pseudo-RZ signal that can be used for optical clock recovery. The pseudo-RZ signal is not an accurate transformation of the NRZ data, but only modifies the NRZ so that the resultant output RZ signal has the proper optical frequency components to allow for injection locking. To produce the required temporal delay, the format conversion uses a Mach-Zehnder interferometer that has an extra optical path in one arm. The interferometer is adjusted so that the output port is destructively interfering. Thus, when both signals are combined at the output, the output signal is zero. In contrast, when one signal or the other is present, a pulse exits the interferometer. This action nearly mimics the exclusive OR logic function. An example of how the format conversion is performed is schematically shown in Fig. 27; two optical configurations of its implementation are displayed in Fig. 28a and b. The benefit of this format conversion is that it employs simple linear optics; however, the interferometer is required to be stabilized for robust performance.

All-Optical Switching for Demultiplexing

In an all-optical switch, light controls light with the aid of a nonlinear optical material. It should be noted here that all materials will exhibit a nonlinear optical response, but the strength of the response will vary widely depending on the specific material. One important effect in an all-optical switch is the optical Kerr effect, whereby the refractive index of a medium is proportional to the square of the incident electric field. Since light is inducing the nonlinearity, or in other words providing the incident electric field, the refractive index becomes proportional to the light intensity. Since the intensity of a light beam can change the refractive index, the speed of a second, weaker beam can be modified owing to the presence of the intense beam. This effect is used extensively in combination with an optical interferometer to realize all-optical switching (see the section on electrooptic modulation using a Mach-Zehnder interferometer). Consider for example a Mach-Zehnder interferometer that

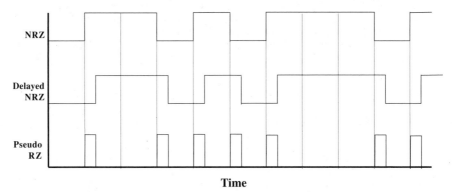

Time

FIGURE 27 Illustration of format conversion between NRZ and RZ line codes. The resultant RZ code is not a representation of the NRZ data, but a pseudo-RZ code that has RZ pulses located at all NRZ transitions.

includes a nonlinear optical material that possesses the optical Kerr effect (see Fig. 29). If data to be demultiplexed is injected into the interferometer, the relative phase delay in each area can be adjusted so that the entire injected data signal is present only at one output port. If an intense optical control beam is injected into the nonlinear optical medium and synchronized with a single data bit passing through the nonlinear medium, that bit can be slowed down such that destructive interference occurs at the original output port and constructive interference

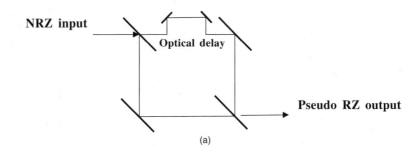

(a)

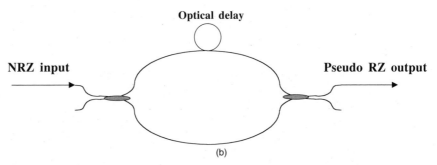

(b)

FIGURE 28 An optical implementation of an NRZ-to-RZ format converter, based on optical interference. (*a*) A simple interferometer demonstrating the operating principle. (*b*) A fiber-optic implementation of the format converter.

occurs at the secondary output port. In this case, the single bit has been switched out of the interferometer, while all other bits are transmitted.

Optical switches have been realized using optical fiber in the form of a Sagnac interferometer, and the fiber itself is used as the nonlinear medium. These devices are usually referred to as *nonlinear loop mirrors*. Other versions of all-optical switches may use semiconductor optical amplifiers as the nonlinear optical element. In this case, it is the change in gain induced by the control pulse that changes the refractive index owing to the Kramers-Kronig relations. Devices such as these are referred to as terahertz optical asymmetric demultiplexers (TOADs), semiconductor laser amplifier loop optical mirrors (SLALOMs), and unbalanced nonlinear interferometers (UNIs).

Receiver Systems

For high-speed optical time-division-multiplexed systems, key components are the optical receiver that detects the optical radiation, the associated electronic/optical circuitry that provides pre- or postamplification of the received signal, and the required clock recovery synchro-

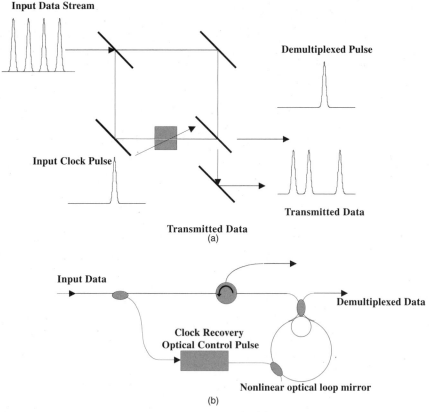

FIGURE 29 Schematic diagram of an all-optical switch. (*a*) A simple configuration based on a Mach-Zehnder interferometer and a separate nonlinear material activated by an independent control pulse. (*b*) An optical fiber implementation of an all-optical switch. This implementation relies on the inherent nonlinearity of the fiber that is induced by an independent control pulse.

nization for demultiplexing. In Fig. 30 is a typical arrangement for an optical receiver system. The incident lightwave signal is converted to an electrical signal by the optical receiver front end, which contains a photodetector and a preamplifier circuit. Usually, to enhance the receiver sensitivity, some means of increasing the average number of photoelectrons generated by the photodetector per incident photon is also included in the receiver setup. Schematically, this process is represented as a gain block G as shown in Fig. 30. This preamplification process can be accomplished in several ways. For example, in a direct detection system (i.e., one that directly detects the incident light), the most commonly adopted method is to use an avalanche photodiode (APD) as the photodetector. This type of detector provides a mechanism for electron multiplication that directly amplifies the detected signal electrically. Statistically, for every incident photon, the average number of photoelectrons generated by the APD is ηM, where η is the quantum efficiency of detection and M is the multiplication factor or avalanche gain, which is typically between 8 and 12 for most common receivers used in telecommunication systems.

An alternate method of realizing the preamplification process is to employ an optical amplifier, such as a semiconductor optical amplifier or an optical fiber amplifier, for example the erbium-doped fiber amplifier (EDFA). Owing to the large gain (which can be greater than 30 dB), the low noise characteristics, and the low insertion loss, the EDFA has been the predominant choice for implementing the optical preamplifier receiver in long-haul system experiments, particularly at high bit rates (>5 Gbit/s). The main disadvantage of employing EDFAs as optical preamplifiers are the high cost, high power consumption, and large size as compared to avalanche photodiodes. For moderate link lengths of less than 50 km, APDs are primarily employed.

In general, the preamplifier in the optical front end is an analog circuit. With a fixed-gain preamplifier, the front-end output signal level will follow the variation of the input optical power. This kind of signal level variation will impair the performance of the clock recovery and decision circuit subsystem, shown as the clock-data recovery block in Fig. 28. In addition, at low input power levels, the output signal level from the front end is usually not sufficiently high to be processed by the decision circuit, which typically requires an input peak-to-peak signal level of a few hundred millivolts. Therefore, a postamplifier is needed after the optical front end to minimize additional degradation in the performance of the CDR section. The main functions of this postamplifier are to provide an adequate signal amplification and to maintain a stable output signal level.

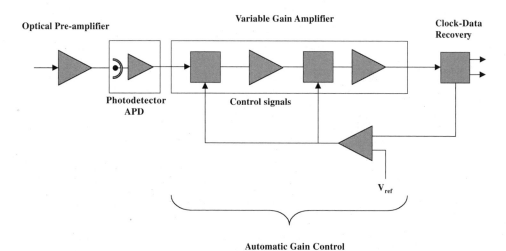

FIGURE 30 High-speed optical receiver showing input optical preamplification and control signals to achieve gain control to compensate for drift in received optical power.

There are two primary methods of creating an amplifier capable of providing the appropriate signal amplification and quantizing the required output signal level: (1) the use of a variable-gain amplifier (VGA) as a postamplifier and (2) the use of a limiting amplifier. When the variable-gain amplifier is employed as the postamplifier, its gain is adjusted according to the input signal power. This method is referred to as *automatic gain control* (AGC). An example of a variable-gain amplifier is created by cascading a chain of variable attenuators and fixed-gain amplifiers. A downstream power detector or peak detector is used to monitor the output signal level from the variable-gain amplifier. The power detector output is then compared with a predetermined reference voltage to generate the control signal, which will then adjust the amount of attenuation in the VGA. Therefore, under the closed-loop condition, the VGA automatically adjusts its overall gain to maintain a constant output signal level.

The second form of quantization amplifier is a limiting amplifier. The simplest form of the limiting amplifier can be pictured as a high-gain amplifier followed by a digital flip-flop. An ideal automatic-gain-controlled amplifier is, by definition, an analog circuit. The signal spectrum of an AGC output should be a scaled replica of that of the input signal. The limiting amplifier, on the other hand, is inherently a device with digital outputs. In practice, there are subtle differences in the rates at which errors may occur (*bit error rate*) and the distortion of the received bits (*eye margin*) between systems that use automatic-gain-controlled amplifiers and systems that employ limiting amplifiers.

While very high-speed optical fiber transmission experiments have been demonstrated, with data rates in excess of 100 Gbit/s (640 Gbit/s max for OTDM links), it should be noted that the key technologies that have made this possible are the use of optical amplifiers and all-optical multiplexers and demultiplexers. Nonetheless, electronic devices continue to have specific advantages over optical devices, such as high functionality, small size, low cost, and high reliability. In addition, as optics continues to push the limits of data transmission, integrated electronic technologies progress and have achieved integrated circuit performance in excess of 50 Gbit/s.

Ultra-High-Speed Optical Time-Division Multiplexed Optical Link—A Tutorial Example

To show how ultra-high-speed optoelectronic device technology can realize state-of-the-art performance in OTDM systems, an example is shown here that incorporates the system and device technology just discussed. While there are several groups that have created OTDM links operating in excess of 100 Gbit/s, generally using different device technology for pulse generation and subsequent demultiplexing, the basic concepts are consistent between each demonstration and are reproduced here to bring together the concepts of OTDM. In Fig. 31 is an ultra-high-speed optical data transmitter and demultiplexing system based on current research in Japan at NTT. To demonstrate the system performance, the transmitter was created by employing a 10-GHz mode-locked laser that uses an erbium-doped fiber as the gain medium and generates a 3-ps optical pulse. In this case, the laser is mode-locked by using a technique referred to as *regenerative mode-locking*. In regenerative mode-locking, the laser initiates mode-locking by using passive mode-locking techniques. A small portion of the resultant pulse train is detected and the subsequent electrical signal is then used to drive an active mode-locking device such as a $LiNbO_3$ modulator within the laser cavity. This approach derives the benefit of obtaining very short pulses that are typical of passively mode-locked lasers, but derives added temporal stability in the optical pulse train from the electrical drive signal. The data pulses are then modulated using a pseudorandom non-return-to-zero electrical signal that drives a 10-Gbit/s $LiNbO_2$ modulator. The optical pulses are then reduced in temporal duration using optical nonlinear effects in fiber (e.g., adiabatic soliton pulse compression), resulting in optical pulses of 250 fs in duration. The pulses are then temporally interleaved by a factor of 64 using a planar lightwave multiplexer. This device is simply a cascade of integrated Mach-Zehnder interferometers that employ a fixed delay equal to half of

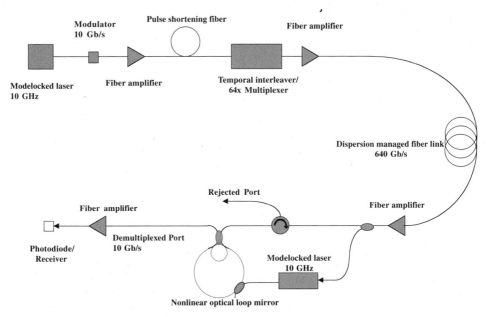

FIGURE 31 Schematic diagram of a 640-Gbit/s optical fiber link, using all the critical components described in this chapter—for example, mode-locked laser, high-speed modulator, temporal interleaver, optical clock recovery, all-optical switching for demultiplexing, and an optical photoreceiver.

the input repetition rate for each Mach-Zender stage. The resulting output from the planar lightwave circuit multiplexer (PLC MUX) is a pseudorandom modulated pulse train at 640 Gbit/s. It should be noted that, in a real system, the data from 64 individual users would need to be interleaved accurately for this scheme to work. However, this can be achieved by distributing or broadcasting the initial 10-Gbit/s pulse train to each user for modulation (recall Fig. 18a). Since the propagation distance to each user is fixed and deterministic, the return signals can be interleaved easily.

After the data has been generated and amplified in an EDFA to compensate for losses, the pulses are launched into a fiber span of approximately 60 km. It should be noted that the typical distance between central switching stations or central offices in conventional telecommunication networks is 40 to 80 km. The fiber span is composed of fiber with varying dispersion to reduce the effects of pulse broadening. The pulses at the demultiplexing end are initially amplified using an EDFA to again compensate for losses encountered in transmission. A portion of the data is extracted from the data line and is used to generate an optical clock signal that will be used as a control pulse in an all-optical switch/demultiplexer. The clock recovery unit is composed of a mode-locked laser that is electrically driven at a nominal data rate of 10 GHz. This clock signal could also be generated by injection locking of a passively mode-locked laser. Once the clock pulses are generated, they are injected into a nonlinear loop mirror, along with the data to provide all-optical demultiplexing at 10 Gbit/s. It should be noted that the clock and data pulses are at 1.533 and 1.556 μm, respectively. Owing to the wavelength difference between the clock data, a simple bandpass filter is used to pass the data stream and filter out the clock pulses after demultiplexing. Finally, after the data pulses are spectrally filtered, detection is performed with a photodetector, with the subsequent electrical signal analyzed by a bit-error-rate measurement system. The resulting performance of this system showed a bit error rate of 10^{-10}, that is, less than one error for every 10 billion bits received, with a received power of −23 dBm or 5 mW of average power.

12.5   *SUMMARY AND FUTURE OUTLOOK*

This chapter has reviewed the fundamental basics of optical time-division multiplexed communication networks, starting from an elementary perspective of digital sampling. The core basics of multiplexing and demultiplexing were then reviewed, with an emphasis on the difference between conventional time-division multiplexing for voice/circuit-switched networks versus data/packet-switched networks. Given this as an underlying background, specific device technology was introduced to show how the system functionality can be realized using ultra-high-speed optics and photonic technologies. Finally, as an example of how these system and device technologies are incorporated into a functioning ultra-high-speed optical time-division multiplexed system, a 640-Gbit/s link was discussed.

As we look toward the future and consider how the current state of the art will evolve to realize faster optical time-division networks and signal processing architectures, new concepts and approaches will be required. Several possible approaches may make use of optical solitons, which are optical pulses that can propagate without the detrimental effects of chromatic dispersion. An alternative approach for high-speed networking may incorporate a unique combination of both time-division multiplexing (TDM) and wavelength-division multiplexing (WDM), ushering in a new generation of hybrid WDM-TDM optical networking and signal processing devices and architectures. Generally speaking, however, in the present competitive industrial arena of telecommunications, there is always a technology-versus-cost trade-off. As a result, the technology that is ultimately deployed will be application specific to maximize performance and minimize cost.

12.6   *FURTHER READING*

Joseph E. Berthold, "Sonet and ATM," in Ivan P. Kaminow and Thomas L. Koch (eds.), *Optical Fiber Telecommunications IIIA,* Academic Press, San Diego, 1997, pp. 2-13 to 2-41.

D. Cotter and A. D. Ellis, "Asynchronous Digital Optical Regeneration and Networks," *J. Lightwave Technol. IEEE* **16**(12):2068–2080 (1998).

J. Das, "Fundamentals of Digital Communication," in Bishnu P. Pal (ed.), *Fundamentals of Fiber Optics in Telecommunication and Sensor Systems,* Wiley Eastern, New Delhi, 1992, pp. 7-415 to 7-451.

S. Kawanishi, "Ultrahigh-Speed Optical Time-Division-Multiplexed Transmission Technology Based on Optical Signal Processing," *J. Quantum Electron. IEEE* **34**(11):2064–2078 (1998).

H. K. Lee, J. T. Ahn, M. Y. Jeon, K. H. Kim, D. S. Lim, and C. H. Lee, "All-Optical Clock Recovery from NRZ Data of 10 Gb/s," *Photon. Technol. Lett. IEEE* **11**(6):730–732 (1999).

Max Ming-Kang Liu, *Principles and Applications of Optical Communications,* Irwin, Chicago, 1996.

K. Ogawa, L. D. Tzeng, Y. K. Park, and E. Sano, "Advances in High Bit-Rate Transmission Systems," in Ivan P. Kaminow and Thomas L. Koch (eds.), *Optical Fiber Telecommunications IIIA,* Academic Press, San Diego, 1997, pp. 11-336 to 11-372.

CHAPTER 13
WAVELENGTH DOMAIN MULTIPLEXED (WDM) FIBER-OPTIC COMMUNICATION NETWORKS

Alan E. Willner and Yong Xie
Department of EE Systems
University of Southern California
Los Angeles, California

13.1 INTRODUCTION

Optical communications have experienced many revolutionary changes since the days of short-distance multimode transmission at 0.8 μm.[1] We have seen, with the advent of erbium-doped fiber amplifiers (EDFAs), single-channel repeaterless transmission at 10 Gb/s across over 8000 km.[2] We may consider single-channel point-to-point links to be state-of-the-art and an accomplished fact, albeit with many improvements possible. (Soliton transmission, which has the potential for much higher speeds and longer distances, is discussed in Chapter 7.) Although single-channel results are quite impressive, they nonetheless have two disadvantages: (1) they take advantage of only a very small fraction of the enormous bandwidth available in an optical fiber, and (2) they connect two distinct end points, not allowing for a multiuser environment. Since the required rates of data transmission among many users have been increasing at an impressive pace for the past several years, it is a highly desirable goal to eventually connect many users with a high-bandwidth optical communication system. By employing wavelength-division multiplexing (WDM) technology, a simple multiuser system may be a point-to-point link with many simultaneous channels, and a more complicated system can take the form of a local, metropolitan, or wide-area network with either high bi-directional connectivity or simple unidirectional distribution.[3]

Much technological progress has been achieved in WDM optical systems since the emergence of EDFAs, a key enabling technology for WDM. Some potential applications of WDM technology include a multiplexed high-bandwidth library resource system, simultaneous information sharing, supercomputer data and processor interaction, and a myriad of multimedia services, video applications, and additional previously undreamed-of services. As demands increase for network bandwidth, the need will become apparent for WDM optical networks, with issues such as functionality, compatibility, and cost determining which systems will eventually be implemented. This chapter will deal with the many technical issues, possible solutions, and recent progress in the exciting area of WDM fiber-optic communication systems.

Fiber Bandwidth

The driving force motivating the use of multichannel optical systems is the enormous bandwidth available in the optical fiber. The attenuation curve as a function of optical carrier wavelength is shown in Fig. 1.[4] There are two low-loss windows, one near 1.3 μm and an even lower-loss one near 1.55 μm. Consider the window at 1.55 μm, which is approximately 25,000 GHz wide. (Note that due to the extremely desirable characteristics of the EDFA, which amplifies only near 1.55 μm, most systems would use EDFAs and therefore not use the dispersion-zero 1.3-μm band of the existing embedded conventional fiber base.) The high-bandwidth characteristic of the optical fiber implies that a single optical carrier at 1.55 μm can be baseband-modulated at ~25,000 Gb/s, occupying 25,000 GHz surrounding 1.55 μm, before transmission losses of the optical fiber would limit transmission. Obviously, this bit rate is impossible for present-day electrical and optical devices to achieve, given that even heroic lasers, external modulators, switches, and detectors all have bandwidths <100 GHz. Practical data links today are significantly slower, perhaps no more than 10 Gb/s per channel. Since a single high-speed channel only takes advantage of an extremely small portion of the available fiber bandwidth, an efficient multiplexing method is needed to take full advantage of the huge bandwidth offered by optical fibers. As we will see in this chapter, WDM has been proven to be the most appropriate approach.

Introduction to WDM Technology

In real systems, even a single channel will probably be a combination of many lower-speed signals, since very few individual applications today utilize this high bandwidth. These lower-speed channels are multiplexed together in time to form a higher-speed channel. This time-division multiplexing (TDM) can be accomplished in either the electrical or optical domain. In TDM, each lower-speed channel transmits a bit (or a collection of bits, known as a *packet*) in a given time slot and then waits its turn to transmit another bit (or packet) after all the other channels have had their opportunity to transmit. TDM is quite popular with today's electrical networks and is fairly straightforward to implement in an optical network at 10 Gb/s speeds. However, as was previously mentioned, this scheme by itself cannot hope to utilize

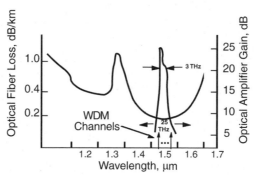

FIGURE 1   Fiber loss as a function of wavelength in conventional single-mode silica fiber. The gain spectrum of the EDFA is also shown.[4]

the available bandwidth because it is limited by the speed of the time-multiplexing and time-demultiplexing components. Moreover, ultra-high-speed transmission becomes severely limited by fiber dispersion and nonlinearities.[5]

To exploit more of the fiber's THz bandwidth, we seek solutions that complement or replace TDM. One obvious choice is WDM, in which several baseband-modulated channels are transmitted along a single fiber but with each channel located at a different wavelength (see Fig. 2.)[6–9] Each of N different-wavelength lasers is operating at the slower Gb/s speeds, but the aggregate system is transmitting at N times the individual laser speed, providing a significant capacity enhancement. The WDM channels are separated in wavelength in order to avoid crosstalk when they are (de)multiplexed by a nonideal optical filter. The wavelengths can be individually routed through a network or individually recovered by wavelength-selective components. WDM allows us to use much of the available fiber bandwidth, although various device, systems, and network issues will limit utilization of the full fiber bandwidth.

Figure 3 shows the continuous capacity growth in optical fiber systems over the past 15 years, where ETDM and OTDM represent electronic time-division multiplexing and optical time-division multiplexing, respectively. The highest demonstrated capacity has been achieved using WDM. One interesting point about this trend, predicted a decade ago by T. Li of AT&T Bell Labs, deserves mention. The trend, which has existed for more than a decade, is that the transmission capacity doubles every two years. WDM technology has provided the means for this trend to continue, and there is no reason to assume that WDM won't continue to produce dramatic progress. In fact, capacity of hundreds of Gb/s has recently been realized in commercial WDM systems.

13.2   FIBER IMPAIRMENTS

As was mentioned in Section 13.1, prior to the widespread use of EDFAs,[10] long-distance transmission required periodic regeneration of a signal. Such detection and retransmission compensate for both attenuation and fiber chromatic dispersion effects. However, when an EDFA is employed in the transmission line, it only compensates for attenuation, allowing other effects such as dispersion and nonlinearities to accumulate unimpeded along a transmission link. Although dispersion-shifted fiber (DSF) can be used in EDFA-based systems to minimize the effect of dispersion when channels are operated near the dispersion-zero wavelength, various nonlinear effects will also accumulate along the link and cause severe limitations in WDM transmission. These nonlinear effects include stimulated scattering processes (i.e., Raman and Brillouin scat-

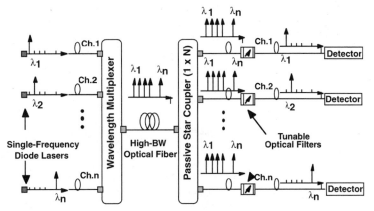

FIGURE 2   Many WDM channels propagating in a single optical fiber.

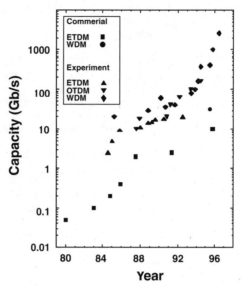

FIGURE 3 Continuous capacity growth in optical fiber transmission systems.

tering) and material refractive-index interactions [i.e., self- and cross-phase modulation (SPM and XPM) and four-wave mixing (FWM)].[11,12] The management of both fiber dispersion and nonlinearities is a key issue in optically amplified long-haul systems. In this section, we will describe these phenomena and discuss various techniques for compensating for their deleterious effects.

Chromatic Dispersion

Without fiber nonlinearity, the electric field of a linearly polarized signal propagating in single-mode fiber (SMF) can be described by

$$E(x, y, z, t) = \tfrac{1}{2}E(x, y, z_0, t_0)e^{i[[\beta(\omega) - \frac{\alpha}{2}]z - \omega t]} + \text{C.C.} \qquad (1)$$

where z is the propagation direction, t is the time, α is the loss of the fiber, $\beta(\omega)$ is the propagation constant, ω is the angular frequency, and C.C. is the complex constant. The mode propagation constant $\beta(\omega)$ can be expanded in a Taylor series about the center frequency ω_0,[11]

$$\beta(\omega) = \beta_0 + \beta_1(\omega - \omega_0) + \tfrac{1}{2}\beta_2(\omega - \omega_0)^2 + \cdots \qquad (2)$$

where

$$\beta_m = \left(\frac{d^m\beta}{d\omega^m}\right)_{\omega = \omega_0} \qquad (m = 0, 1, 2, \cdots) \qquad (3)$$

ω_0/β_0 is the phase velocity, whereas $1/\beta_1$ is the group velocity, which is the speed of the energy (optical pulse) propagation. The term β_2 describing the frequency dependence of the group velocity is the chromatic dispersion (or group velocity dispersion) of the fiber.

Since an optical pulse consists of different frequency components that travel at different speeds in the fiber due to chromatic dispersion, the pulse will be broadened as it propagates through the fiber, as shown in Fig. 4.

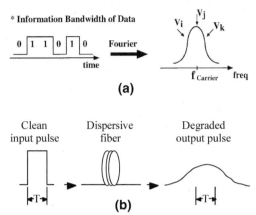

FIGURE 4 (*a*) Information bandwidth of data. (*b*) Broadening of an optical pulse propagating through a single-mode fiber.

In lightwave systems, it is more convenient to use

$$D = \frac{d}{d\lambda}\left(\frac{1}{v_g}\right) = \frac{d^2\beta}{d\lambda d\omega} \approx -\frac{2\pi c}{\lambda^2}\beta_2 \qquad (4)$$

in units of ps/nm/km, which is the amount of broadening of a pulse with a bandwidth of 1 nm after propagating through 1 km of fiber, instead of β_2, to represent chromatic dispersion. For conventional SMF D is typically 17 ps/nm/km.

In fact, chromatic dispersion is also wavelength dependent. We define

$$\frac{dD}{d\lambda} = \frac{2\pi c}{\lambda^3}\left(2\beta_2 - \frac{2\pi c}{\lambda}\beta_3\right) \qquad (5)$$

as the chromatic dispersion slope, which plays an important role in wide-band long-haul WDM systems.[13,14]

In a communication system, when a data stream composed of multiple optical pulses in series propagates through a dispersive fiber, each of the broadened pulses spreads into other time slots, causing intersymbol-inference after transmission, as shown in the simulation results for 10 Gb/s signal transmission in Fig. 5.

In Fig. 5, we can also see that the distortion due to chromatic dispersion is dependent on the data rates of the signal. Actually, fiber chromatic dispersion places limits on both channel data rates and transmission distance, since the pulse broadening is dependent on both of them. A simple estimation of the dispersion limit is:

$$L_D = \frac{1}{BD\Delta\lambda} \qquad (6)$$

where L_D is the transmission distance at which the pulse broadening exceeds one bit-time, B is the data rate, and $\Delta\lambda$ is the bandwidth of the signal launched into the fiber. Since $\Delta\lambda$ is linearly proportional to the data rate B, the dispersion-limited transmission distance is inversely proportional to B^2. Using 2.5 Gb/s, 17 ps/nm/km, and 0.025 nm for B, D, and $\Delta\lambda$, we estimate the dispersion limit for 2.5 Gb/s chirp-free modulation systems would be around 1000 km. However, when the data rate increases to 10 Gb/s, the limit decreases to 60 to 70 km. Consequently, after 100 km SMF transmission, although there is no degradation at 2.5 Gb/s, 10 Gb/s pulses are seriously distorted.

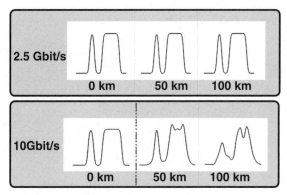

FIGURE 5   Signal waveform distortion due to fiber dispersion in 2.5 Gb/s and 10 Gb/s optical systems after 0, 50, and 100 km SMF transmission (assuming chirp-free modulation).

It is clear that in high data rate systems, fiber chromatic dispersion must be reduced to allow repeaterless long-haul transmission.[15] A straightforward way to overcome these dispersion limits is to shift the dispersion-zero wavelength of fiber to 1.55 μm.[16] The dispersion limit in DSF is dramatically increased compared to SMF. However, as we will discuss in next section, most fiber nonlinear effects would impose critical limitations on system performance when dispersion is low, which makes simultaneous management of both chromatic dispersion and fiber nonlinearities one of the most important issues in system design. Therefore, we will discuss dispersion compensation and management after a brief review of fiber nonlinearities.

Fiber Nonlinearities

Silica optical fiber has certain nonlinearities associated with it. The major nonlinear effects are each dependent on the power (i.e., intensity) of the propagating signals, with weak signals not incurring significant effects. Furthermore, these affects require a certain propagation distance to allow for the interactions to accumulate and become relevant.

A detailed discussion of fiber nonlinearities can be found in Chapter 3. In this section we will briefly describe system impairments due to some of the most important fiber nonlinearities.

Self-Phase Modulation and Cross-Phase Modulation   The nonlinearities discussed in this section and the next all owe their origin to the fact that the index of refraction of an optical fiber varies nonlinearly with signal power:[11]

$$n = n_{\text{linear}} + \overline{n}_2\left(\frac{P}{A_{\text{eff}}}\right) \tag{7}$$

where $\overline{n}_2$ is the nonlinear index coefficient (3.2×10^{-16} cm²/W for silica fiber), n_{linear} is the linear index of refraction, P is the optical power, and A_{eff} is the cross-sectional mode area.

Since the local refractive index is a function of the optical intensity of a propagating signal, a nonrectangular-shaped optical pulse will experience a varying refractive index depending on the optical power at each temporal location. As we know, the speed of an optical wave within a medium is dependent on the refractive index. Therefore, the varying refractive index will then cause the different intensities to propagate at different speeds along the fiber. Thus, optical power fluctuations are converted to phase fluctuations. This phase shift causes the pulse to temporally disperse in the fiber and limits transmission distance and signal speed.

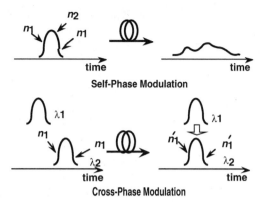

FIGURE 6 Pulse broadening due to SPM and XPM.

This effect for a single optical pulse is known as self-phase modulation (SPM) and will distort a transmitted optical pulse as shown in Fig. 6. Note that both temporal and spectral broadening result from this nonlinearity.

This effect of intensity-dependent propagation speeds is quite important when considering WDM transmission of several channels. The optical power from data pulses on one channel will affect the refractive index, propagation speed, and dispersion-induced distortion of data pulses on another channel if they are temporally colocated. Since each wave is located at a different transmission wavelength, the different channels will propagate at different speeds, and data pulses from one channel will propagate through the data pulses from another channel, causing an overall smearing of the distortion and dispersion from one channel to the next. This cross-channel interference is known as cross-phase modulation (XPM) (see Fig. 6). Note that XPM is not symmetric as two pulses pass through each other since optical power is not constant along a fiber span with lumped amplification. Moreover, the distortion caused by XPM is dependent on the local dispersion of fiber.[17] Increasing local dispersion can suppress the effects of XPM because the pulses on different channels tend to walk away faster, and the XPM can not accumulate.

Figure 7 shows the effects of XPM on the performance of 10 Gb/s systems. The signal-to-noise ratio (SNR) of the received signal is plotted as a function of the local dispersion in a

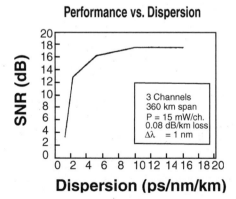

FIGURE 7 System performance in the presence of XPM.[17]

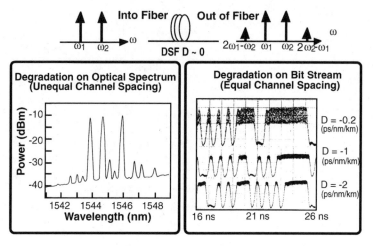

FIGURE 8 Four-wave mixing in WDM systems.[16]

dispersion-compensated system, showing that XPM effects could be suppressed by introducing an appropriate amount of local dispersion in the fiber.

Four-Wave Mixing Figure 8 shows the situation wherein two channels mix with each other, producing optical power at the sum and difference beat frequencies (see Chapter 3). It is important to note that FWM occurs only when (1) the two mixing channels are located near the dispersion-zero wavelength of the fiber, and (2) the channel spacing is less than a few 10s of GHz. These conditions exist since the two waves must maintain certain phase-matching requirements. The less critical effect of FWM is that the original channels suffer a power loss due to power transfer to the newly-generated products, and the more critical problem occurs when there are three or more original WDM signal channels producing many mixing products. It is quite probable that some of the products will appear at the same wavelength as one of the original signals. Then, when an optical filter ultimately recovers this original signal, both the original signal and a FWM product will be detected. The product has some modulation on it based on the combination of the channels that created it, and severe crosstalk results.[18,19] The equation describing the formation of FWM products between two waves A_1 and A_2 is:[20]

$$\frac{dA_1}{dz} = i\lambda_1 A_2^* A_3 A_4 e^{i\Delta k z}, \qquad \Delta k = k_3 + k_4 - k_1 - k_2 \qquad (8)$$

where * denotes the complex conjugate, and k is the phase of the propagating wave.

Figure 8 also shows the dependence of FWM effects on fiber dispersion.[16] Increasing the local dispersion value in transmission fibers helps to suppress the FWM-induced crosstalk, since high dispersion may destroy the phase-matching condition required for FWM generation.

Dispersion Compensation and Management

In the previous section, we saw that although fiber dispersion is generally considered a negative characteristic, it does have the desirable effect of inhibiting FWM and XPM. Therefore, it may not be good to reduce the fiber dispersion to zero by using DSF in systems. As an alternative, we keep the local dispersion along the transmission link high enough to suppress nonlinear effects, while managing the total dispersion of the link to be close to zero, as shown in

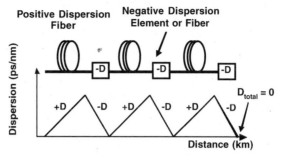

FIGURE 9 Sample dispersion map for a dispersion-managed system.

Fig. 9. We will introduce some of the most important dispersion compensation/management techniques in this section.

Dispersion Compensating Fiber Dispersion compensating fiber (DCF) has been the most successful method for dispersion compensation[21,22] because of its obvious advantage—wideband operation. Furthermore, as shown in Fig. 10, DCF can provide not only a very high negative dispersion, but also a negative dispersion slope,[21] which could mitigate the effect of dispersion accumulation on channels far from the dispersion-zero wavelength in WDM long-haul systems. The negative dispersion value for DCF is typically −80 to −100 ps/nm, so a relatively short length of DCF can compensate much longer lengths of SMF.

Tailoring the fiber's refractive index profile with the addition of germanium to the core generates the high negative dispersion of DCF. Therefore, losses in such fibers (typically 0.3 to 0.5 dB/km) are higher than standard SMF. Another approach for DCF is to guide the fundamental mode into the cladding to increase the waveguide dispersion, which minimizes the intrinsic loss. However, bending loss would be high for such designs, making it difficult to package the fibers for practical use. Moreover, the core size of DCF is usually much smaller than SMF (<4 μm compared to ~9 μm for SMF), leading to higher fiber-induced nonlinearity. Therefore, although DCF is commercially available and used in optical systems, people continue to seek better solutions to manage fiber dispersion.

Specialty Fibers for Dispersion Management A significant disadvantage of DCF is that it can not be employed as transmission fiber. Moreover, if DCF must be used, it would be better if the dispersion value of the transmission fiber were as small as possible to reduce the required DCF length, thus minimizing the induced loss and nonlinearity. The idea of dispersion management is to establish a long fiber link composed of alternating short lengths of two different types of fiber.[23] One type of fiber has the opposite sign but a similar dispersion mag-

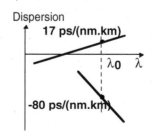

FIGURE 10 Dispersion compensating fiber.

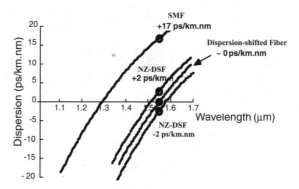

FIGURE 11   Dispersion of single-mode fibers.[15]

nitude as that of the other type of fiber for a given wavelength; note that these two types of fiber are similar except that their dispersion-zero wavelengths are shifted from each other. Each short length of fiber may have a dispersion value of, for instance, either +2 or −2 ps/(nm km), depending on which type of fiber is used. After a given distance of propagation, the total effective dispersion is close to zero, but there is an absolute value of dispersion at any given point along the fiber, reducing the impact of FWM and XPM.

Figure 11 shows the dispersion of non-zero dispersion-shifted fibers (NZ-DSF), which are specialty fibers designed for dispersion management. There are two types of NZ-DSF,[24] one having a dispersion-zero wavelength at about 1520 nm and another having a dispersion-zero wavelength at about 1580 nm. Their dispersion at 1550 nm is about +2 ps/nm/km or −2 ps/nm/km. These amounts of dispersion are lower than in standard SMF to mitigate the distortion due to accumulated dispersion, but higher than DSF to suppress fiber nonlinearities. Normally NZ-DSF has a smaller core area than SMF due to the different waveguide design, resulting in higher nonlinearity. However, with proper design, large-core NZ-DSF is also available.[25]

Some possible periodic dispersion maps for dispersion management include:[26,27]

- Positive NZ-DSF alternated with negative NZ-DSF
- Negative NZ-DSF alternated with standard SMF
- Positive NZ-DSF compensated with DCF

Chirped Fiber Bragg Grating   In the past few years, chirped fiber Bragg gratings (FBGs) have become very attractive devices for dispersion compensation.[28–30] The grating period of a chirped FBG varies linearly along the grating, causing light with different wavelengths to be reflected at different points inside the FBG. Therefore, the optical pulses broadened during transmission can be recompressed using the chirped FBG as shown in Fig. 12.

Figure 13 shows the reflection and time delay of a chirped FBG. The slope of the time delay curve represents the amount of dispersion provided by the grating. Chirped FBGs are advantageous due to their low loss, compact size, and polarization insensitivity. Additionally, they do not induce additional optical nonlinearities, which is a main drawback of DCF.

Achieving wideband operation has been one of the most important issues for chirped FBGs. Let's consider a 10-cm-long grating. The maximum time delay that can be provided by the grating is ~1000 ps across the whole grating bandwidth. Assuming the grating bandwidth to be 0.2 nm for single-channel operation, the dispersion provided by the grating could be as high as ~5000 ps/nm. However, if the grating bandwidth increases to, say, 6 nm for multichannel operation, then the dispersion would be <200 ps/nm, which is too low for dispersion compensation. It is possible to make gratings as long as 1 m to obtain wideband operation.[31]

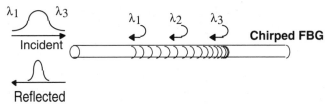

FIGURE 12   Dispersion compensation by a chirped FBG.

However, the fabrication process is difficult, and grating uniformity is hard to maintain. Another approach for wideband operation involves sampling the grating to generate multiple replicas of the grating reflection band.[32,33] Each of the replicas will have the same time-delay characteristics so that simultaneous dispersion compensation for multiple channels can be realized.

Tunable Dispersion Compensation   All of the dispersion compensators previously discussed share one common feature—they are not tunable. However, the accumulated dispersion for a data channel may vary in time for (1) dynamically reconfigurable optical networks in which a given channel may originate locally or far away, and (2) transmission systems having changing operating conditions (i.e., due to laser or modulator chirp). For such dynamic systems, especially at data rates 10 Gb/s, dispersion compensation must be adjustable in order to track dynamically the accumulated dispersion. For DCF, there is no method at present for tuning *in situ* its dispersion. For linearly-chirped FBGs, simply stretching the grating only produces a shift in the resonant wavelength range of the reflected band but does nothing to change the induced dispersion.

A recent report used 21 separate stretching segments to asymmetrically stretch a uniform FBG to change the induced dispersion,[34] but in this case the grating bandwidth varied significantly during dispersion tuning.

Another approach is to use a nonlinearly-chirped FBG.[35] As shown in Fig. 14, the time delay induced by this nonlinearly-chirped FBG changes with wavelength in a nonlinear fashion. The time-delay characteristics, bandwidth, and spectral shape of the grating do not change when the grating is stretched by a linear stretcher, but the signal channel will experience a varying dispersion value during stretching. The tuning speed is on the order of a ms, which is fast enough for circuit-switched systems.

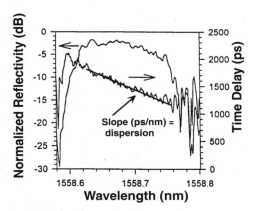

FIGURE 13   Reflection and time delay characteristics of a chirped FBG.

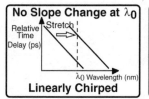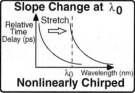

FIGURE 14   Concept of tunable dispersion compensation using a nonlinearly-chirped FBG.

Demonstration of dynamic dispersion compensation has also been accomplished by using a nonlinearly chirped FBG.[35] The eye diagrams before and after the dynamic dispersion compensation at both 50 and 104 km are shown in Fig. 15.

13.3   BASIC ARCHITECTURE OF WDM NETWORKS

We have explained how WDM enables the utilization of a significant portion of the available fiber bandwidth by allowing many independent signals to be transmitted simultaneously in one fiber, as well as the basic fiber impairments in WDM systems. Actually, the WDM channels can be routed and detected independently, with the wavelength determining the communication path by acting as the signature address of the origin, destination, or routing. Therefore, the basic system architecture that can take the full advantage of WDM technology is an important issue, and will be discussed in this section.

Point-to-Point Links

Figure 16 shows a simple point-to-point WDM system in which several channels are multiplexed at one node; the combined signals are transmitted across some distance of fiber; and the channels are demultiplexed at a destination node. This facilitates high-bandwidth fiber transmission. Add-drop multiplexers (ADMs) can be inserted in the transmission line to

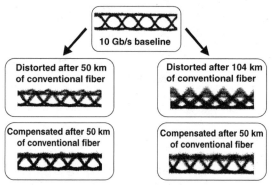

FIGURE 15   The results of tunable chromatic dispersion compensation with a nonlinearly-chirped FBG.

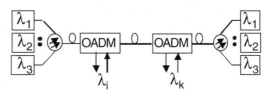

FIGURE 16 Point-to-point WDM links with optical ADMs.

allow the adding and dropping of one or more channels at specific nodes, forming a more efficient WDM link.

Wavelength-Routed Networks

Figure 17 shows a more complex multiuser WDM network structure, where the wavelength is used as the signature address for either the transmitters or the receivers, and determines the routing path through an optical network. In order for each node to be able to communicate with any other node and facilitate proper link setup, either the transmitters or the receivers must be wavelength tunable; we have arbitrarily chosen the transmitters to be tunable in this network example. Note that the wavelengths are routed passively in wavelength-routed networks.

WDM Stars, Rings, and Meshes

In this section, we will discuss three common network topologies that can use WDM, namely the *star, ring,* and *mesh* networks.[36–38]

In the star topology, each node has a transmitter and receiver, with the transmitter connected to one of the passive central star's inputs and the receiver connected to one of the star's outputs, as is shown in Fig. 18(*a*). Rings, as shown in Fig. 18(*b*), are also popular because (1) many electrical networks use this topology, and (2) rings are easy to implement for any geographical network configuration. In this example, each node in the unidirectional ring can transmit on a specific signature wavelength, and each node can recover any other node's wavelength signal by means of a wavelength-tunable receiver. Although not depicted in the figure, each node must recover a specific channel. This can be performed (1) where a small portion of the combined traffic is tapped off by a passive optical coupler, thereby allowing a tunable filter to recover a specific channel, or (2) in which a channel-dropping filter completely removes only the desired signal and allows all other channels to continue propagating around the ring. Furthermore, a synchronous optical network (SONET) dual-ring architecture, with one ring providing service and the other protection, can provide automatic fault detection and protection switching.[39]

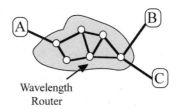

Wavelength
Router

FIGURE 17 A wavelength-routed network.

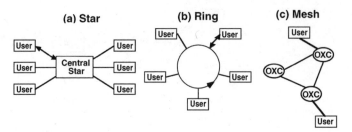

FIGURE 18   WDM stars, rings, and meshes.

In both the star and ring topologies, each node has a signature wavelength, and any two nodes can communicate with each other by transmitting and recovering that wavelength. This implies that N wavelengths are required to connect N nodes. The obvious advantage of this configuration, known as a *single-hop network,* is that data transfer occurs with an uninterrupted optical path between the origin and destination; the optical data starts at the originating node and reaches the destination node without stopping at any other intermediate node. A disadvantage of this single-hop WDM network is that the network and all its components must accommodate N wavelengths, which may be difficult (or impossible) to achieve in a large network.

An alternative is to have a mesh network, shown in Fig. 18(c), in which the nodes are connected by *reconfigurable optical crossconnects* (OXCs).[40] The wavelength can be dynamically switched and routed by controlling the OXCs. Therefore, the required number of wavelengths and the wavelength-tunable range of the components can be reduced in this topology. Moreover, the mesh topology can also provide multiple paths between two nodes to make network protection and restoration easier to realize. If a failure occurs in one of the paths, the system can automatically find another path and restore communications between any two nodes. However, OXCs with large numbers of ports are extremely difficult to obtain, which limits the scalability of the mesh network.

Network Reconfigurability

As mentioned in the previous section, a reconfigurable network will be highly desirable in future networks, in order to meet the requirements of high bandwidth and bursty traffic.[41]

A network is reconfigurable if it can provide the following functionality for multichannel operations:

- Channel add/drop
- Path reconfiguration for bandwidth allocation or restoration

These functions could be provided by using a reconfigurable optical crossconnect system as shown in Fig. 19.

A reconfigurable network allows dynamic network optimization to accommodate changing traffic patterns, which provides more efficient use of network resources. Figure 20 shows blocking probability as a function of call arrival rate in a WDM ring network with 20 nodes.[42] A configurable topology can support six times the traffic of a fixed WDM topology for the same blocking probability.

The key component technologies enabling network reconfigurability include:

- Wavelength-tunable lasers and laser arrays
- Wavelength routers

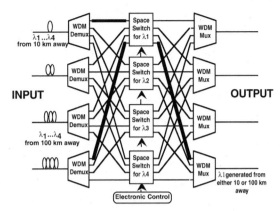

FIGURE 19 An optical crossconnect system in reconfigurable optical networks.

- Optical switches
- OXCs
- ADMs
- Tunable optical filters
- EDFAs

Although huge benefits are possible with a reconfigurable topology, the path to reconfigurability is paved with various degrading effects. As shown in Fig. 19, the signal may pass through different lengths of fiber links due to the dynamic routing, causing some degrading effects in reconfigurable networks to be more critical than in static networks, such as:

- Nonstatic dispersion and nonlinearity accumulation due to reconfigurable paths
- EDFA gain transients
- Channel power nonuniformity
- Crosstalk in optical switching and crossconnects
- Wavelength drift of components

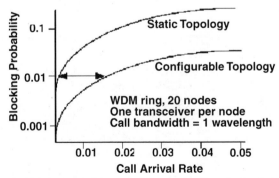

FIGURE 20 Blocking probability as a function of call arrival rate in a WDM ring.[42]

A detailed discussion on network reconfigurability can be found in Ref. 41. We note that dynamic optical routing and switching networks are still in the test-bed phase, for example, in the Multiwavelength Optical Network (MONET), the Optical Networks Technology Consortium (ONTC), and the wideband All-Optical Network (AON) Consortium.

Circuit and Packet Switching

Optical switching is one of the key components in reconfigurable optical networks. The two common schemes are (1) *circuit* and (2) *packet* switching.[43] Figure 21 shows the scenario for both these switching schemes in a simple 6-node network.

To begin with, it is important to realize that circuit switching is widely used and is the method typically used in the public switched telephone network. In circuit switching, if user A wishes to communicate with user E, user A sends a request signal to user E in order to initiate data transmission. If user E is available for interconnection, then it will acknowledge user A's original request, and then user A can begin data transmission. This process of request and acknowledgment is known as *handshaking*. Once transmission commences, a fixed path or circuit along the network is created between the two users until such time that the communications link is terminated by either user. The fiber medium on which the A-E circuit resides has a certain total bandwidth potential, and only a portion of this bandwidth is dedicated (i.e., reserved) for these two users. However, the medium bandwidth is reserved for the A-E circuit independent of whether a stream of bits is actually being transmitted (e.g., consider a pause in the conversation).

A circuit-switched network is quite easy to implement and control. Circuit switching emphasizes a transparent data pipe with a long set-up and tear-down time, and is more appropriate for low-speed transmission and long transmission time, such as a telephone call. Therefore, it has low component switching speed requirements (~ms), since the communications path is intended for minutes of transmission, as in a telephone call. In this case, the handshaking process between widely separated nodes may take only a short time compared to the slow data transfer speed. However, if the handshaking process takes a long time compared to the data transfer rate, then the link overhead due to handshaking is undesirably large. (Note that the switching speed discussed is the time it takes for a switch to change the routing of a data stream or packet, i.e., determining which output port of the switch is enabled for data arriving at a given input port.)

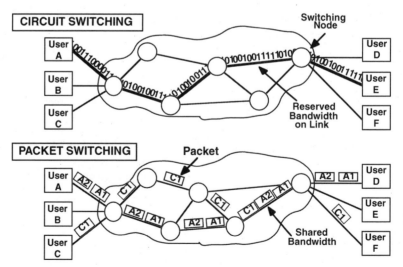

FIGURE 21   Circuit and packet switching systems for a simple 6-node network.

However, for high-speed (Gb/s) optical transmission, in which a user may require only a few seconds of transmission time, the more likely scenario to maximize system throughput is to implement a packet-switched network that does not require a link with dedicated bandwidth. Each user in such a network transmits data in the form of packets, as is depicted in Fig. 21. A data packet in a generic network is composed of (1) a flag (i.e., a unique set of bits) which is typically at the beginning of the packet in order to synchronize data recovery and alert a receiver that a packet is arriving, (2) header bits which contain destination and routing information for switching through the network, and (3) the user-generated end-to-end data payload.[44] A packet may be required to traverse many intermediate switching nodes before it is successfully routed to its destination. Depending on the network and on the available bandwidth on different lines, any packet may take several possible routes to reach a destination. In such a case, a sequence number is included in the packet to ensure that the packets will not be misinterpreted if they arrive out of order. Each transmitted packet uses some fraction of the bandwidth available in the fiber links. If no packets are transmitted, the bandwidth in the fiber links can be allocated to transmission of different packets, such as between users B and D. Therefore, packet switching efficiently uses the available bandwidth, since bandwidth is only used for transmitting high-speed data and the line will not be forced idle if a given user is not transmitting data (as is the case for circuit switching). Thus, bandwidth can be freely allocated in this system to where it is needed and is used quite efficiently, making it ideal for either bursty or uniform high-speed traffic. However, packet switching has much more stringent requirements on the switching technologies since switching must occur on the time scale of an individual packet. The component switching speed requirement is severe, with the range being μs to ns, depending on the specific system being implemented. Other critical issues in packet switching include:

- Header processing[45]
- Bit synchronization
- Variable bit rate processing
- Wavelength translation

13.4 *ERBIUM-DOPED FIBER AMPLIFIERS IN WDM NETWORKS*

Erbium-doped fiber amplifiers (EDFAs) are discussed in great detail in Chapter 5. In this section, we will consider some important issues about EDFAs with regard to their implementation in WDM systems. We emphasize that the EDFA has been perhaps the most important recent key enabling technology for WDM systems, because it can simultaneously amplify signals over a very wide bandwidth (~35 nm). EDFAs can be used in multichannel WDM systems to compensate for (1) fiber attenuation losses in transmission, (2) component excess losses, and (3) optical network splitting losses. These optical splitting losses can occur in a passive star, in which the optical power is divided by the number of users (N), or in a ring/bus, in which there may possibly be optical tapping losses at each node.

Figure 22(a) shows WDM transmission in a conventional electrically regenerated system. Regenerators can correct for fiber attenuation and chromatic dispersion by detecting an optical signal and then retransmitting it as a new signal using an internal laser. However, regenerators (being a hybrid of optics and electronics) are expensive, bit-rate and modulation-format specific, and waste much power and time in converting from photons to electrons and back again to photons. In contrast, as shown in Fig. 22(b), the EDFA is ideally a transparent box which is insensitive to the bit-rate, modulation-format, power, and wavelengths of the signal(s) passing through it, and, most important, provides gain for all the WDM channels simultaneously. Since all the channels remain in optical form during amplification, optically-amplified WDM systems are potentially cheaper and more reliable than electrically regenerated systems.

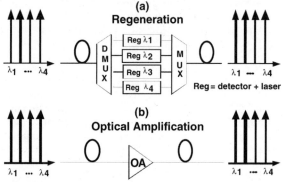

FIGURE 22   Wideband amplifiers enable WDM.

Gain Peaking in EDFA Cascades

The EDFA is an almost ideal optical amplifier for WDM systems, except for one major flaw: The gain is not uniform with wavelength, whereas the interamplifier losses are nearly wavelength independent.[46–49] For a single amplifier, as shown in Fig. 23, the gain exhibits a peak at 1530 nm and a relatively flat region near 1555 nm. Moreover, the gain shape of an EDFA is dependent on the inversion of Er^{3+} in the erbium-doped fiber.[16] When the inversion is low, which can be achieved by operating the amplifier in deep saturation, the gain peak at 1530 nm can be suppressed, and the gain flatness around 1555 nm would become quite flat.

If several channels are located on the relatively flat shoulder region of the gain spectrum, then the gain differential after a single amplifier will be within a few dB. However, when a cascade of EDFAs is used to periodically compensate for losses, the differential in gain and resultant SNR can become quite severe. A large differential in SNR among many channels can be deleterious for proper system performance. Figure 24 shows the gain spectrum after a single amplifier and after 13 cascaded amplifiers. The gain does not accumulate linearly from stage to stage, and the resultant wavelength-dependent gain shape dramatically changes in a cascade. Along the cascade, gain is gradually pulled away from the shorter wavelengths and made available at the longer wavelengths, resulting in a usable bandwidth of only several nm.

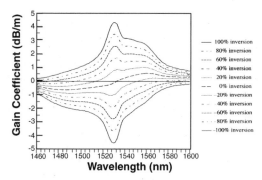

FIGURE 23   Nonuniform gain spectrum of an EDFA for different values of inversion.[16]

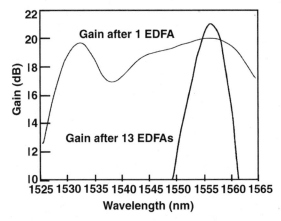

FIGURE 24   EDFA gain nonuniformity accumulation.

EDFA Gain Flattening

We have shown the bandwidth reduction due to nonuniform gain in a cascade of EDFAs. It is clear that gain flattening is an important issue in optically-amplified networks. Several methods have been reported for equalizing nonuniform EDFA gain. These methods include:

- *Long-period grating filters.*   A long-period grating (LPG) with an index-varying period of ~100 μm provides coupling between the core modes and the cladding modes, creating ·a wavelength-dependent loss to equalize the EDFA gain shape.[50–53] The results in Reference 50 are shown in Fig. 25. (Note that in a recent report, similar bandwidth is obtained after more than 9000 km transmission.)

- *Mach-Zehnder filters.*   The wavelength-dependent transmission characteristics of cascaded Mach-Zehnder filters can be tailored to compensate for the gain nonuniformity of EDFAs.[54–55]

- *Fluoride-based EDFAs.*   A fluoride-based EDFA[56] can provide an intrinsically flat gain over a wide wavelength range from 1530 to 1560 nm, which is a much wider bandwidth than for a silica-based EDFA. However, the fluoride-based fiber is extremely difficult to splice to normal fibers, and mechanical connections result in high connection loss and instability.

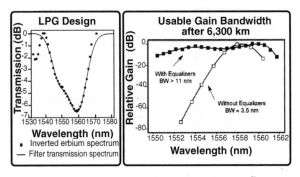

FIGURE 25   LPG design and gain equalization results.[50]

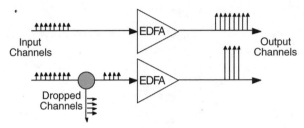

FIGURE 26 EDFA gain transients.

Fast Power Transients

The lifetime of a stimulated erbium ion is generally ~10 ms, which seems to be long enough to be transparent to signals modulated by data at the rates of several Gb/s or higher. However, the EDFAs could be critically affected by the adding/dropping of WDM channels, network reconfiguration, or link failures, as illustrated in Fig. 26. To achieve optimal channel SNRs, the EDFAs are typically operated in the gain-saturation regime where all channels must share the available gain.[57] Therefore, when channels are added or dropped, the power of the remaining channels will increase resulting transient effects.

The transients can be very fast in EDFA cascades.[58,59] As shown in Fig. 27, with an increase in the number of cascaded EDFAs, the transients can occur in ~2 μs. These fast power transients in chain-amplifier systems should be controlled dynamically, and the response time required scales as the size of the network. For large-scale networks, response times shorter than 100 ns may be necessary.

From a system point of view, fiber nonlinearity may become a problem when too much channel power exists, and a small SNR at the receiver may arise when too little power remains.[60] The corresponding fiber transmission penalty of the surviving channel is shown in Fig. 28 in terms of the Q-factor, for varying numbers of cascaded EDFAs. When 15 channels are dropped or added, the penalties are quite severe. Note that this degradation increases with the number of channels N simply because of enhanced SPM due to a large power excursion as a result of dropping $N - 1$ channels.

In order to maintain the quality of service, the surviving channels must be protected when channel add/drop or network reconfiguration occurs. The techniques include:

- Optical attenuation, by adjusting optical attenuators between the gain stages in the amplifier to control the amplifier gain[61]

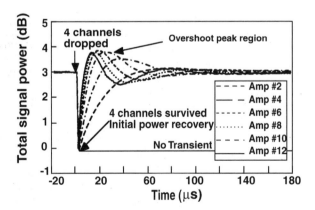

FIGURE 27 Fast power transients in EDFA cascades.[58]

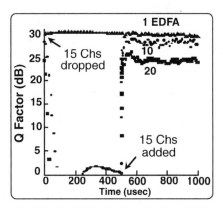

FIGURE 28 Q factor versus time for adding and dropping 15 channels of a 16-channel system at a bit rate of 10 Gb/sec.

- Pump control, by adjusting the drive current of the pump lasers to control the amplifier gain[62–64]
- Link control, using a power-variable control channel propagating with the signal channels to balance the amplifier gain[65]
- Gain clamping, with an automatic optical feedback control scheme to achieve all-optical gain clamping[66,67]

Ultrawideband EDFA

In Fig. 1 we saw that the bandwidth provided by conventional EDFAs is narrow compared to the low-loss window of the optical fiber. It is extremely important to broaden the gain spectrum of the optical amplifier to increase the transmission capacity of WDM systems.

Figure 29 shows a very promising scheme utilizing a parallel-type amplifier with both a 1550 nm-band and a 1580 nm-band EDFA.[68,69] The gain in the 1580 nm-band is provided by an underpumped EDFA, sometimes using an erbium-doped fiber of as long as ~200 m. The combined amplifier can offer a bandwidth of up to 80 nm, which is twice the bandwidth provided by conventional EDFAs.

In fact, the available bandwidth of amplified WDM systems can be further extended by combining this parallel configuration with other types of optical amplifiers, such as Raman amplifiers, as shown in Fig. 30. Moreover, the available bandwidth may exceed 25 THz (the low-loss window around 1.55 μm) by suppressing the high-loss peak around 1400 nm in the optical fiber. An additional bandwidth of more than 100 nm could be opened up by using this new type of fiber.[70]

13.5 *DYNAMIC CHANNEL POWER EQUALIZATION*

In Section 13.4 we discussed EDFA gain-flattening, which is a passive channel power equalization scheme effective only for a static link. However, as mentioned in Section 13.3, all optical networks are nonstatic, since the power in each channel suffers from dynamic network changes, including:

- Wavelength drift of components
- Changes in span loss
- Channel add/drop

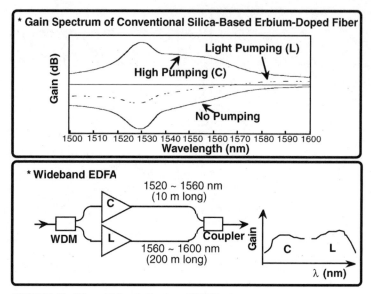

FIGURE 29 Ultrawideband silica-based EDFA.

As an example, consider Fig. 31, which shows how the gain shape of a cascaded EDFA chain varies significantly with link loss changes due to environmental problems. This is because the EDFA gain spectra are dependent on the saturation level of the amplifiers (see Fig. 23). The simulation results in Fig. 31 are for a cascade of 10 gain-flattened EDFAs, each with 20-dB gain, saturated by 16 input channels with −18 dBm per channel.

As illustrated in Fig. 32, system performance can be degraded due to unequalized WDM channel power. These degrading effects include:

- SNR differential (reduced system dynamic range)
- Widely varying channel crosstalk
- Nonlinear effects
- Low signal power at the receiver

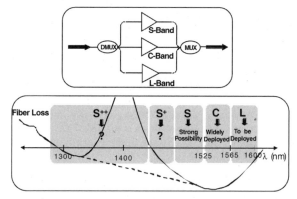

FIGURE 30 Potential wavelength regions for future WDM systems.

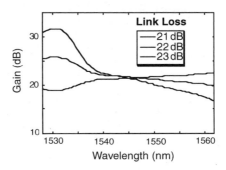

FIGURE 31 Gain spectra variation due to link loss changes for a cascade of 10 EDFAs.

Therefore, channel power needs to be equalized dynamically in WDM networks to ensure stable system performance. To obtain feedback for control purposes, a channel power monitoring scheme is very important. A simple way to accomplish this is to demultiplex all the channels and detect the power in each channel using different photodetectors or detector arrays. To avoid the high cost of many discrete components in WDM systems with large numbers of channels, other monitoring techniques that take advantage of wavelength-to-time mapping have also been proposed, including the use of concatenated FBGs[71] or swept acousto-optic tunable filters (AOTFs).[72]

Various techniques have been proposed for dynamic channel power equalization, including:

- Parallel loss elements[73]
- Individual bulk devices (e.g., AOTFs)[74]
- Serial filters[75]
- Micro-opto-mechanics (MEMS)[76]
- Integrated devices[77]

As an example, Fig. 33 shows the parallel loss element scheme, where the channels are demultiplexed and attenuated by separate loss elements. An additional advantage of this scheme is that ASE noise is reduced by the WDM multiplexer and demultiplexer. Possible candidates for the loss elements in this scheme include:

- Opto-mechanical attenuators
- Acousto-optic modulators
- FBG

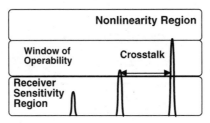

FIGURE 32 Degradation due to unequalized WDM channel power.

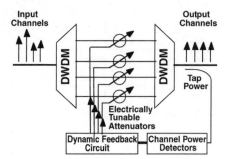

FIGURE 33 Parallel loss element scheme for dynamic channel power equalization.

13.6   *CROSSTALK IN WDM NETWORKS*

Incoherent and coherent crosstalk between adjacent channels are major problems that limit the density of WDM channels. In particular, coherent crosstalk in WDM ADMs and switches places severe requirements on the acceptable crosstalk-suppression levels provided by these components. In this section, we will introduce these two types of crosstalk and discuss their impact on system performance.

Incoherent Crosstalk

WDM channel crosstalk involves the most basic function of demultiplexing many WDM channels using an optical filter.[78] Figure 34 shows the demultiplexing of WDM channels by an optical filter so that only one channel is received and the other channel is blocked. However, the slowly diminishing spectral shape of the filter with long tails will induce crosstalk to the demultiplexed channel when the signals are detected at the photodetector.

In the case of incoherent crosstalk, the crosstalk level is determined by the channel spacing and channel powers. This interchannel crosstalk power P_{cr} will act to reduce the power extinction ratio of the selected recovered signal:

$$S = P_S - P_{cr} \qquad (9)$$

where S is the effective recovered signal power, and P_S is the power in the selected wavelength that passes through the filter. The crosstalk due to power leakage from the adjacent channels may raise the level of a "zero," increasing the probability of errors. Although the tolerable amount of crosstalk depends on the specific WDM system, crosstalk should not exceed a few percent of the selected channel power in order to maintain good system performance.[79–81]

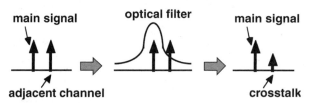

FIGURE 34   Incoherent crosstalk from adjacent channels.

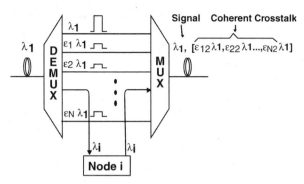

FIGURE 35   Coherent crosstalk in an ADM.

Coherent Crosstalk

Add-drop multiplexers are key subsystems in WDM networks.[82] They are typically composed of a wavelength demultiplexer followed by a wavelength multiplexer. Ideally, they allow a network node to have access to a single wavelength while allowing all other wavelengths to pass through the node unaffected (see Fig. 35). Unfortunately, wavelength (de)multiplexers[83] are not ideal and allow unwanted crosstalk of any input wavelength onto unintended output ports. For example, a small replica of the signal at wavelength λ_1 will appear at all the demultiplexer output ports, as shown in Fig. 35. If a nonideal demultiplexer and nonideal multiplexer are placed back-to-back, small replicas of the signal at λ_1 will recombine with the main signal at λ_1 at the multiplexer output.

This ADM structure with crosstalk represents an interferometer, since it consists of input and output ports connected by optical paths. Due to coherent interaction of the fields, the amplitude of the output signal will fluctuate as the signals' relative phases and polarization states change.[84,85]

Coherent crosstalk has been shown to be a serious limiting factor in scaleable WDM networks,[86,87] since the crosstalk signal has the same frequency as the main signal and the crosstalk is added to the main signal in the electric field. Let's consider the case shown in Fig. 36, where two signals with identical wavelengths enter an OXC and then are routed to different paths. Assuming the two channels have the same input power P and the power leakage of the OXC is ε, then the optical power at one of the output ports can be approximated by

$$P_0 \propto (\sqrt{P} + \sqrt{\varepsilon\, P})^2 = P \pm 2\sqrt{\varepsilon}P + \varepsilon P \tag{10}$$

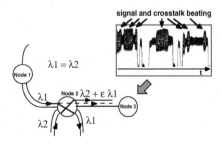

FIGURE 36   Power fluctuation due to coherent crosstalk.[86]

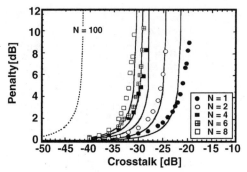

FIGURE 37   Scaling limitations due to coherent crosstalk.[86]

Therefore, the crosstalk power can be as high as 20 percent even when the power leakage ratio of the OXC is only 1 percent.

Figure 37 shows scaling limitations due to coherent crosstalk.[86] For the case of a single crosstalk channel, a coherent crosstalk of −25 dB will cause a power penalty of 1 dB, which is much more severe than the effects of incoherent crosstalk (1 dB power penalty at −8 dB crosstalk).[87] Because of the severity of this penalty, coherent crosstalk critically limits the scalability of an optical network employing ADMs and OXCs. If 100 channels are required, the crosstalk should be lower than −45 dB. However, the typical crosstalk of current components (WDM arrayed-waveguide gratings, switches, and modulators) is only −25 dB.

To mitigate the effects of coherent crosstalk in WDM networks, the following techniques have been proposed:

- *Architecture dilation.*[88]   Purposely introducing fiber delays (on the order of a bit time) between different crosstalk paths decorrelates the bit pattern of the main signal from the interfering signals and places some of the crosstalk power in the "0" bits of the main signal instead of the "1" bits.

- *Low-coherence light sources.*   Using *light-emitting diodes* (LEDs) or *chirped distributed feedback lasers* (DFBs) with coherence times less than a bit time can reduce coherent crosstalk, but the impact of dispersion would become more critical.

- *Polarization scrambling or modulation.*   Polarization scrambling reduces the effect of coherent crosstalk, since interference between the crosstalk and signal decreases when their polarization states are not matched.

- *Phase modulation.*   Modulating the phase of the signal at a rate greater than the bit rate averages out the crosstalk and improves performance, since signal degradation depends on the relative phase between the signal and crosstalk.

13.7   SUMMARY

In this chapter we have covered many different aspects of high-speed WDM fiber-optic communication networks. We have endeavored to treat the most important topics—those that will likely impact these networks for years to come. The enormous growth of these systems is all due to the revolutionary introduction of the EDFA. With increased knowledge, more development, higher data rates, and increasing channel count, WDM network limitations are being continually redefined. Network reconfigurability can offer great benefits for future WDM networks. However, a number of new degrading effects must be solved before reconfigurable networks become a reality. Yet the push for more bandwidth in WDM systems continues due to the enormous inherent potential of the optical fiber.

13.8 ACKNOWLEDGMENTS

We are indebted to Steve Havstad for his generous help and valuable contributions to this chapter.

13.9 REFERENCES

1. F. P. Kapron, "Fiber-Optic System Tradeoffs," *IEEE Spectrum Magazine,* 60–75 (March 1985).
2. Y. K. Park, T. V. Nguyen, O. Mizuhara, C. D. Chen, L. D. Tzeng, P. D. Yeates, F. Heismann, Y. C. Chen, D. G. Ehrenbergh, and J. C. Feggeler, "Field Demonstration of 10-Gb/s Line-Rate Transmission on an Installed Transoceanic Submarine Lightwave Cable," *IEEE Photonics Technology Letters* **8**:425–427 (1996).
3. D. P. Berztsekas and R. G. Gallager, *Data Networks,* Prentice Hall, Englewood Cliffs, New Jersey, 1987.
4. P. Kaiser and D. B. Keck, "Fiber Types and Their Status," *Optical Fiber Telecommunications II,* S. E. Miller and I. P. Kaminow (eds.), Academic Press, New York, 1988.
5. A. R. Chraplyvy, "Limitations on Lightwave Communications Imposed by Optical Fiber Nonlinearities," *IEEE Journal of Lightwave Technology* **8**:1548–1557 (1990).
6. P. E. Green, Jr., *Fiber Optic Networks,* Prentice Hall, Englewood Cliffs, New Jersey, 1993.
7. Special Issue on Wavelength Division Multiplexing, *IEEE Journal on Selected Areas in Communications,* **8** (1990).
8. C. A. Brackett, "Dense Wavelength Division Multiplexing: Principles and Applications," *IEEE Journal on Selected Areas in Communications* **8**:948 (1990).
9. I. P. Kaminow, "FSK with Direct Detection in Optical Multiple-Access FDM Networks," *IEEE Journal on Selected Areas in Communications* **8**:1005 (1990).
10. T. Li, "The Impact of Optical Amplifiers on Long-Distance Lightwave Telecommunications," *Proceedings of the IEEE* **81**:1568–1579 (1993).
11. G. P. Agrawal, *Nonlinear Fiber Optics,* Academic Press, New York, 1990.
12. A. R. Chraplyvy and R. W. Tkach, "What Is the Actual Capacity of Single-Mode Fibers In Amplified Lightwave Systems?," *IEEE Photonics Technology Letters,* **5**:666–668 (1987).
13. N. S. Bergano, and C. R. Davidson, "Wavelength Division Multiplexing in Long-Haul Transmission Systems," *IEEE J. Lightwave Technology,* **14**:1299–1308 (1996).
14. H. Taga, "Technologies for Hundreds of Gb/s Capacities in Undersea Lightwave Systems," *Conference on Optical Fiber Communication,* paper TuD3 (1999).
15. T. Li, "The Impact of Optical Amplifiers on Long-Distance Lightwave Telecommunications," Proceedings of the IEEE **81**:1568, (1993).
16. I. Kaminow and T. Koch, *Optical Fiber Communications,* IIIA, Academic Press, 1997.
17. D. Marcuse, A. R. Chraplyvy, and R. W. Tkach, "Dependence of Cross-Phase Modulation on Channel Number in Fiber WDM Systems," *IEEE J. Lightwave Technology,* **12**:885–890 (1994).
18. R. G. Waarts and R. P. Braun, "System Limitations Due to Four-Wave Mixing In Single-Mode Optical Fibers," *Electronics Letters,* **22**:873 (1986).
19. M. W. Maeda, W. B. Sessa, W. I. Way, A. Yi-Yan, L. Curtis, R. Spicer, and R. I. Laming, "The Effect of Four Wave Mixing in Fibers on Optical Frequency-Division Multiplexed Systems," *IEEE J. Lightwave Technology* **8**:1402–1408 (1990).
20. K. O. Hill, D. C. Johnson, B. S. Kawasaki, and R. I. MacDonald, "CW Three-Wave Mixing In Single-Mode Optical Fibers," *J. Appl. Phys.,* **49**:5098 (1978).

21. A. M. Vengsarkar, "Dispersion Compensating Fibers," *Conference on Optical Fiber Communication,* paper ThA2 (1997).

22. A. J. Antos and D. K. Smith, "Design and Characterization of Dispersion Compensating Fiber Based on the LP_{01} Mode," *IEEE J. Lightwave Technology,* **12**:1739–1745 (1994).

23. C. Kurtzke, "Suppression of Fiber Nonlinearities By Appropriate Dispersion Management," *IEEE Photonics Technology Letters,* **5**:1250–1252 (1993).

24. A. R. Chraplyvy, "High-Capacity Lightwave Transmission Experiments," *Bell Labs Technical Journal* (January–March 1999).

25. Y. Liu and A. J. Antos, "Dispersion-Shifted Large-Effective-Area Fiber for Amplified High-Capacity Long-Distance Systems," *Conference on Optical Fiber Communication,* paper TuN5 (1997).

26. R. W. Tkach, A. R. Chraplyvy, F. Forghieri, A. H. Gnauck, and R. M. Derosier, "Four-Photon Mixing and High-Speed WDM Systems," *IEEE J. Lightwave Technology* **13**:841–849 (1995).

27. A. R. Chraplyvy, A. H. Gnauck, R. W. Tkach, and R. M. Derosier, "8 × 10 Gb/s Transmission Through 280 km of Dispersion Management," *IEEE Photonics Technology Letters,* **5**:1233–1235 (1993).

28. K. O. Hill, S. Theriault, B. Malo, F. Bilodeau, T. Kitagawa, D. C. Johnson, J. Albert, K. Takiguchi, T. Kataoka, and K. Hagimoto, "Chirped In-Fibre Bragg Grating Dispersion Compensators: Linearisation of Dispersion Characteristic and Demonstration of Dispersion Compensation in 100 km, 10 Gb/s Optical Fibre Link," *Electronics Letters,* **30**:1755–1756 (1994).

29. W. H. Loh, R. I. Laming, N. Robinson, A. Cavaciuti, F. Vaninetti, C. J. Anderson, M. N. Zervas, and M. J. Cole, "Dispersion Compensation Over Distances in Excess of 500 km for 10-Gb/s Systems Using Chirped Fiber Gratings," *IEEE Photonics Technology Letters,* **8**:944–946 (1996).

30. L. Dong, M. J. Cole, A. D. Ellis, N. Durkin, M. Ibsen, V. Gusmeroli, and R. I. Laming, "40 Gb/s 1.55 μm Transmission Over 109 km of Non-Dispersion Shifted Fiber with Long Continuous Chirped Fiber Gratings," *Conference on Optical Fiber Communication,* paper PD6 (1997).

31. M. Dukin, M. Ibsen, M. J. Cole, and R. I. Laming, "1m Long Continuously-Written Fiber Bragg Gratings for Combined Second- and Third-Order Dispersion Compensation," *Electron. Letters,* **33**:1891–1893 (1997).

32. M. Ibsen, M. K. Durkin, M. J. Cole, and R. I. Laming, "Sinc-Sampled Fiber Bragg Gratings for Identical Multiple Wavelength Operation," *IEEE Photonics Technology Letters,* **10**:842–845 (1998).

33. J.-X. Cai, K.-M. Feng, A. E. Willner, V. Grubsky, D. S. Starodubov, and J. Feinberg, "Simultaneous Tunable Dispersion Compensation of Many WDM Channels Using a Sampled Nonlinearly-Chirped Fiber-Bragg-Grating," *IEEE Photonics Technology Letters,* **11**:1455–1457, (1999).

34. M. M. Ohn, A. T. Alavie, R. Maaskant, M. G. Xu, F. Bilodeau, and K. O. Hill, "Tunable Fiber Grating Dispersion Using a Piezoelectric Stack," *Conference on Optical Fiber Communication,* paper WJ3 (1997).

35. K.-M. Feng, J.-X. Cai, V. Grubsky, D. S. Starodubov, M. I. Hayee, S. Lee, X. Jiang, A. E. Willner, and J. Feinberg, "Dynamic Dispersion Compensation in a 10-Gb/s Optical System Using a Novel Voltage Tuned Nonlinearly Chirped Fiber Bragg Grating," *IEEE Photonics Technology Letters,* **11**:373–375 (1999).

36. A. E. Willner, I. P. Kaminow, M. Kuznetsov, J. Stone, and L. W. Stulz, "1.2Gb/s Closely-Spaced FDMA-FSK Direct-Detection Star Network," *IEEE Photonics. Technology Letters,* **2**:223–226 (1990).

37. N. R. Dono, P. E. Green, K. Liu, R. Ramaswami, and F. F. Tong, "A Wavelength Division Multiple Access Network for Computer Communication," *IEEE Journal on Selected Areas in Communications,* **8**:983–994 (1990).

38. W. I. Way, D. A. Smith, J. J. Johnson, and H. Izadpanah, "A Self-Routing WDM High-Capacity SONET Ring Network," *IEEE Photonics Technology Letters* **4**:402–405 (1992).

39. T-H. Wu, *Fiber Network Service Survivability,* Boston: Artech House.

40. A. S. Acampora, M. J. Karol, and M. G. Hluchyj, "Terabit Lightwave Networks: The Multihop Approach," *AT&T Technical Journal,* **66**:21–34 (1987).

41. A. E. Willner, "Combating Degrading Effects in Non-Static and Reconfigurable WDM Systems and Networks," *Conference on Optical Fiber Communication,* Short Course SC114 (1999).

42. V. W. S. Chan, K. L. Hall, E. Modiano, and K. A. Rauschenbach, "Architectures and Technologies for High-Speed Optical Data Networks," *IEEE J. Lightwave Technology,* **16**:2146–2168 (1998).

43. J. Y. Hui, *Switching and Traffic Theory for Integrated Broadband Networks,* Kluwer Academic Publishers, Boston, 1990.

44. J. A. McEachern, "Gigabit Networking on the Public Transmission Network," *IEEE Communications Magazine,* 70–78 (April 1992).

45. M. C. Cardakli, S. Lee, A. E. Willner, V. Grubsky, D. Starodubov, and J. Feinberg, "All-Optical Packet Header Recognition and Switching in a Reconfigurable Network Using Fiber Bragg Gratings for Time-to-Wavelength Mapping and Decoding," *Conference on Optical Fiber Communications,* paper ThM4 (1999).

46. E. L. Goldstein, A. F. Elrefaie, N. Jackman, and S. Zaidi, "Multiwavelength Fiber-Amplifier Cascades in Unidirectional Interoffice Ring Networks," *Conference on Optical Fiber Communications,* paper TuJ3 (1993).

47. J. P. Blondel, A. Pitel, and J. F. Marcerou, "Gain-Filtering Stability in Ultralong-Distance Links," *Conference on Optical Fiber Communications '93,* paper TuI3 (1993).

48. H. Taga, N. Edagawa, Y. Yoshida, S. Yamamoto, and H. Wakabayashi, "IM-DD Four-Channel Transmission Experiment Over 1500 km Employing 22 Cascaded Optical Amplifiers," *Electron. Letters,* **29**:485 (1993).

49. A. E. Willner and S.-M. Hwang, "Transmission of Many WDM Channels Through a Cascade of EDFAs in Long-Distance Link and Ring Networks," *IEEE Journal of Lightwave Technology* **13**:802–816 (1995).

50. A. M. Vengsarkar, P. J. Lemaire, J. B. Judkins, V. Bhatia, T. Erdogan, and J. E. Sipe, "Long-Period Fiber Gratings as Band-Rejection Filters," *IEEE Journal of Lightwave Technology* **14**:58–65 (1996).

51. Y. Sun, J. B. Judkins, A. K. Srivastava, L. Garrett, J. L. Zyskind, J. W. Sulhoff, C. Wolf, R. M. Derosier, A. P. Gnauck, R. W. Tkach, J. Zhou, R. P. Espindola, A. M. Vengsarkar, and A. R. Chraplyvy, "Transmission of 32-WDM 10-Gb/s Channels Over 640 km Using Broad-Band, Gain-Flattened Erbium-Doped Silica Fiber Amplifiers," *IEEE Photonics Technology Letters* **9**:1652–1654 (1997).

52. A. M. Vengsarkar, P. J. Lemaire, J. B. Judkins, J. E. Sipe, and T. Erdogan, "Long-Period Fiber Gratings as Band-Rejection Filters," *Conference on Optical Fiber Communications,* paper PD4 (1995).

53. P. F. Wysocki, J. B. Judkins, R. P. Espindola, M. Andrejco, and A. M. Vengsarkar, "Broad-Band Erbium-Doped Fiber Amplifier Flattened Beyond 40 nm Using Long-Period Grating Filter," *IEEE Photonics Technology Letters* **9**:1343–1345 (1997).

54. K. Inoue, T. Kominato, and H. Toba, "Tunable Gain Equalization Using a Mach-Zehnder Optical Filter in Multistage Fiber Amplifiers," *IEEE Photonics Technology Letters,* **3**:718–720 (1991).

55. J. Y. Pan, M. A. Ali, A. F. Elrefaie, and R. E. Wagner, "Multiwavelength Fiber-Amplifier Cascades with Equalization Employing Mach-Zehnder Optical Filter," *IEEE Photonics Technology Letters* **7**:1501–1503 (1995).

56. M. Yamada, T. Kanamori, Y. Terunuma, K. Oikawa, M. Shimizu, S. Sudo, and K. Sagawa, "Fluoride-Based Erbium-Doped Fiber Amplifier with Inherently Flat Gain Spectrum," *IEEE Photonics Technology Letters* **8**:882–884 (1996).

57. E. Desurvire, R. Giles, and J. Simpson, "Gain Saturation Effects in High-Speed, Multi Channel Erbium-Doped Fiber Amplifiers at 1 = 1.53 mm," *IEEE Journal of Lightwave Technology* **7**:2095–2104 (1989).

58. J. Zyskind, Y. Sun, A. Srivastava, J. Sulhoff, A. Lucero, C. Wolf, and R. Tkach, "Fast Power Transients in Optically Amplified Multiwavelength Optical Networks," *Conference on Optical Fiber Communications,* paper PD-31 (1996).

59. Y. Sun, A. Saleh, J. Zyskind, D. Wilson, A. Srivastava, and J. Sulhoff, "Time Dependent Perturbation Theory and Tones in Cascaded Erbium-Doped Fiber Amplifier System," *IEEE Journal of Lightwave Technology* **15**:1083–1087 (1997).

60. M. Hayee and A. Willner, "Fiber Transmission Penalties Due to EDFA Power Transients Resulting from Fiber Nonlinearity and ASE Noise in Add/Drop Multiplexed WDM Networks," *IEEE Photonics Tech. Lett.* **11**:889–891 (1999).

61. J.-X. Cai, K.-M. Feng, and A. E. Willner, "Simultaneous Compensation of Fast Add/Drop Power-Transients and Equalization of Inter-Channel Power Differentials for Robust WDM Systems with EDFAs," *Conference on Optical Amplifiers and Their Applications, Victoria, Canada,* paper MC6, (July 1997).

62. K. Motoshima, L. Leba, D. Chen, M. Downs, T. Li and E. Desurvire, "Dynamic Compensation of Transient Gain Saturation in Erbium-Doped Fiber Amplifiers by Pump Feedback Control," *IEEE Photonics Technology Letters* **5**:1423–1426 (1993).

63. A. Srivastava, Y. Sun, J. Zyskind, J. Sulhoff, C. Wolf, and R. Tkach, "Fast Gain Control in an Erbium-Doped Fiber Amplifier," *Conference on Optical Amplifiers and Their Applications,* paper PD4 (1996).

64. C. Konishi, T. Yoshida, S. Hamada, K. Asahi, and S. Fujita, "Dynamic Gain-Controlled Erbium-Doped Fiber Amplifier Repeater for WDM Network," *Conference on Optical Fiber Communications,* paper TuE1 (1997).

65. A. Srivastava, J. Zyskind, Y. Sun, J. Ellson, G. Newsome, R. Tkach, A. Chraplyvy, J. Sulhoff, T. Strasser, C. Wolf, and J. Pedrazzani, "Fast-Link Control Protection of Surviving Channels in Multiwavelength Optical Networks," *IEEE Photonics Technology Letters* **9**:1667–1669 (1997).

66. B. Landousies, T. Georges, E. Delevaque, R. Lebref, and M. Monerie, "Low Power Transient in Multichannel Equalized and Stabilised Gain Amplifier Using Passive Gain Control," *Electronics Letters* **32**:1912–1913 (1996).

67. G. Luo, J. Zyskind, Y. Sun, A. Srivastava, J. Sulhoff, and M. Ali, "Performance Degradation of All-Optical Gain-Clamped EDFAs Due to Relaxation-Oscillations and Spectral-Hole Burning in Amplified WDM Networks," *IEEE Photonics Technology Letters* **9**:1346–1348 (1997).

68. Y. Sun, J. W. Sulhoff, A. K. Srivastava, J. L. Zyskind, T. A. Strasser, J. R. Pedrazzani, C. Wolf, J. Zhou, J. B. Jukins, R. P. Espindola, and A. M. Vengsarkar, "80 nm Ultra-Wideband Erbium-Doped Silica Fibre Amplifier," *Electronics Letters* **33**:1965–1967 (1997).

69. M. Yamada, H. Ono, T. Kanamori, S. Sudo, and Y. Ohishi, "Broadband and Gain-Flattened Amplifier Composed of a 1.55 μm-band and a 1.58 μm-band Er^{3+}-Doped Fibre Amplifier in a Parallel Configuration," *Electronics Letters* **33**:710–711 (1997).

70. Allwave Single-Mode Optical Fiber," *Bell Labs Technical Document* (June 1, 1999).

71. R. Giles, T. Strasser, K. Dryer, and C. Doerr, "Concatenated Fiber Grating Optical Monitor," *IEEE Photonics Technology Letters* **10**:1452–1454 (1998).

72. St. Schmid, S. Morasca, D. Scarano, and H. Herrmann, "High-Performance Integrated Acousto-Optic Channel Analyzer," *Conference on Optical Fiber Communications,* paper TuC3 (1997).

73. J. Cai, K. Feng, X. Chen, and A. Willner, "Experimental Demonstration of Dynamic High-Speed Equalization of Three WDM Channels Using Acousto-Optic Modulators and a Wavelength Demultiplexer," *IEEE Photonics Technology Letters* **9**:678–680 (1997).

74. S. Huang, X. Zou, A. Willner, Z. Bao, and D. Smith, "Experimental Demonstration of Active Equalization and ASE Suppression of Three 2.5 Gb/s WDM-Network Channels Over 2500 km Using AOTF as Transmission Filters," *IEEE Photonics Technology Letters* **9**:389–391 (1997).

75. D. Starodubov, V. Grubsky, J. Feinberg, J. Cai, K. Feng, and A. Willner, "Novel Fiber Amplitude Modulators for Dynamic Channel Power Equalization in WDM Systems," *Conference on Optical Fiber Communication,* paper PD-8 (1998).

76. J. Ford and J. Walker, "Dynamic Spectral Power Equalization Using Micro-Opto-Mechanics," *IEEE Photonics Technology Letters* **10**:1440–1442 (1998).

77. C. Doerr, C. Joyner, and L. Stulz, "Integrated WDM Dynamic Power Equalizer with Low Insertion Loss," *IEEE Photonics Technology Letters* **10**:1443–1445 (1998).

78. P. A. Humblet and W. M. Hamdy, "Crosstalk Analysis and Filter Optimization of Single- and Double-Cavity Fabry-Perot Filters," *IEEE Journal on Selected Areas of Communications* **8**:1095–1107 (1990).

79. I. P. Kaminow, P. P. Iannone, J. Stone, and L. W. Stulz, "FDMA-FSK Star Network with a Tunable Optical Fiber Demultiplexer," *J. Lightwave Technology* **6**:1406–1414 (1988).

80. A. E. Willner, "Simplified Model of a FSK-to-ASK Direct-Detection System Using a Fabry-Perot Demodulator," *IEEE Photonics Technology Letters* **2**:363–366 (1990).

81. A. E. Willner, I. P. Kaminow, M. Kuznetsov, J. Stone, and L. W. Stulz, "FDMA-FSK Noncoherent Star Network Operated at 600 Mb/s Using Two-Electrode DFB Lasers and a Fiber Optical Filter Demultiplexer," *Electronics Letters* **25**:1600–1601 (1989).

82. R. E. Wagner, R. C. Alferness, A. A. M. Saleh, and M. S. Goodman, "MONET: Multiwavelength Optical Networking," *J. Lightwave Technology* **14**:1349–1355 (1996).

83. C. Dragone, C. A. Edwards, and R. C. Kistler, "Integrated Optics N X N Multiplexer on Silicon," *IEEE Photonics Technology Letters* **3**:896–899 (1991).

84. A. Arie, M. Tur, and E. L. Goldstein, "Probability-Density Function of Noise at the Output of a Two-Beam Interferometer," *J. Opt. Soc. Am. A* **8**:1936–1941 (1991).

85. P. J. Legg, D. K. Hunter, I. Andonovic and P. E. Barnsley, "Inter-Channel Crosstalk Phenomena in Optical Time Division Multiplexed Switching Networks," *IEEE Photonics Technology Letters* **6**:661–663 (1994).

86. E. L. Goldstein and L. Eskildsen, "Scaling Limitations in Transparent Optical Networks Due to Low-Level Crosstalk," *IEEE Photonics Technology Letters* **7**:93–94 (1995).

87. E. L. Goldstein, L. Eskildsen, and A. F. Elrefaie, "Performance Implications of Component Crosstalk in Transparent Lightwave Networks," *IEEE Photonics Technology Letters* **6**:657–660 (1994).

88. R. Khosravani, M. I. Hayee, B. Hoanca, and A. E. Willner, "Reduction of Coherent Crosstalk in WDM Add/Drop Multiplexing Nodes by Bit Pattern Misalignment," *IEEE Photonics Technology Letters* **11**:134–136 (1999).

CHAPTER 14
INFRARED FIBERS

James A. Harrington
Rutgers University
Piscataway, New Jersey

14.1 INTRODUCTION

Infrared (IR) optical fibers may be defined as fiber optics transmitting radiation with wavelengths greater than approximately 2 μm. The first IR fibers were fabricated in the mid-1960s from chalcogenide glasses such as arsenic trisulfide and had losses in excess of 10 dB/m.[1] During the mid-1970s, the interest in developing an efficient and reliable IR fiber for short-haul applications increased, partly in response to the need for a fiber to link broadband, long-wavelength radiation to remote photodetectors in military sensor applications. In addition, there was an ever increasing need for a flexible fiber delivery system for transmitting CO_2 laser radiation in surgical applications. Around 1975, a variety of IR materials and fibers were developed to meet these needs. These included the heavy metal fluoride glass (HMFG) and polycrystalline fibers as well as hollow rectangular waveguides. While none of these fibers had physical properties even approaching those of conventional silica fibers, they were nevertheless useful in lengths less than 2 to 3 m for a variety of IR sensor and power delivery applications.[2]

IR fiber optics may logically be divided into three broad categories: glass, crystalline, and hollow waveguides. These categories may be further subdivided based on fiber material, structure, or both, as shown in Table 1. Over the past 25 years many novel IR fibers have been made in an effort to fabricate a fiber optic with properties as close as possible to those of silica, but only a relatively small number have survived. A good source of general information on these various IR fiber types may be found in the literature.[3-6] In this review only the best, most viable, and, in most cases, commercially available IR fibers are discussed. In general, both the optical and mechanical properties of IR fibers remain inferior to those of silica fibers, and therefore the use of IR fibers is still limited primarily to nontelecommunication, short-haul applications requiring only tens of meters of fiber rather than the kilometer lengths common to telecommunication applications. The short-haul nature of IR fibers results from the fact that most IR fibers have losses in the range of a few decibels per meter. An exception is fluoride glass fibers, which can have losses as low as a few decibels per kilometer. In addition, IR fibers are much weaker than silica fiber and, therefore, more fragile. These deleterious features have slowed the acceptance of IR fibers and restricted their use today to applications in chemical sensing, thermometry, and laser power delivery.

TABLE 1   Categories of IR Fibers with a Common Example to Illustrate Each Subcategory

Main	Subcategory	Examples
Glass	Heavy metal fluoride (HMFG)	ZrF_4-BaF_2-LaF_3-AlF_3-NaF (ZBLAN)
	Germanate	GeO_2-PbO
	Chalcogenide	As_2S_3 and AsGeTeSe
Crystal	Polycrystalline (PC)	AgBrCl
	Single crystal (SC)	Sapphire
Hollow waveguide	Metal/dielectric film	Hollow glass waveguide
	Refractive index < 1	Hollow sapphire at 10.6 µm

A key feature of current IR fibers is their ability to transmit longer wavelengths than most oxide glass fibers can. In some cases the transmittance of the fiber can extend well beyond 20 µm, but most applications do not require the delivery of radiation longer than about 12 µm. In Fig. 1 we give the attenuation values for some of the most common IR fibers as listed in Table 1. From the data it is clear that there is a wide variation in range of transmission for the different IR fibers and that there is significant extrinsic absorption that degrades the overall optical response. Most of these extrinsic bands can be attributed to various impurities, but, in the case of the hollow waveguides, they are due to interference effects resulting from the thin film coatings used to make the guides.

Some of the other physical properties of IR fibers are listed in Table 2. For comparison, the properties of silica fibers are also listed. The data in Table 2 and in Fig. 1 reveal that, compared to silica, IR fibers usually have higher losses, larger refractive indices and dn/dT values, lower melting or softening points, and greater thermal expansion. For example, chalcogenide and polycrystalline Ag halide fibers have refractive indices greater than 2. This means that the Fresnel loss exceeds 20 percent for two fiber ends. The higher dn/dT and low melting or softening point lead to thermal lensing and low laser-induced damage thresholds for some of the

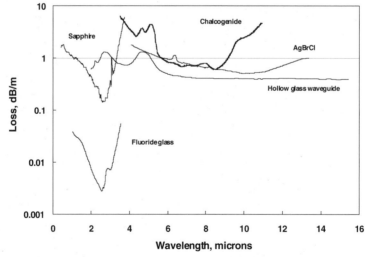

FIGURE 1   Composite loss spectra for some common IR fiber optics: ZBLAN fluoride glass,[7] SC sapphire,[8] chalcogenide glass,[9] PC AgBrCl,[10] and hollow glass waveguide.[11]

TABLE 2   Selected Physical Properties of Key IR Fibers Compared to Conventional Silica Fiber

Property	Glass			Crystal		Hollow
	Silica	HMFG ZBLAN	Chalcogenide AsGeSeTe	PC AgBrCl	SC Sapphire	Hollow Silica Waveguide
Glass transition or melting point, °C	1175	265	245	412	2030	150 (usable T)
Thermal conductivity, W/m °C	1.38	0.628	0.2	1.1	36	1.38
Thermal expansion coefficient, 10^{-6} °C^{-1}	0.55	17.2	15	30	5	0.55
Young's modulus, GPa	70.0	58.3	21.5	0.14	430	70.0
Density, g/cm^3	2.20	4.33	4.88	6.39	3.97	2.20
Refractive index (λ, μm)	1.455 (0.70)	1.499 (0.589)	2.9 (10.6)	2.2 (10.6)	1.71 (3.0)	NA
dn/dT, 10^{-5} °C^{-1} (λ, μm)	+1.2 (1.06)	−1.5 (1.06)	+10 (10.6)	−1.5 (10.6)	+1.4 (1.06)	NA
Fiber transmission range, μm	0.24–2.0	0.25–4.0	4–11	3–16	0.5–3.1	0.9–25
Loss* at 2.94 μm, dB/m	~800	0.08	5	3	0.4	0.5
Loss* at 10.6 μm, dB/m	NA	NA	2	0.5	NA	0.4

* Typical measured loss.
NA = not applicable.

fibers. Finally, a number of these fibers do not have cladding analogous to clad oxide glass fibers. Nevertheless, core-only IR fibers such as sapphire and chalcogenide fibers can still be useful because their refractive indices are sufficiently high. For these high-index fibers, the energy is largely confined to the core of the fiber as long as the unprotected fiber core does not come in contact with an absorbing medium.[12]

The motivation to develop a viable IR fiber stems from many proposed applications. A summary of the most important current and future applications and the associated candidate IR fiber that will best meet each need is given in Table 3. We may note several trends from this table. The first is that hollow waveguides are an ideal candidate for laser power delivery at all IR laser wavelengths. The air core of these special fibers or waveguides gives an inherent advantage over solid-core fibers, whose damage threshold is frequently very low for these IR-transmissive materials. The high refractive index of chalcogenide fibers is ideal for chem-

TABLE 3   Examples of IR Fiber Candidates for Various Sensor and Power Delivery Applications

Application	Comments	Suitable IR fibers
Fiber-optic chemical sensors	Evanescent wave principle—liquids	AgBrCl, sapphire, chalcogenide, HMFG
	Hollow core waveguides—gases	Hollow glass waveguides
Radiometry	Blackbody radiation, temperature measurements	Hollow glass waveguides, AgBrCl, chalcogenide, sapphire
Er:YAG laser power delivery	3-μm transmitting fibers with high damage threshold	Hollow glass waveguides, sapphire, germanate glass
CO_2 laser power delivery	10-μm transmitting fibers with high damage threshold	Hollow glass waveguides
Thermal imaging	Coherent bundles	HMFG, chalcogenide
Fiber amplifiers and lasers	Doped IR glass fibers	HMFG, chalcogenide

ical sensing via evanescent wave coupling of a small portion of the light from the core into an IR-absorbing medium. For the measurement of temperature through the simple transmission of blackbody radiation, IR fibers that transmit beyond about 8 μm, such as the Ag halide, chalcogenide, and hollow waveguides, are excellent candidates for use in measuring temperatures below 50°C. This is because the peak for room-temperature blackbody radiation is about 10 μm.

14.2   NONOXIDE AND HEAVY-METAL OXIDE GLASS IR FIBERS

There are two IR-transmitting glass fiber systems that are relatively similar to conventional silica-containing glass fibers. One is the HMFG and the other is heavy-metal germanate glass fibers based on GeO_2. The germanate glass fibers generally do not contain fluoride compounds; instead, they contain heavy metal oxides to shift the IR absorption edge to longer wavelengths. The advantage of germanate fibers over HMFG fibers is that germanate glass has a higher glass transition temperature and, therefore, higher laser damage thresholds. But the level of loss for the HMFG fibers is lower. Finally, chalcogenide glass fibers made from chalcogen elements such as As, Ge, S, and Te contain no oxides or halides, making them a good choice for nonlaser power delivery applications.

HMFG Fibers

Poulain et al.[13] discovered HMFGs or fluoride glasses accidentally in 1975 at the University of Rennes. In general, the typical fluoride glass has a glass transition temperature T_g four times less than that of silica, is considerably less stable than silica, and has failure strains of only a few percent compared to greater than 5 percent for silica. While an enormous number of multicomponent fluoride glass compositions have been fabricated, comparably few have been drawn into fiber. This is because the temperature range for fiber drawing is normally too small in most HMFGs to permit fiberization of the glass. The most popular HMFGs for fabrication into fibers are the fluorozirconate and fluoroaluminate glasses, of which the most common are ZrF_4-BaF_2-LaF_3-AlF_3-NaF (ZBLAN) and AlF_3-ZrF_4-BaF_2-CaF_2-YF_3, respectively. The key physical properties of these glasses are summarized in Table 4. An important feature of the fluoroaluminate glass is its higher T_g, which largely accounts for the higher laser damage threshold for the fluoroaluminate glasses compared to ZBLAN at the Er:YAG laser wavelength of 2.94 μm.

The fabrication of HMFG fiber is similar to any glass fiber drawing technology except that the preforms are made using some type of melt-forming method rather than by a vapor deposition process as is common with silica fibers. Specifically, a casting method based on first forming a clad glass tube and then adding the molten core glass is used to form either multimode or single-mode fluorozirconate fiber preforms. The cladding tube is made either by a

TABLE 4   Fluorozirconate vs. Fluoroaluminate Glasses

Property	Fluorozirconate (ZBLAN)	Fluoroaluminate (AlF_3-ZrF_4-BaF_2-CaF_2-YF_3)
Glass transition temperature, °C	265	400
Durability	Medium	Excellent
Loss at 2.94 μm, dB/m	0.01	0.1
Er:YAG laser peak output energy, mJ	300 (300-μm core)	850 (500-μm core)

Comparison between fluorozirconate and fluoroaluminate glasses of some key properties that relate to laser power transmission and durability of the two HMFG fibers. Other physical properties are relatively similar.

rotational casting technique in which the tube is spun in a metal mold or by merely inverting and pouring out most of the molten cladding glass contained in a metal mold to form a tube.[14] The cladding tube is then filled with a higher-index core glass. Other preform fabrication techniques include rod-in-tube and crucible techniques. The fluoroaluminate fiber preforms have been made using an unusual extrusion technique in which core and cladding glass plates are extruded into a core-clad preform.[15] All methods, however, involve fabrication from the melted glass rather than from the more pristine technique of vapor deposition used to form SiO_2-based fibers. This process creates inherent problems such as the formation of bubbles, core-cladding interface irregularities, and small preform sizes. Most HMFG fiber drawing is done using preforms rather than the crucible method. A ZBLAN preform is drawn at about 310°C in a controlled atmosphere (to minimize contamination by moisture or oxygen impurities, which can significantly weaken the fiber) using a narrow heat zone compared to silica. Either ultraviolet (UV) acrylate or Teflon coatings are applied to the fiber. In the case of Teflon, heat-shrink Teflon (FEP) is generally applied to the glass preform prior to the draw.

The attenuation in HMFG fibers is predicted to be about 10 times less than that for silica fibers.[16] Based on extrapolations of the intrinsic losses resulting from Rayleigh scattering and multiphonon absorption, the minimum in the loss curves or V-curves is projected to be about 0.01 dB/km at 2.55 μm. Recent refinements of the scattering loss have modified this value slightly to be 0.024 dB/km, or about eight times less than that for silica fiber.[7] In practice, however, extrinsic loss mechanisms still dominate fiber loss. In Fig. 2, losses for two ZBLAN fibers are shown. The data from British Telecom (BTRL) represents state-of-the-art fiber 110 m in length.[7] The other curve is more typical of commercially available (Infrared Fiber Systems, Silver Spring, Maryland) ZBLAN fiber. The lowest measured loss for a BTRL 60-m-long fiber is 0.45 dB/km at 2.3 μm. Some of the extrinsic absorption bands that contribute to the total loss shown in Fig. 2 for the BTRL fiber are Ho^{3+} (0.64 and 1.95 μm), Nd^{3+} (0.74 and 0.81 μm), Cu^{2+} (0.97 μm), and OH^- (2.87 μm). Scattering centers such as crystals, oxides, and bubbles have also been found in the HMFG fibers. In their analysis of the data in Fig. 2, the BTRL group separated the total minimum attenuation coefficient (0.65 dB/km at 2.59 μm)

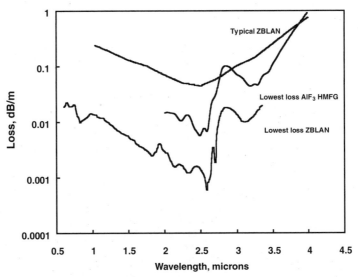

FIGURE 2   Losses in the best BTRL[7] and typical (Infrared Fiber Systems, Silver Spring, Maryland) ZBLAN fluoride glass fibers compared to those for fluoroaluminate glass fibers.[15]

into an absorptive loss component equal to 0.3 dB/km and a scattering loss component equal to 0.35 dB/km. The losses for the fluoroaluminate glass fibers are also shown for comparison in Fig. 2.[15] Clearly, the losses are not as low as for the BTRL-ZBLAN fiber, but the AlF₃-based fluoride fibers do have the advantage of higher glass transition temperatures and therefore are better candidates for laser power delivery.

The reliability of HMFG fibers depends on protecting the fiber from attack by moisture and on pretreatment of the preform to reduce surface crystallization. In general, the HMFGs are much less durable than oxide glasses. The leach rates for ZBLAN glass range between 10^{-3} and 10^{-2} g/cm²/day. This is about five orders of magnitude higher than the leach rate for Pyrex glass. The fluoroaluminate glasses are more durable, with leach rates that are more than three times lower than those for the fluorozirconate glasses. The strength of HMFG fibers is less than that of silica fibers. From Table 2 we see that Young's modulus E for fluoride glass is 51 GPa compared to 73 GPa for silica glass. Taking the theoretical strength to be about one-fifth that of Young's modulus gives a theoretical value of strength of 11 GPa for fluoride glass. The largest bending strength measured has been about 1.4 Gpa, well below the theoretical value. To estimate the bending radius R, we may use the approximate expression $R = 1.198r(E/\sigma_{max})$, where σ_{max} is the maximum fracture stress and r is the fiber radius.[17]

Germanate Fibers

Heavy metal oxide glass fibers based on GeO₂ have recently shown great promise as an alternative to HMFG fibers for 3-μm laser power delivery.[18] Today, GeO₂-based glass fibers are composed of GeO₂ (30–76 percent)–RO (15–43 percent)–XO (3–20 percent), where R represents an alkaline earth metal and X represents an element of Group IIIA.[19] In addition, small amounts of heavy metal fluorides may be added to the oxide mixture. The oxide-only germanate glasses have glass transition temperatures as high as 680°C, excellent durability, and a relatively high refractive index of 1.84. In Fig. 3, loss data is given for a typical germanate glass fiber. While the losses are not as low as they are for the fluoride glasses shown in Fig. 2, these fibers have an exceptionally high damage threshold at 3 μm. Specifically, over 20 W (2 J at 10 Hz) of Er:YAG laser power has been launched into these fibers.

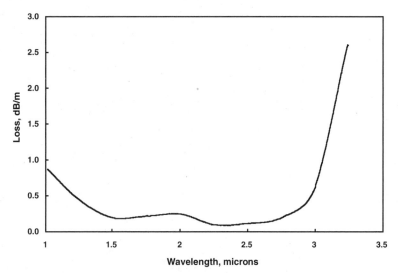

FIGURE 3 Germanate glass fiber manufactured by Infrared Fiber Systems, Silver Spring, Maryland.

Chalcogenide Fibers

Chalcogenide glass fibers were drawn into essentially the first IR fiber in the mid-1960s.[1] Chalcogenide fibers fall into three categories: sulfide, selenide, and telluride.[20] One or more chalcogen elements are mixed with one or more elements such as As, Ge, P, Sb, Ga, Al, Si, and so on to form a glass having two or more components. From the data in Table 2 we see that the glasses have low softening temperatures more comparable to those of fluoride glass than to those of oxide glasses. Chalcogenide glasses are very stable, durable, and insensitive to moisture. A distinctive difference between these glasses and the other IR fiber glasses is that they do not transmit well in the visible region and their refractive indices are quite high. Additionally, most of the chalcogenide glasses, except for As_2S_3, have a rather large value of dn/dT.[9] This fact limits the laser power handling capability of the fibers. In general, chalcogenide glass fibers have proven to be an excellent candidate for evanescent wave fiber sensors and for IR fiber image bundles.[21]

Chalcogenide glass is made by combining highly purified (>6 nines purity) raw elements in an ampoule that is heated in a rocking furnace for about 10 h. After melting and mixing, the glass is quenched and a glass preform is fabricated using rod-in-tube or rotational casting methods. Preform fiber draws involve drawing a core-clad preform or a core-only preform. For the core-only preform draw, either a soft chalcogenide cladding can be extruded over the fiber as it is drawn or the preform can be Teflon clad. Crucible drawing is also possible.

The losses for the most important chalcogenide fibers are given in Fig. 4. Arsenic trisulfide (As_2S_3) fiber, one of the simplest and oldest chalcogenide fibers, has a transmission range from 0.7 to about 6 μm.[20] This fiber is red in color and therefore transmits furthest into the visible region but cuts off in the long-wavelength end well before the heavier chalcogenide fibers.[9] Longer wavelengths are transmitted through the addition of heavier elements like Te, Ge, and Se, as shown in Fig. 4. A key feature of essentially all chalcogenide glasses is the strong extrinsic absorption resulting from the bonding of contaminants such as hydrogen, H_2O, and OH⁻ to the elemental cations. In particular, absorption peaks between 4.0 and 4.6 μm are due to S-H or Se-H bonds, and those at 2.78 and 6.3 μm are due to OH⁻ (2.78 μm) and/or molecular water. The hydride impurities are often especially strong and can be deleterious when these fibers are used in chemical sensing applications where the desired chemical signature falls in the region of extrinsic absorption. Another important feature of most of the chalcogenide fibers is that their

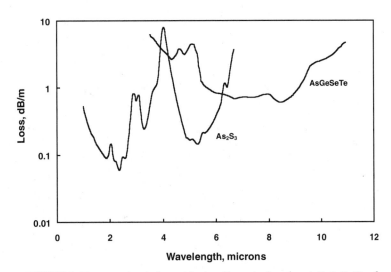

FIGURE 4   Two common chalcogenide glass fibers: As_2S_3 and an AsGeSeTe fiber.[9] Note the many impurity bands pervasive in these fiber systems.

losses are in general much higher than those of the fluoride glasses. In fact, at the important CO_2 laser wavelength of 10.6 μm, the lowest loss is still above 1 dB/m for the Se-based fibers.[20]

14.3   CRYSTALLINE FIBERS

Crystalline IR fibers are an attractive alternative to glass IR fibers because most nonoxide crystalline materials can transmit longer-wavelength radiation than IR glasses and, in the case of sapphire, exhibit some superior physical properties as well.[2] The disadvantage is that crystalline fibers are difficult to fabricate. There are two types of crystalline fiber: single-crystal (SC)[8] and polycrystalline (PC).[22,23] Historically, the first crystalline fiber made was hot-extruded KRS-5 fiber fabricated at Hughes Research Labs in 1975.[24] KRS-5 or TlBrI was chosen because it is very ductile and because it can transmit beyond the 20-μm range required for the intended military surveillance satellite application. In fact, crystalline fibers such as KRS-5 and other halide crystals were initially thought to hold great potential as next-generation ultra-low-loss fibers because their intrinsic loss was predicted to be as low as 10^{-3} dB/m.[24] Unfortunately, this loss was not only never achieved but not even approached experimentally.

PC Fibers

There are many halide crystals that have excellent IR transmission, but only a few have been fabricated into fiber optics. The technique used to make PC fibers is hot extrusion. As a result, only the silver and thallium halides have the requisite physical properties (such as ductility, low melting point, and independent slip systems) to be successfully extruded into fiber. In the hot extrusion process, a single-crystal billet or preform is placed in a heated chamber and the fiber is extruded to net shape through a diamond or tungsten carbide die at a temperature equal to about half the melting point. The final PC fibers are usually from 500 to 900 μm in diameter with no buffer jacket. The polycrystalline structure of the fiber consists of grains on the order of 10 μm or larger in size. The billet may be clad using the rod-in-tube method. In this method, a mixed silver halide such as AgBrCl is used as the core and then a lower-index tube is formed using a Cl^--rich AgBrCl crystal. The extrusion of a core-clad fiber is not as easy to achieve as it is in glass drawing, but Artjushenko et al.[10] at the General Physics Institute (GPI) in Moscow have achieved clad Ag halide fibers with losses nearly as low as those for the core-only Ag halide fiber. Today, the PC Ag halide fibers represent the best PC fibers. KRS-5 is no longer a viable candidate due largely to the toxicity of Tl and the greater flexibility of the Ag halide fibers.

The losses for the Ag halide fibers are shown in Fig. 5. Both the core-only and core-clad fibers are shown, and, as with the other IR fibers, we again see that there are several extrinsic absorption bands. Water is often present at 3 and 6.3 μm and there is sometimes an SO_4^- absorption near 9.6 μm. Furthermore, we note the decreasing attenuation as the wavelength increases. This is a result of λ^{-2} scattering from strain-induced defects in the extruded fiber. An important feature of the data is that the loss at 10.6 μm can be as low as 0.2 dB/m for the core-only fiber and these fibers will transmit to almost 20 μm. These fibers have been used to transmit about 100 W of CO_2 laser power, but the safe limit seems to be 20 to 25 W.[25] This is due to the low melting point of the fibers.

There are several difficulties in handling and working with PC fibers. One is an unfortunate aging effect in which the fiber transmission is observed to decrease over time.[26] Normally the aging loss, which increases uniformly over the entire IR region, is a result of strain relaxation and possible grain growth as the fiber is stored. Another problem is that Ag halides are photosensitive; exposure to visible or UV radiation creates colloidal Ag, which in turn leads to increased losses in the IR. Finally, AgBrCl is corrosive to many metals. Therefore, the fibers should be packaged in dark jackets and connectorized with materials such as Ti, Au, or ceramics.

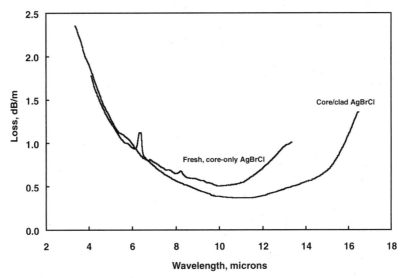

FIGURE 5 Losses in a typical PC silver halide fiber[22] compared to those in recently developed core-clad silver halide fiber.[10]

The mechanical properties of these ductile fibers are quite different from those of glass fibers. The fibers are weak, with ultimate tensile strengths of about 80 MPa for a 50-50 mixture of AgBrCl. However, the main difference between the PC and glass fibers is that the PC fibers plastically deform well before fracture. This plastic deformation leads to increased loss as a result of increased scattering from separated grain boundaries. Therefore, in use, the fibers should not be bent beyond their yield point; too much bending can lead to permanent damage and a region of high loss in the fiber.

SC Fibers

Meter-long lengths of SC fibers have been made from only a small number of the over 80 IR transmissive crystalline materials. Initially some SC fibers were grown by zone-refining methods from the same metal halides used to extrude PC fibers. The idea was that removal of the grain boundaries in the PC fibers would improve the optical properties of the fiber. This did not occur, so most of the crystalline materials chosen today for SC fiber fabrication have been oxides. Compared to halides, oxide materials like Al_2O_3 (sapphire) have the advantage of high melting points, chemical inertness, and the ability to be conveniently melted and grown in air. Currently, sapphire is the most popular SC fiber.[8,27,28]

Sapphire is an insoluble, uniaxial crystal (trigonal structure) with a melting point of over 2000°C. It is an extremely hard and robust material with a usable fiber transmission from about 0.5 to 3.2 μm. Other important physical properties shown in Table 2 include a refractive index equal to 1.75 at 3 μm, a thermal expansion about 10 times higher than that of silica, and a Young's modulus approximately six times greater than that of silica. These properties make sapphire an almost ideal IR fiber candidate for applications less than about 3.2 μm. In particular, this fiber has been used to deliver over 10 W of average power from an Er:YAG laser operating at 2.94 μm.[29]

Sapphire fibers are fabricated using either the edge-defined, film-fed growth (EFG) or the laser-heated pedestal growth (LHPG) techniques.[30] In either method, some or all of the starting sapphire material is melted and an SC fiber is pulled from the melt. In the EFG method, a capillary tube is used to conduct the molten sapphire to a seed fiber, which is drawn slowly into a long fiber. Multiple capillary tubes, which also serve to define the shape and diameter of the fiber, may be placed in one crucible of molten sapphire so that many fibers can be drawn at one time. The LHPG process is a crucibleless technique in which a small molten zone at the tip of an SC sapphire source rod (<2 mm diameter) is created using a CO_2 laser. A seed fiber slowly pulls the SC fiber as the source rod continuously moves into the molten zone to replenish the molten material. Both SC fiber growth methods are very slow (several millimeters per minute) compared to glass fiber drawing. The EFG method, however, has an advantage over LHPG methods because more than one fiber can be continuously pulled at a time. LHPG methods, however, have produced the cleanest and lowest-loss fibers owing to the fact that no crucible is used that can contaminate the fiber. The sapphire fibers grown by these techniques are unclad, pure Al_2O_3 with the C-axis usually aligned along the fiber axis. Fiber diameters range from 100 to 300 μm and lengths are generally less than 2 m. Postcladding techniques mostly involve a Teflon coating using heat-shrink tubing.

The optical properties of the as-grown sapphire fibers are normally inferior to those of the bulk starting material. This is particularly evident in the visible region and is a result of color-center-type defect formation during the fiber drawing. These defects and the resulting absorption can be greatly reduced if the fibers are postannealed in air or oxygen at about 1000°C. In Fig. 6, the losses for LHPG fiber grown at Rutgers University[8] and EFG fiber grown by Saphikon, Inc. (Milford, New Hampshire) are shown. Both fibers have been annealed at 1,000°C to reduce short-wavelength losses. We see that the LHPG fiber has the lowest overall loss. In particular, LHPG fiber loss at the important Er:YAG laser wavelength of 2.94 μm is less than 0.3 dB/m, compared to the intrinsic value of 0.15 dB/m. There are also several impurity absorptions beyond 3 μm that are believed to be due to transition metals like Ti or Fe. Sapphire fibers have been used at temperatures of up to 1400°C without any change in their transmission.

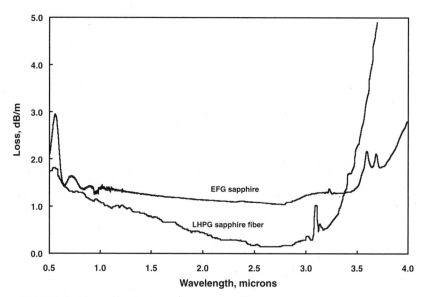

FIGURE 6   SC sapphire fibers grown by the EFG[30] (Saphikon, Inc., Milford, New Hampshire) and LHPG[29] methods.

14.4 HOLLOW WAVEGUIDES

The first optical-frequency hollow waveguides were similar in design to microwave guides. Garmire et al.[31] made a simple rectangular waveguide using aluminum strips spaced 0.5 mm apart by bronze shim stock. Even when the aluminum was not well polished, these guides worked surprisingly well. Losses at 10.6 μm were well below 1 dB/m, and Garmire early demonstrated the high power handling capability of an air-core guide by delivering over 1 kW of CO_2 laser power through this simple structure. These rectangular waveguides, however, never gained much popularity, primarily because their overall dimensions (about 0.5×10 mm) were quite large in comparison to circular-cross-section guides and also because the rectangular guides cannot be bent uniformly in any direction. As a result, hollow circular waveguides with diameters of 1 mm or less fabricated using metal, glass, or plastic tubing are the most common guides today. In general, hollow waveguides are an attractive alternative to conventional solid-core IR fibers for laser power delivery because of the inherent advantage of their air core. Hollow waveguides not only enjoy the advantage of high laser power thresholds but also low insertion loss, no end reflection, ruggedness, and small beam divergence. A disadvantage, however, is a loss on bending, which varies as $1/R$ where R is the bending radius. In addition, the losses for these guides vary as $1/a^3$ where a is the radius of the bore; therefore the loss can be arbitrarily small for a sufficiently large core. The bore size and bending radius dependence of all hollow waveguides are characteristics of these guides not shared by solid-core fibers. Initially these waveguides were developed for medical and industrial applications involving the delivery of CO_2 laser radiation, but more recently they have been used to transmit incoherent light for broadband spectroscopic and radiometric applications.[32–34] Today they are one of the best alternatives for power delivery in IR laser surgery and industrial laser delivery systems, with losses as low as 0.1 dB/m and transmitted CW laser powers as high as 2.7 kW.[35]

Hollow core waveguides may be grouped into two categories: (1) those whose inner core materials have refractive indices greater than 1 (leaky guides) and (2) those whose inner wall materials have refractive indexes less than 1 [attenuated total reflectance (ATR) guides]. Leaky or $n > 1$ guides have metallic and dielectric films deposited on the inside of metallic,[36] plastic,[37] or glass[11] tubing. ATR guides are made from dielectric materials with refractive indices of less than 1 in the wavelength region of interest.[38] Therefore, $n < 1$ guides are fiber-like in that the core index ($n \approx 1$) is greater than the clad index. Hollow sapphire fibers operating at 10.6 μm ($n = 0.67$) are an example of this class of hollow guide.[39]

Hollow Metal and Plastic Waveguides

The earliest circular-cross-section hollow guides were formed using metallic and plastic tubing as the structural members. Miyagi and colleagues in Japan used sputtering methods to deposit Ge,[40] ZnSe, and ZnS[36] coatings on aluminum mandrels. Then a final layer of Ni was electroplated over these coatings before the aluminum mandrel was removed by chemical leaching. The final structure was then a flexible Ni tube with optically thick dielectric layers on the inner wall to enhance the reflectivity in the infrared. Croitoru and colleagues[41] at Tel Aviv University applied Ag followed by AgI coatings on the inside of polyethylene and Teflon tubing to make a very flexible waveguide. Similar Ag and Ag halide coatings were deposited inside Ag tubes by Morrow and colleagues.[42]

Hollow Glass Waveguides

The most popular structure today is the hollow glass waveguide (HGW) developed initially at Rutgers University.[43] The advantage of glass tubing is that it is much smoother than either metal and plastic tubing and, therefore, the scattering losses are less. HGWs are fabricated using wet chemistry methods to first deposit a layer of Ag on the inside of silica glass tubing and then to form a dielectric layer of AgI over the metallic film by converting some of the Ag to AgI. The silica tubing used has a polymer coating of UV acrylate or polyimide on the outside surface to preserve the mechanical strength. The thickness of the AgI is optimized to give high reflectivity at a particular laser wavelength or range of wavelengths. Using these techniques, HGWs have been fabricated with lengths as long as 13 m and bore sizes ranging from 250 to 1300 μm.

The spectral loss for an HGW with a 530-μm bore is given in Fig. 2. This HGW was designed for an optimal response at 10 μm. The peaks at about 3 and 5 μm are not absorption peaks but rather interference bands due to thin-film optical effects. For broadband applications and shorter-wavelength applications, a thinner AgI coating would be used to shift the interference peaks to shorter wavelengths. For such HGWs the optical response will be nearly flat without interference bands in the far IR fiber region of the spectrum. The data in Fig. 7 shows the straight loss measured using a CO_2 and Er:YAG laser for different bore sizes. An important feature of this data is the $1/a^3$ dependence of loss on bore size predicted by the theory of Marcatili and Schmeltzer.[44] In general, the losses are less than 0.5 dB/m at 10 μm for bore sizes larger than ~400 μm. Furthermore, the data at 10.6 μm agrees well with the calculated values, but at 3 μm the measured losses are somewhat above those predicted by Marcatili and Schmeltzer. This is a result of increased scattering at the shorter wavelengths from the metallic and dielectric films. The bending loss depends on many factors such as the quality of the films, the bore size, and the uniformity of the silica tubing. A typical bending loss curve for an HGW with a 530-μm bore measured with a CO_2 laser is given in Fig. 8. The losses are seen to increase linearly with increasing curvature as predicted. It is important to note that while there is an additional loss on bending for any hollow guide, it does not necessarily mean that this restricts the use of hollow guides in power delivery or sensor applications. Normally most fiber delivery systems have rather large bend radii and therefore a minimal amount of the guide is under tight bending conditions and the bending loss is low. From the

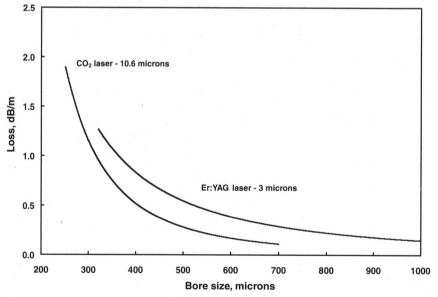

FIGURE 7 Straight losses measured in hollow glass waveguides with Ag/AgI films. The guide labeled "CO_2 laser" was designed for optimal transmission at 10.6 μm while that labeled "Er:YAG laser" was designed for optimal transmission at 3 μm. Note that the loss varies approximately as $1/a^3$.

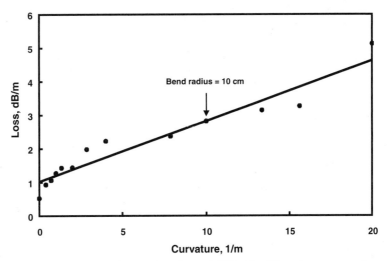

FIGURE 8   Additional loss on bending an HGW with a 530-μm bore, measured at 10.6 μm. The loss is seen to increase as the curvature increases.

data in Fig. 8 one can calculate the bending loss contribution for an HGW link by assuming some modest bends over a small section of guide length. An additional important feature of hollow waveguides is that they are nearly single mode. This is a result of the strong dependence of loss on the fiber mode parameter. That is, the loss of high-order modes increases as the square of the mode parameter, so even though the guides are very multimode, in practice only the lowest-order modes propagate. This is particularly true for the small-bore (<300 μm) guides, in which virtually only the lowest-order HE_{11} mode is propagated.

HGWs have been used quite successfully in IR laser power delivery and, more recently, in some sensor applications. Modest CO_2 and Er:YAG laser powers below about 80 W can be delivered without difficulty. At higher powers, water-cooling jackets have been placed around the guides to prevent laser damage. The highest CO_2 laser power delivered through a water-cooled hollow metallic waveguide with a bore of 1800 μm was 2700 W, and the highest power through a water-cooled HGW with a 700-μm bore was 1040 W.[45] Sensor applications include gas and temperature measurements. A coiled HGW filled with gas can be used in place of a more complex and costly White cell to provide an effective means for gas analysis. Unlike evanescent wave spectroscopy, in which light is coupled out of a solid-core-only fiber into media in contact with the core, all of the light is passing through the gas in the hollow guide cell, making this a sensitive, quick-response fiber sensor. Temperature measurements may be aided by using an HGW to transmit blackbody radiation from a remote site to an IR detector. Such an arrangement has been used to measure jet engine temperatures.

14.5   *SUMMARY AND CONCLUSIONS*

During the past 25 years of the development of IR fibers, there has been a great deal of fundamental research designed to produce a fiber with optical and mechanical properties close to those of silica. We can see that today we are still far from that Holy Grail, but some viable IR fibers have emerged that, as a class, can be used to address some of the needs for a fiber that can transmit greater than 2 μm. Yet we are still limited with the current IR fiber technology by high loss and low strength. Nevertheless, more applications are being found for IR fibers as users become aware of their limitations and, more importantly, how to design around their properties.

There are two near-term applications of IR fibers: laser power delivery and sensors. An important future application for these fibers, however, may be more in active fiber systems like the Er- and Pr-doped fluoride fibers and emerging doped chalcogenide fibers. In regards to power delivery fibers, the best choice seems to be hollow waveguides for CO_2 lasers and SC sapphire, germanate glass, or HGWs for Er:YAG laser delivery. Chemical, temperature, and imaging bundles make use mostly of solid-core fibers. Evanescent wave spectroscopy (EWS) using chalcogenide and fluoride fibers is quite successful. A distinct advantage of an IR-fiber EWS sensor is that the signature of the analyte is often very strong in the infrared or finger-print region of the spectrum. Temperature sensing generally involves the transmission of blackbody radiation. IR fibers can be very advantageous at low temperatures, especially near room temperature, where the peak in the blackbody radiation is near 10 μm. Finally, there is an emerging interest in IR imaging using coherent bundles of IR fibers. Several thousand chalcogenide fibers have been bundled by Amorphous Materials (Garland, Texas) to make an image bundle for the 3- to 10-μm region.

14.6   REFERENCES

1. N. S. Kapany and R. J. Simms, "Recent Developments of Infrared Fiber Optics," *Infrared Phys.* **5**:69 (1965).

2. J. A. Harrington, *Selected Papers on Infrared Fiber Optics,* Milestone Series, vol. MS-9, SPIE Press, Bellingham, Washington, 1990.

3. T. Katsuyama and H. Matsumura, *Infrared Optical Fibers,* Adam Hilger, Bristol, UK, 1989.

4. I. Aggarwal and G. Lu, *Fluoride Glass Optical Fiber,* Academic Press, San Diego, 1991.

5. P. France, M. G. Drexhage, J. M. Parker, M. W. Moore, S. F. Carter, and J. V. Wright, *Fluoride Glass Optical Fibers,* CRC Press, Boca Raton, Florida, 1990.

6. J. Sanghera and I. Aggarwal, *Infrared Fiber Optics,* CRC Press, Boca Raton, Florida, 1998.

7. S. F. Carter, M. W. Moore, D. Szebesta, D. Ransom, and P. W. France, "Low Loss Fluoride Fibre by Reduced Pressure Casting," *Electron. Lett.* **26**:2115–2117 (1990).

8. R. Nubling and J. A. Harrington, "Optical Properties of Single-Crystal Sapphire Fibers," *Appl. Opt.* **36**:5934–5940 (1997).

9. J. Nishii, S. Morimoto, I. Inagawa, R. Iizuka, T. Yamashita, and T. Yamagishi, "Recent Advances and Trends in Chalcogenide Glass Fiber Technology: A Review," *J. Non-Cryst. Sol.* **140**:199–208 (1992).

10. V. Artjushenko, V. Ionov, K. I. Kalaidjian, A. P. Kryukov, E. F. Kuzin, A. A. Lerman, A. S. Prokhorov, E. V. Stepanov, K. Bakhshpour, K. B. Moran, and W. Neuberger, "Infrared Fibers: Power Delivery and Medical Applications," *Proc. SPIE* **2396**:25–36 (1995).

11. Y. Matsuura, T. Abel, and J. A. Harrington, "Optical Properties of Small-Bore Hollow Glass Wave-guides," *Appl. Opt.* **34**:6842–6847 (1995).

12. P. Kaiser, A. C. Hart Jr., and L. L. Blyler, "Low Loss FEP-clad Silica Fibers," *Appl. Opt.* **14**:156 (1975).

13. M. Poulain, M. Chanthanasinh, and J. Lucas, "New Fluoride Glasses," *Mat. Res. Bull.* **12**:151–156 (1977).

14. D. Tran, G. H. Sigel, and B. Bendow, "Heavy Metal Fluoride Glasses and Fibers: A Review," *J. Light-wave Technol.* **LT-2**:566–586 (1984).

15. K. Itoh, K. Miura, M. Masuda, M. Iwakura, and T. Yamagishi, "Low-Loss Fluorozirco-Aluminate Glass Fiber," in *Proceedings of 7th International Symposium on Halide Glass,* Center for Advanced Materials Technology, Monash University, Lorne, Australia, 1991, pp. 2.7–2.12.

16. P. W. France, S. F. Carter, M. W. Moore, and C. R. Day, "Progress in Fluoride Fibres for Optical Communications," *Brit. Telecom Tech. J.* **5**:28–44 (1987).

17. M. J. Matthewson, C. R. Kurkjian, and S. T. Gulati, "Strength Measurement of Optical Fibers by Bending," *J. Am. Cer. Soc.* **69**:815–821 (1986).

18. S. Kobayashi, N. Shibata, S. Shibata, and T. Izawa, "Characteristics of Optical Fibers in Infrared Wavelength Region," *Rev. Electrical Comm. Lab.* **26**:453–467, 1978.

19. D. Tran, Heavy Metal-Oxide Glass Optical Fibers for Use in Laser Medical Surgery, U.S. Patent 5,274,728, December 28, 1993.

20. Y. Kanamori, Y. Terunuma, and T. Miyashita, "Preparation of Chalcogenide Optical Fiber," *Rev. Electrical Comm. Lab.* **32**:469–477 (1984).

21. J. Nishii, T. Yamashita, T. Tamagishi, C. Tanaka, and H. Sone, "As$_2$S$_3$ Fibre for Infrared Image Bundle," *Int. J. Optoelectron.* **7**:209–216 (1992).

22. V. G. Artjushenko, L. N. Butvina, V. V. Vojtsekhovsky, E. M. Dianov, and J. G. Kolesnikov, "Mechanisms of Optical Losses in Polycrystalline KRS-5 Fibers," *J. Lightwave Technol.* **LT-4**:461–465 (1986).

23. A. Sa'ar, F. Moser, S. Akselrod, and A. Katzir, "Infrared Optical Properties of Polycrystalline Silver Halide Fibers," *Appl. Phys. Lett.* **49**:305–307 (1986).

24. D. A. Pinnow, A. L. Gentile, A. G. Standlee, A. J. Timper, and L. M. Hobrock, "Polycrystalline Fiber Optical Waveguides for Infrared Transmission," *Appl. Phys. Lett.* **33**:28–29 (1978).

25. K. Takahashi, N. Yoshida, and M. Yokota, "Optical Fibers for Transmitting High-power CO$_2$ Laser Beam," *Sumitomo Electric Tech. Rev.* **23**:203–210 (1984).

26. J. A. Wysocki, R. G. Wilson, A. G. Standlee, A. C. Pastor, R. N. Schwartz, A. R. Williams, G.-D. Lei, and L. Kevan, "Aging Effects in Bulk and Fiber TlBr-TlI," *J. Appl. Phys.* **63**:4365–4371 (1988).

27. D. H. Jundt, M. M. Fejer, and R. L. Byer, "Characterization of Single-Crystal Sapphire Fibers for Optical Power Delivery Systems," *Appl. Phys. Lett.* **55**:2170–2172 (1989).

28. R. S. F. Chang, V. Phomsakha, and N. Djeu, "Recent Advances in Sapphire Fibers," *Proc. SPIE* **2396**:48–53 (1995).

29. R. Nubling and J. A. Harrington, "Single-Crystal LHPG Sapphire Fibers for Er:YAG Laser Power Delivery," *Appl. Opt.* **37**:4777–4781 (1998).

30. H. E. LaBelle, "EFG, the Invention and Application to Sapphire Growth," *J. Cryst. Growth* **50**:8–17 (1980).

31. E. Garmire, T. McMahon, and M. Bass, "Flexible Infrared Waveguides for High-Power Transmission," *J. Quant. Elect.* **QE-16**:23–32 (1980).

32. S. J. Saggese, J. A. Harrington, and G. H. Sigel Jr., "Attenuation of Incoherent Infrared Radiation in Hollow Sapphire and Silica Waveguides," *Opt. Lett.* **16**:27–29 (1991).

33. M. Saito, Y. Matsuura, M. Kawamura, and M. Miyagi, "Bending Losses of Incoherent Light in Circular Hollow Waveguides," *J. Opt. Soc. Am. A* **7**:2063–2068 (1990).

34. M. Saito and K. Kikuchi, "Infrared Optical Fiber Sensors," *Opt. Rev.* **4**:527–538 (1997).

35. A. Hongo, K. Morosawa, K. Matsumoto, T. Shiota, and T. Hashimoto, "Transmission of Kilowatt-Class CO$_2$ Laser Light Through Dielectric-Coated Metallic Hollow Waveguides for Material Processing," *Appl. Opt.* **31**:5114–5120 (1992).

36. Y. Matsuura, M. Miyagi, and A. Hongo, "Fabrication of Low-Loss Zinc-Selenide Coated Silver Hollow Waveguides for CO$_2$ Laser Light," *J. Appl. Phys.* **68**:5463–5466 (1990).

37. M. Alaluf, J. Dror, R. Dahan, and N. Croitoru, "Plastic Hollow Fibers as a Selective Infrared Radiation Transmitting Medium," *J. Appl. Phys.* **72**:3878–3883 (1992).

38. C. C. Gregory and J. A. Harrington, "Attenuation, Modal, Polarization Properties of $n < 1$, Hollow Dielectric Waveguides," *Appl. Opt.* **32**:5302–5309 (1993).

39. J. A. Harrington and C. C. Gregory, "Hollow Sapphire Fibers for the Delivery of CO$_2$ Laser Energy," *Opt. Lett.* **15**:541–543 (1990).

40. M. Miyagi, Y. Shimada, A. Hongo, K. Sakamoto, and S. Nishida, "Fabrication and Transmission Properties of Electrically Deposited Germanium-Coated Waveguides for Infrared Radiation," *J. Appl. Phys.* **60**:454–456 (1986).

41. O. Morhaim, D. Mendlovic, I. Gannot, J. Dror, and N. Croitoru, "Ray Model for Transmission of Infrared Radiation Through Multibent Cylindrical Waveguides," *Opt. Eng.* **30**:1886–1891 (1991).

42. P. Bhardwaj, O. J. Gregory, C. Morrow, G. Gu, and K. Burbank, "Performance of a Dielectric-Coated Monolithic Hollow Metallic Waveguide," *Mat. Lett.* **16**:150–156 (1993).

43. T. Abel, J. Hirsch, and J. A. Harrington, "Hollow Glass Waveguides for Broadband Infrared Transmission," *Opt. Lett.* **19**:1034–1036 (1994).

44. E. A. J. Marcatili and R. A. Schmeltzer, "Hollow Metallic and Dielectric Waveguides for Long Distance Optical Transmission and Lasers," *Bell Syst. Tech. J.* **43**:1783–1809 (1964).

45. R. K. Nubling and J. A. Harrington, "Hollow-Waveguide Delivery Systems for High-Power, Industrial CO_2 Lasers," *Appl. Opt.* **34**:372–380 (1996).

CHAPTER 15

OPTICAL FIBER SENSORS

Richard O. Claus
Virginia Tech
Blacksburg, Virginia

Ignacio Matias and Francisco Arregui
Public University Navarra
Pamplona, Spain

15.1 INTRODUCTION

Optical fiber sensors are a broad topic. The objective of this chapter is to briefly summarize the fundamental properties of representative types of optical fiber sensors and how they operate. Four different types of sensors are evaluated systematically on the basis of performance criteria such as resolution, dynamic range, cross-sensitivity to multiple ambient perturbations, fabrication, and demodulation processes. The optical fiber sensing methods that will be investigated include well-established technologies such as fiber Bragg grating (FBG)–based sensors, and rapidly evolving measurement techniques such as those involving long-period gratings (LPGs). Additionally, two popular versions of Fabry-Perot interferometric sensors (intrinsic and extrinsic) are evaluated.

The outline of this chapter is as follows. The principles of operation and fabrication processes of each of the four sensors are discussed separately. The sensitivity of the sensors to displacement and simultaneous perturbations such as temperature is analyzed. The overall complexity and performance of a sensing technique depends heavily on the signal demodulation process. Thus, the detection schemes for all four sensors are discussed and compared on the basis of their complexity. Finally, a theoretical analysis of the cross-sensitivities of the four sensing schemes is presented and their performance is compared.

Measurements of a wide range of physical measurands by optical fiber sensors have been investigated for more than 20 years. Displacement measurements using optical fiber sensors are typical of these, and both embedded and surface-mounted configurations have been reported by researchers in the past.[1] Fiber-optic sensors are small in size, are immune to electromagnetic interference, and can be easily integrated with existing optical fiber communication links. Such sensors can typically be easily multiplexed, resulting in distributed networks that can be used for health monitoring of integrated, high-performance materials and structures.

Optical fiber sensors of displacement are perhaps the most basic of all fiber sensor types because they may be configured to measure many other related environmental factors. They should possess certain important characteristics. First, they should either be insensitive to ambient fluctuations in temperature and pressure or should employ demodulation techniques that compensate for changes in the output signal due to these additional perturbations. In an embedded configuration, the sensors for axial strain measurements should have minimum cross-sensitivity to other strain states. The sensor signal should itself be simple and easy to demodulate. Nonlinearities in the output require expensive decoding procedures or necessitate pre-calibration and sensor-to-sensor incompatibility. The sensor should ideally provide an absolute and real-time displacement or strain measurement in a form that can be easily processed. For environments where large strain magnitudes are expected, the sensor should have a large dynamic range while at the same time maintaining the desired sensitivity. We now discuss each of the four sensing schemes individually and present their relative advantages and shortcomings.

15.2 EXTRINSIC FABRY-PEROT INTERFEROMETRIC SENSORS

The extrinsic Fabry-Perot interferometric (EFPI) sensor, proposed by a number of groups and authors, is one of the most popular fiber-optic sensors used for applications in health monitoring of smart materials and structures.[3] As the name suggests, the EFPI is an interferometric sensor in which the detected intensity is modulated by the parameter under measurement. The simplest configuration of an EFPI is shown in Fig. 1.

The EFPI system consists of a single-mode laser diode that illuminates a Fabry-Perot cavity through a fused biconical tapered coupler. The cavity is formed between an input single-mode fiber and a reflecting target element that may be a fiber. Since the cavity is external to the lead-in/lead-out fiber, the EFPI sensor is independent of transverse strain and small ambient temperature fluctuations. The input fiber and the reflecting fiber are typically aligned using a hollow core tube as shown in Fig. 10. For optical fibers with uncoated ends, Fresnel reflection of approximately 4 percent results at the glass-to-air and air-to-glass interfaces that

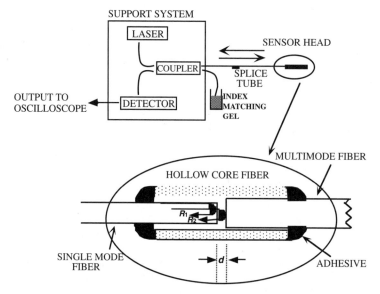

FIGURE 1 Extrinsic Fabry-Perot interferometric (EFPI) sensor and system.

define the cavity. The first reflection at the glass-air interface R_1, called the *reference reflection,* is independent of the applied perturbation. The second reflection at the air-glass interface R_2, termed the *sensing reflection,* is dependent on the length of the cavity d, which in turn is modulated by the applied perturbation. These two reflections interfere (provided $2d < L_c$, the coherence length of the light source) and the intensity I at the detector varies as a function of the cavity length,

$$I = I_0 \cos\left(\frac{4\pi}{\lambda}d\right) \tag{1}$$

where I_0 is the maximum value of the output intensity and λ is the center wavelength of the light source, here assumed to be a laser diode.

The typical intensity-versus-displacement transfer function curve [Eq.(1)] for an EFPI sensor is shown in Fig. 2. Small perturbations that result in operation around the quiescent or Q point of the sensor lead to an approximately linear variation in output intensity versus applied displacement. For larger displacements, the output signal is not a linear function of the input signal, and the output signal may vary over several sinusoidal periods. In this case, a *fringe* in the output signal is defined as the change in intensity from a maximum to a maximum, or from a minimum to a minimum, so each fringe corresponds to a change in the cavity length by half of the operating wavelength λ. The change in the cavity length Δd is then employed to calculate the strain using the expression

$$\varepsilon = \frac{\Delta d}{L} \tag{2}$$

where L is defined as the gauge length of the sensor and is typically the distance between two points where the input and reflecting fibers are bonded to the hollow-core support tube.

The EFPI sensor has been used for the analysis of materials and structures.[1,3] The relatively low temperature sensitivity of the sensor element, due to the opposite directional expansion of the fiber and tube elements, makes it attractive for the measurement of strain and displacement in environments where the temperature is not anticipated to change over a wide range.

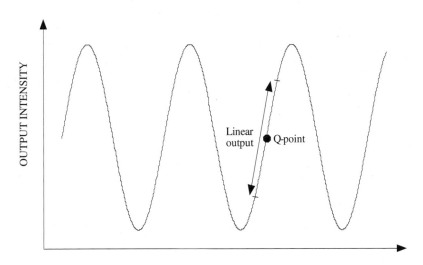

FIGURE 2 EFPI transfer function curve.

The EFPI sensor is capable of measuring subangstrom displacements with strain resolution better than 1 με and a dynamic range greater than 10,000 με. Moreover, the large bandwidth simplifies the measurement of highly cyclical strain. The sensor also allows single-ended operation and is hence suitable for applications where ingress to and egress from the sensor location are important. The sensor requires simple and inexpensive fabrication equipment and an assembly time of a few minutes. Additionally, since the cavity is external to the fibers, transverse strain components that tend to influence the response of similar intrinsic sensors through Poisson-effect cross-coupling have negligible effect on the EFPI sensor output.

15.3   *INTRINSIC FABRY-PEROT INTERFEROMETRIC SENSORS*

The intrinsic Fabry-Perot interferometric (IFPI) sensor is similar in operation to its extrinsic counterpart, but significant differences exist in the configurations of the two sensors.[4] The basic IFPI sensor is shown in Fig. 3. An optically isolated laser diode is used as the optical source to one of the input arms of a bidirectional 2×2 coupler. The Fabry-Perot cavity is formed internally by fusing a small length of single-mode fiber to one of the output legs of the coupler. As shown in Fig. 3, the reference (R) and sensing (S) reflections interfere at the detector to again provide a sinusoidal intensity variation versus cavity path length modulation. The cavity can also be implemented by introducing two Fresnel or other reflectors along the length of a single fiber. The photosensitivity effect in germanosilicate fibers has been used in the past to fabricate broadband grating-based reflector elements to define such an IFPI cavity.[5] Since the cavity is formed within an optical fiber, changes in the refractive index of the fiber due to the applied perturbation can significantly alter the phase of the sensing signal S. Thus the intrinsic cavity results in the sensor being sensitive to ambient temperature fluctuations and all states of strain.

The IFPI sensor, like all other interferometric signals, has a nonlinear output that complicates the measurement of large-magnitude strain. This can again be overcome by operating the sensor in the linear regime around the Q point of the sinusoidal transfer function curve. The main limitation of the IFPI strain sensor is that the photoelastic-effect-induced change in index of refraction results in a nonlinear relationship between the applied perturbation and the change in cavity length. For most IFPI sensors, the change in the propagation constant of the fundamental mode dominates the change in cavity length. Thus IFPIs are highly susceptible to temperature changes and transverse strain components.[6] In embedded applications, the sensitivity to all of the strain components can result in complex signal output. The process of fabricating an IFPI strain sensor is more complicated than that for the EFPI sensor since the sensing cavity of the IFPI sensor must be formed within the optical fiber by some special procedure. The strain resolution of IFPI sensors is approximately 1 με with an operating range

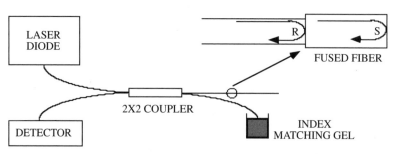

FIGURE 3   The intrinsic Fabry-Perot interferometric (IFPI) sensor.

greater than 10,000 με. IFPI sensors also suffer from drift in the output signal due to variations in the polarization state of the input light.

Thus the preliminary analysis shows that the extrinsic version of the Fabry-Perot optical fiber sensor seems to have an overall advantage over its intrinsic counterpart. The extrinsic sensor has negligible cross-sensitivity to temperature and transverse strain. Although the strain sensitivities, dynamic ranges, and bandwidths of the two sensors are comparable, the IFPIs can be expensive and cumbersome to fabricate due to the intrinsic nature of the sensing cavity.

The extrinsic and intrinsic Fabry-Perot interferometric sensors possess nonlinear sinusoidal outputs that complicate signal processing at the detector. Although intensity-based sensors have a simple output variation, they suffer from limited sensitivity to strain or other perturbations of interest. Grating-based sensors have recently become popular as transducers that provide wavelength-encoded output signals that can typically be easily demodulated to derive information about the perturbation under investigation. We next discuss the advantages and drawbacks of Bragg grating sensing technology. The basic operating mechanism of Bragg grating-based strain sensors is then reviewed and the expressions for strain resolution are obtained. These sensors are then compared to the recently developed long-period grating devices in terms of fabrication process, cross-sensitivity to multiple measurands, and simplicity of signal demodulation.

15.4   FIBER BRAGG GRATING SENSORS

The phenomenon of photosensitivity was discovered by Hill and coworkers in 1978.[7] It was found that permanent refractive index changes could be induced in optical fibers by exposing the germanium-doped core of a fiber to intense light at 488 or 514 nm. Hill found that a sinusoidal modulation of index of refraction in the core created by the spatial variation of such an index-modifying beam gives rise to refractive index grating that can be used to couple the energy in the fundamental guided mode to various guided and lossy modes. Later Meltz et al.[8] suggested that photosensitivity is more efficient if the fiber is side-exposed to a writing beam at wavelengths close to the absorption wavelength (242 nm) of the germanium defects in the fiber. The side-writing process simplified the fabrication of Bragg gratings, and these devices have recently emerged as highly versatile components for optical fiber communication and sensing systems. Recently, loading of the fibers with hydrogen prior to writing has been used to produce order-of-magnitude larger changes in index in germanosilicate fibers.[9]

Principle of Operation

Bragg gratings in optical fibers are based on a phase-matching condition between propagating optical modes. This phase-matching condition is given by

$$k_g + k_c = k_B \tag{3}$$

where k_g, k_c, and k_B are, respectively, the wave vectors of the coupled guided mode, the resulting coupling mode, and the grating. For a first-order interaction, $k_B = 2\pi/\Lambda$, where Λ is the spatial period of the grating. In terms of propagation constants, this condition reduces to the general form of interaction for mode coupling due to a periodic perturbation

$$\Delta\beta = \frac{2\pi}{\Lambda} \tag{4}$$

where $\Delta\beta$ is the difference in the propagation constants of the two modes involved in mode coupling, where both modes are assumed to travel in the same direction.

Fiber Bragg gratings (FBGs) involve the coupling of the forward-propagating fundamental LP_{01} in a single-mode fiber to the reverse-propagating LP_{01} mode.[10] Here, consider a single-mode fiber with β_{01} and $-\beta_{01}$ as the propagation constants of the forward- and reverse-propagating fundamental LP_{01} modes. To satisfy the phase-matching condition,

$$\Delta\beta = \beta_{01} - (-\beta_{01}) = \frac{2\pi}{\Lambda} \tag{5}$$

where $\beta_{01} = 2\pi n_{eff}/\lambda$, n_{eff} is the effective index of the fundamental mode, and λ is the free-space wavelength of the source. Equation (5) reduces to[10]

$$\lambda_B = 2\Lambda n_{eff} \tag{6}$$

where λ_B is termed the *Bragg wavelength*—the wavelength at which the forward-propagating LP_{01} mode couples to the reverse-propagating LP_{01} mode. Such coupling is wavelength dependent, since the propagation constants of the two modes are a function of the wavelength. Hence, if an FBG element is interrogated using a broadband optical source, the wavelength at which phase matching occurs is back-reflected. This back-reflected wavelength is a function of the grating period Λ and the effective index n_{eff} of the fundamental mode as shown in Eq. (6). Since strain and temperature effects can modulate both of these parameters, the Bragg wavelength is modulated by both of these external perturbations. The resulting spectral shifts are utilized to implement FBGs for sensing applications.

Figure 4 shows the mode-coupling mechanism in fiber Bragg gratings using a β-plot. Since the difference in propagation constants ($\Delta\beta$) between the modes involved in coupling is large, we see from Eq. (4) that only a small period, Λ, is needed to induce this mode coupling. Typically for optical fiber communication system applications the value of λ_B is approximately 1.5 μm. From Eq. (6), Λ is thus approximately 0.5 μm for $n_{eff} = 1.5$, the approximate index of refraction of the glass in a fiber. Due to the small period, on the order of 1 μm, FBGs are typically classified as short-period gratings (SPGs).

Bragg Grating Sensor Fabrication

Fiber Bragg gratings have commonly been manufactured using two side-exposure techniques, namely *interferometric* and *phase mask* methods. The interferometric method, shown in Fig. 5, uses an ultraviolet (UV) writing beam at 244 or 248 nm, split into two parts of approximately the same intensity by a beam splitter.[8] The two beams are focused on a portion of the Ge-doped fiber, whose protective coating has been removed using cylindrical lenses. The period of the resulting interference pattern, and hence the period of the Bragg grating element to be written, is varied by altering the mutual angle θ. A limitation of this method is that any relative vibration of the pairs of mirrors and lenses can lead to the degradation of the

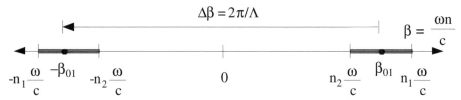

FIGURE 4   Mode-coupling mechanism in fiber Bragg gratings. The large value of $\Delta\beta$ in FBGs requires a small value of the grating periodicity Λ. The hatched regions represent the guided modes in the forward ($\beta > 0$) and reverse ($\beta > 0$) directions.

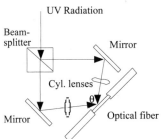
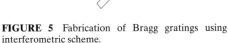

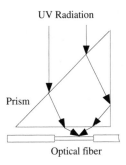

FIGURE 5 Fabrication of Bragg gratings using interferometric scheme.

FIGURE 6 Bragg grating fabrication using prism method.

quality of the fringe pattern and the fabricated grating; thus the entire system has a stringent stability requirement. To overcome this drawback, Kashyap[10] proposed a novel interferometer technique in which the path difference between the interfering UV beams is produced by propagation through a right-angled prism, as shown in Fig. 6. This geometry is inherently stable because both beams are perturbed similarly by any prism vibration.

The phase mask technique has gained popularity as an efficient holographic side-writing procedure for grating fabrication.[11] In this method, shown in Fig. 7, an incident UV beam is diffracted into −1, 0, and +1 orders by a relief grating typically generated on a silica plate by electron beam exposure and plasma etching. The two first diffraction orders undergo total internal reflection at the glass-air interface of a rectangular prism and interfere at the location of the fiber placed directly behind the mask. This technique is wavelength specific, since the period of the resulting two-beam interference pattern is uniquely determined by the diffraction angle of −1 and +1 orders and thus the properties of the phase mask. Obviously, different phase masks are required for the fabrication of gratings at different Bragg wavelengths. A setup for actively monitoring the growth of a grating in the transmission mode during fabrication is shown in Fig. 8.

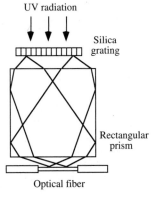

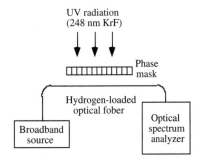

FIGURE 7 Phase mask method of fabricating Bragg gratings.

FIGURE 8 Setup to write Bragg gratings in germanosilicate fibers.

Bragg Grating Sensors

From Eq. (6) we see that a change in the value of n_{eff} and/or Λ can cause the Bragg wavelength λ to shift. This fractional change in the resonance wavelength $\Delta\lambda/\lambda$ is given by

$$\frac{\Delta\lambda}{\lambda} = \frac{\Delta\Lambda}{\Lambda} + \frac{\Delta n_{eff}}{n_{eff}} \qquad (7)$$

where $\Delta\Lambda/\Lambda$ and $\Delta n_{eff}/n_{eff}$ are the fractional changes in the period and the effective index, respectively. The relative magnitudes of the two changes depend on the type of perturbation to which the grating is subjected. For most applications the effect due to change in effective index is the dominating mechanism.

An axial strain ε in the grating changes the grating period and the effective index and results in a shift in the Bragg wavelength, given by

$$\frac{1}{\lambda}\frac{\Delta\lambda}{\varepsilon} = \frac{1}{\Lambda}\frac{\Delta\Lambda}{\varepsilon} + \frac{1}{n_{eff}}\frac{\Delta n_{eff}}{\varepsilon} \qquad (8)$$

The first term on the right side of Eq. (8) is unity, while the second term has its origin in the photoelastic effect. An axial strain in the fiber serves to change the refractive index of both the core and the cladding. This results in a variation in the value of the effective index of glass. The photoelastic or strain-optic coefficient is approximately −0.27. Thus, the variations in n_{eff} and Λ due to strain have contrasting effects on the Bragg peak. The fractional change in the Bragg wavelength due to axial strain is 0.73ε, or 73 percent of the applied strain. At 1550 and 1300 nm, the shifts in the resonance wavelength are 11 nm/%ε and 9 nm/%ε, respectively. An FBG at 1500 nm shifts by 1.6 nm for every 100°C rise in temperature.[7]

Limitations of Bragg Grating Strain Sensors

The primary limitation of Bragg grating sensors is the complex and expensive fabrication technique. Although side-writing is commonly being used to manufacture these gratings, the requirement of expensive phase masks increases the cost of the sensing system. In the interferometric technique, stability of the setup is a critical factor in obtaining high-quality gratings. Since index changes of the order of 10^{-3} are required to fabricate these gratings, laser pulses of high energy levels are necessary.

The second primary limitation of Bragg gratings is their limited bandwidth. The typical value of the full width at half-maximum (FWHM) is between 0.1 and 1 nm. Although higher bandwidths can be obtained by chirping the index or period along the grating length, this adds to the cost of the grating fabrication. The limited bandwidth requires high-resolution spectrum analysis to monitor the grating spectrum. Kersey and Berkoff[12] have proposed an unbalanced Mach-Zender interferometer to detect the perturbation-induced wavelength shift. Two unequal arms of the Mach-Zender interferometer are excited by the back reflection from a Bragg grating sensor element. Any change in the input optical wavelength modulates the phase difference between the two arms and results in a time-varying sinusoidal intensity at the output. This interference signal can be related to the shift in the Bragg peak and the magnitude of the perturbation can be obtained. Recently, modal interferometers have also been proposed to demodulate the output of a Bragg grating sensor.[13] The unbalanced interferometers are also susceptible to external perturbations and hence need to be isolated from the parameter under investigation. Moreover, the nonlinear output may require fringe counting, which can be complicated and expensive. Additionally, a change in the perturbation polarity at the maxima or minima of the transfer function curve will not be detected by this demodulation scheme. To overcome this limitation, two unbalanced interferometers may be employed for dynamic measurements.

Cross-sensitivity to temperature leads to erroneous displacement measurements in applications where the ambient temperature has a temporal variation. So a reference grating used to measure temperature change may be utilized to compensate for the output of the strain sensor. Recently, temperature-independent sensing has been demonstrated using chirped gratings written in tapered optical fibers.[14]

Finally, the sensitivity of fiber Bragg grating strain sensors may not be adequate for certain applications. This sensitivity of the sensor depends on the minimum detectable wavelength shift at the receiver. Although excellent wavelength resolution can be obtained with unbalanced interferometric detection techniques, standard spectrum analysis systems typically provide a resolution of 0.1 nm. At 1300 nm, this minimum detectable change in wavelength corresponds to a strain resolution of 111 $\mu\varepsilon$. Hence, in applications where strains smaller than 100 $\mu\varepsilon$ are anticipated, Bragg grating sensors may not be practical. The dynamic range of strain measurement can be as much as 15,000 $\mu\varepsilon$.

15.5   *LONG-PERIOD GRATING SENSORS*

This section discusses the use of novel long-period gratings (LPGs) as displacement-sensing devices. We analyze the principle of operation of these gratings, their fabrication process, typical experimental evaluation, their demodulation process, and their cross-sensitivity to ambient temperature.

Principle of Operation

Long-period gratings that couple the fundamental guided mode to different guided modes have been demonstrated.[15,16] Gratings with longer periodicities that involve coupling of a guided mode to forward-propagating cladding modes were recently proposed by Vengsarkar et al.[17,18] As discussed previously, fiber gratings satisfy the Bragg phase-matching condition between the guided and cladding or radiation modes or another guided mode. This wavelength-dependent phase-matching condition is given by

$$\beta_{01} - \beta = \Delta\beta = \frac{2\pi}{\Lambda} \tag{9}$$

where Λ is the period of the grating and β_{01} and β are the propagation constant of the fundamental guided mode and the mode to which coupling occurs, respectively.

For conventional fiber Bragg gratings, the coupling of the forward-propagating LP_{01} mode occurs to the reverse-propagating LP_{01} mode ($\beta = -\beta_{01}$). Since $\Delta\beta$ is large in this case, as shown in Fig. 9a, the grating periodicity is small, typically on the order of 1 μm. Unblazed long-period gratings couple the fundamental mode to the discrete and circularly symmetric, forward-propagating cladding modes ($\beta = \beta^n$), resulting in smaller values of $\Delta\beta$, as shown in Fig. 9b, and hence periodicities ranging in the hundreds of micrometers.[17] The cladding modes attenuate rapidly as they propagate along the length of the fiber, due to the lossy cladding-coating interface and bends in the fiber. Since $\Delta\beta$ is discrete and a function of the wavelength, this coupling to the cladding modes is highly selective, leading to a wavelength-dependent loss. As a result, any modulation of the core and cladding guiding properties modifies the spectral response of long-period gratings, and this phenomenon can be utilized for sensing purposes. Moreover, since the cladding modes interact with the fiber jacket or any other material surrounding the cladding, changes in the index of refraction or other properties of these effective coatings materials can also be detected.

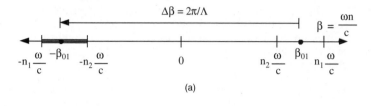

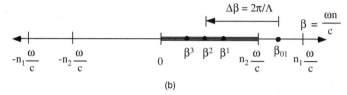

FIGURE 9 Depiction of mode coupling in (*a*) Bragg gratings and (*b*) long-period gratings. The differential propagation constant $\Delta\beta$ determines the grating periodicity.

LPG Fabrication Procedure

To fabricate long-period gratings, hydrogen-loaded (3.4 mole%), germanosilicate fibers may be exposed to 248-nm UV radiation from a KrF excimer laser through a chrome-plated amplitude mask possessing a periodic rectangular transmittance function. Figure 10 shows a typical setup used to fabricate such gratings. The laser is pulsed at approximately 20 Hz with a pulse duration of several nanoseconds. The typical writing times for an energy of 100 mJ/cm²/pulse and a 2.5-cm exposed length vary between 6 and 15 min for different fibers. The coupling wavelength λ_p shifts to higher values during exposure due to the photoinduced enhancement of the refractive index of the fiber core and the resulting increase in β_{01}. After writing, the gratings are annealed at 150°C for several hours to remove the unreacted hydrogen. This high-temperature annealing causes λ_p to move to shorter wavelengths due to the decay of UV-induced defects and the diffusion of molecular hydrogen from the fiber. Figure 11 depicts the typical transmittance of a grating. Various attenuation bands correspond to coupling to discrete cladding modes of different orders. A number of gratings can be fabricated at the same time by placing more than one fiber behind the amplitude mask. Due to the relatively long spatial periods, the stability requirements during the writing process are not so severe as those for short-period Bragg gratings.

For coupling to the highest-order cladding mode, the maximum isolation (loss in transmission intensity) is typically in the 5- to 20-dB range on wavelengths depending on fiber param-

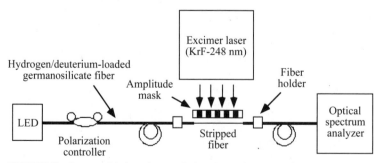

FIGURE 10 Setup to fabricate long-period gratings using an amplitude mask.

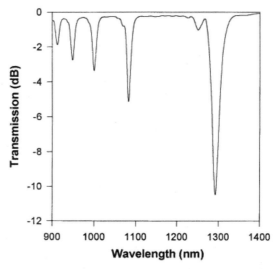

FIGURE 11 Transmission spectrum of a long-period grating written in Corning FLEXCOR fiber with period $\Lambda = 198 \ \mu$m. The discrete, spiky loss bands correspond to the coupling of the fundamental guided mode to discrete cladding modes.

eters, duration of UV exposure, and mask periodicity. The desired fundamental coupling wavelength can easily be varied by using inexpensive amplitude masks of different periodicities. The insertion loss, polarization mode dispersion, backreflection, and polarization-dependent loss of a typical grating are 0.2 dB, 0.01 ps, −80 dB, and 0.02 dB, respectively. The negligible polarization sensitivity and backreflection of these devices eliminates the need for expensive polarizers and isolators.

We now look at representative experiments that have been performed and discussed to examine the displacement sensitivity of long-period gratings written in different fibers.[19,20] For example, gratings have been fabricated in four different types of fibers—standard dispersion-shifted fiber (DSF), standard 1550-nm fiber, and conventional 980- and 1050-nm single-mode fibers. For the sake of brevity, these will be referred to as fibers A, B, C, and D, respectively. The strain sensitivity of gratings written in different fibers was determined by axially straining the gratings between two longitudinally separated translation stages. The shift in the peak loss wavelength of the grating in fiber D as a function of the applied strain is depicted in Fig. 12 along with that for a Bragg grating (about 9 nm/%ε at 1300 nm).[7] The strain coefficients of wavelength shift β for fibers A, B, C, and D are shown in Table 1. Fiber D has a coefficient of 15.2 nm/%ε, which gives it a strain-induced shift that is 50 percent larger than that for a conventional Bragg grating. The strain resolution of this fiber for a 0.1-nm detectable wavelength shift is 65.75 με.

TABLE 1 Strain Sensitivity of Long-Period Gratings Written in Four Different Types of Fibers

Type of fiber	Strain sensitivity (nm/%ε)
A—standard dispersion-shifted fiber (DSF)	−7.27
B—standard 1550-nm communication fiber	4.73
C—conventional 980-nm single-mode fiber	4.29
D—conventional 1060-nm single-mode fiber	15.21

Values correspond to the shift in the highest order resonance wavelength.

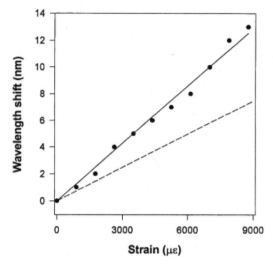

FIGURE 12 Shift in the highest order resonance band with strain for a long-period grating written in fiber D (circles). Also depicted is the shift for a conventional Bragg grating (dashed line).

The demodulation scheme of a sensor determines the overall simplicity and sensitivity of the sensing system. Short-period Bragg grating sensors were shown to possess signal processing techniques that are complex and expensive to implement. We now present a simple demodulation method to extract information from long-period gratings. The wide bandwidth of the resonance bands enables the wavelength shift due to the external perturbation to be converted into an intensity variation that can be easily detected.

Figure 13 shows the shift induced by strain in a grating written in fiber C. The increase in the loss at 1317 nm is about 1.6 dB. A laser diode centered at 1317 nm was used as the optical source, and the change in transmitted intensity was monitored as a function of applied strain. The transmitted intensity is plotted in Fig. 14 for three different trials. The repeatability of the experiment demonstrates the feasibility of using this simple scheme to utilize the high sensitivity of long-period gratings. The transmission of a laser diode centered on the slope of the grating spectrum on either side of the resonance wavelength can be used as a measure of the applied perturbation. A simple detector and amplifier combination at the output can be used to determine the transmission through the detector. On the other hand, a broadband source can also be used to interrogate the grating. At the output an optical bandpass filter can be used to transmit only a fixed bandwidth of the signal to the detector. The bandpass filter should again be centered on either side of the peak loss band of the resonance band. These schemes are easy to implement, and unlike the case for conventional Bragg gratings, complex and expensive interferometric demodulation schemes are not necessary.[20]

Temperature Sensitivity of Long-Period Gratings

Gratings written in different fibers were also tested for their cross-sensitivity to temperature.[20] The temperature coefficients of wavelength shift for different fibers are shown in Table 2. The temperature sensitivity of a fiber Bragg grating is 0.014 nm/°C. Hence the temperature sensitivity of a long-period grating is typically an order of magnitude higher than that of a Bragg grating. This large cross-sensitivity to ambient temperature can degrade the strain sens-

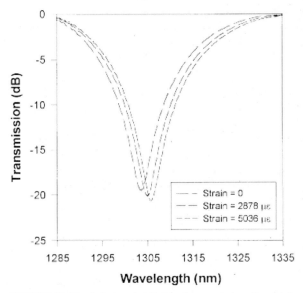

FIGURE 13 Strain-induced shift in a long-period grating fabricated in fiber C. The loss at 1317 nm increases by 1.6 dB due to the applied strain (5036 µε).

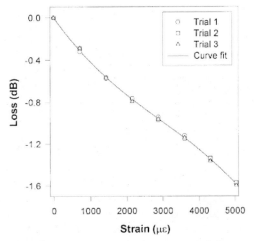

FIGURE 14 The change in the grating transmission at 1317 nm as a function of strain for three different trials. The increase in loss by 1.6 dB at 5036 µε provides evidence of the feasibility of the simple setup used to measure strain.

TABLE 2   Temperature Sensitivity of Long-period Gratings Written in Four Different Types of Fibers

Type of fiber	Temperature sensitivity (nm/°C)
A—standard dispersion-shifted fiber (DSF)	0.062
B—standard 1550-nm communication fiber	0.058
C—conventional 980-nm single mode fiber	0.154
D—conventional 1060 nm single mode fiber	0.111

Values correspond to the shift in the highest order resonance wavelength.

ing performance of the system unless the output signal is adequately compensated. Multi-parameter sensing using long-period gratings has been proposed to obtain precise strain measurements in environments with temperature fluctuations.[19]

In summary, long-period grating sensors are highly versatile. These sensors can easily be used in conjunction with simple and inexpensive detection techniques. Experimental results prove that these methods can be used effectively without sacrificing the enhanced resolution of the sensors. Long-period grating sensors are insensitive to input polarization and do not require coherent optical sources. Cross-sensitivity to temperature is a major concern while using these gratings for strain measurements.

15.6   COMPARISON OF SENSING SCHEMES

Based on these results, interferometric sensors have a high sensitivity and bandwidth but are limited by nonlinearity in their output signals. Conversely, intrinsic sensors are susceptible to ambient temperature changes, while grating-based sensors are simpler to multiplex. Each may be used in specific applications.

15.7   CONCLUSION

We have briefly summarized the performance of four different interferometric and grating-based sensors as representative of the very wide range of possible optical fiber sensor instrumentation and approaches. This analysis was based on the sensor head fabrication and cost, signal processing, cross-sensitivity to temperature, resolution, and operating range. Relative merits and demerits of the various sensing schemes were discussed.

15.8   REFERENCES

1. R. O. Claus, M. F. Gunther, A. Wang, and K. A. Murphy, "Extrinsic Fabry-Perot Sensor for Strain and Crack Opening Displacement Measurements from Minus 200 to 900°C," *Smart Mat. and Struct*, **1**:237–242 (1992).

2. K. A. Murphy, M. F. Gunther, A. M. Vengsarkar, and R. O. Claus, "Fabry-Perot Fiber Optic Sensors in Full-Scale Fatigue Testing on an F-15 Aircraft," *App. Opt.*, **31**:431–433 (1991).

3. V. Bhatia, C. A. Schmid, K. A. Murphy, R. O. Claus, T. A. Tran, J. A. Greene, and M. S. Miller, "Optical Fiber Sensing Technique for Edge-Induced and Internal Delamination Detection in Composites," *J. Smart Mat Strut.*, **4** (1995).

4. C. E. Lee and H. F. Taylor, "Fiber-Optic Fabry-Perot Temperature Sensor Using a Low-Coherence Light Source," *J. Lightwave Technol.*, **9**:129–134 (1991).

5. J. A. Greene, T. A. Tran, K. A. Murphy, A. J. Plante, V. Bhatia, M. B. Sen, and R. O. Claus, "Photo-induced Fresnel Reflectors for Point-Wise and Distributed Sensing Applications," in *Proceedings of the Conference on Smart Structures and Materials,* SPIE'95, paper 2444-05, February 1995.

6. J. Sirkis, "Phase-Strain-Temperature Model for Structurally Embedded Interferometric Optical Fiber Strain Sensors with Applications," *Fiber Opt. Smart Struct. Skins IV,* SPIE **1588** (1991).

7. K. O. Hill, Y. Fuiji, D. C. Johnson, and B. S. Kawasaki, "Photosensitivity in Optical Fiber Waveguides: Applications to Reflection Filter Fabrication," *Appl. Phys. Lett.* **32**:647 (1978).

8. G. Meltz, W. W. Morey, and W. H. Glenn, "Formation of Bragg Gratings in Optical Fibers by Transverse Holographic Method," *Opt. Lett.* **14**:823 (1989).

9. P. J. Lemaire, A. M. Vengsarkar, W. A. Reed, V. Mizrahi, and K. S. Kranz, "Refractive Index Changes in Optical Fibers Sensitized with Molecular Hydrogen," in *Proceedings of the Conference on Optical Fiber Communications,* OFC'94, Technical Digest, paper TuL1, 1994, p. 47.

10. R. Kashyap, "Photosensitive Optical Fibers: Devices and Applications," *Opt. Fiber Technol.* **1**:17–34 (1994).

11. D. Z. Anderson, V. Mizrahi, T. Ergodan, and A. E. White, "Phase-Mask Method for Volume Manufacturing of Fiber Phase Gratings," *in Proceedings of the Conference on Optical Fiber Communication,* post-deadline paper PD16, 1993, p. 68.

12. A. D. Kersey and T. A. Berkoff, "Fiber-Optic Bragg-Grating Differential-Temperature Sensor," *IEEE Phot. Techno. Lett.* **4**:1183–1185 (1992).

13. V. Bhatia, M. B. Sen, K. A. Murphy, A. Wang, R. O. Claus, M. E. Jones, J. L. Grace, and J. A. Greene, "Demodulation of Wavelength-Encoded Optical Fiber Sensor Signals Using Fiber Modal Interferometers," *SPIE Photon. East,* Philadelphia, PA, paper 2594–09, October 1995.

14. M. G. Xu, L. Dong, L. Reekie, J. A. Tucknott, and J. L. Cruz, "Chirped Fiber Gratings for Temperature-Independent Strain Sensing," in *Proceedings of the First OSA Topical Meeting on Photosensitivity and Quadratic Nonlinearity in Glass Waveguides: Fundamentals and Applications,* paper PMB2, 1995.

15. K. O. Hill, B. Malo, K. Vineberg, F. Bilodeau, D. Johnson, and I. Skinner, "Efficient Mode-Conversion in Telecommunication Fiber Using Externally Written Gratings," *Electron. Lett.,* **26**:1270–1272 (1990).

16. F. Bilodeau, K. O. Hill, B. Malo, D. Johnson, and I. Skinner, "Efficient Narrowband $LP_{01} <-> LP_{02}$ Mode Converters Fabricated in Photosensitive Fiber: Spectral Response," *Electron. Lett.* **27**:682–684 (1991).

17. A. M. Vengsarkar, P. J. Lemaire, J. B. Judkins, V. Bhatia, J. E. Sipe, and T. E. Ergodan, "Long-Period Fiber Gratings as Band-Rejection Filters", in *Proceedings of Conference on Optical Fiber Communications,* OFC'95, post-deadline paper, PD4-2, 1995.

18. A. M. Vengsarkar, P. J. Lemaire, J. B. Judkins, V. Bhatia, J. E. Sipe, and T. E. Ergodan, "Long-Period Fiber Gratings as Band-Rejection Filters," *J. Lightwave Technol.,* accepted for publication, 1995.

19. V. Bhatia, M. B. Burford, K. A. Murphy, and A. M. Vengsarkar, "Long-Period Fiber Grating Sensors," in *Proceedings of the Conference on Optical Fiber Communication,* paper ThP1, February 1996.

20. V. Bhatia and A. M. Vengsarkar, "Optical Fiber Long-Period Grating Sensors," *Opt. Lett.* 1995, submitted.

15.9 FURTHER READING

Bhatia, V., M. J. de Vries, K. A. Murphy, R. O. Claus, T. A. Tran, and J. A. Greene, "Extrinsic Fabry-Perot Interferometers for Absolute Measurements," *Fiberoptic Prod. News* **9**:12–3, (December 1994).

Bhatia, V., M. B. Sen, K. A. Murphy, and R. O. Claus, "Wavelength-Tracked White Light Interferometry for Highly Sensitive Strain and Temperature Measurements", *Electron. Lett.,* 1995, submitted.

Butter, C. D., and G. B. Hocker, "Fiber Optics Strain Gage," *Appl. Opt.,* **17**:2867–2869 (1978).

J. S. Sirkis and H. W. Haslach, "Interferometric Strain Measurement by Arbitrarily Configured, Surface Mounted, Optical Fiber," *Lightwave Technol.,* **8**:1497–1503 (1990).

CHAPTER 16
FIBER-OPTIC COMMUNICATION STANDARDS

Casimer DeCusatis
IBM Corporation
Poughkeepsie, New York

16.1 INTRODUCTION

In the past 10 years several international standards have been adopted for optical communications. This chapter presents a brief overview of several major industry standards, including the following:

- ESCON/SBCON (Enterprise System Connection / Serial Byte Connection)
- FDDI (Fiber Distributed Data Interface)
- Fibre Channel Standard
- ATM (Asynchronous Transfer Mode) / SONET (Synchronous Optical Network)
- Gigabit Ethernet

16.2 ESCON

The Enterprise System Connection (ESCON)* architecture was introduced on the IBM System/390 family of mainframe computers in 1990 as an alternative high-speed I/O channel attachment.[1,2] The ESCON interface specifications were adopted in 1996 by the ANSI X3T1 committee as the serial byte connection (SBCON) standard.[3]

The ESCON/SBCON channel is a bidirectional, point-to-point 1300-nm fiber-optic data link with a maximum data rate of 17 Mbytes/s (200 Mbit/s). ESCON supports a maximum unrepeated distance of 3 km using 62.5-micron multimode fiber and LED transmitters with an 8-dB link budget, or a maximum unrepeated distance of 20 km using single-mode fiber and laser transmitters with a 14-dB link budget. The laser channels are also known as the ESCON

---

* ESCON is a registered trademark of IBM Corporation, 1991

extended distance feature (XDF). Physical connection is provided by an ESCON duplex connector, illustrated in Fig. 1. Recently, the single-mode ESCON links have adopted the SC duplex connector as standardized by Fibre Channel. With the use of repeaters or switches, an ESCON link can be extended up to 3 to 5 times these distances; however, performance of the attached devices typically falls off quickly at longer distances due to the longer round-trip latency of the link, making this approach suitable only for applications that can tolerate a lower effective throughput, such as remote backup of data for disaster recovery. ESCON devices and CPUs may communicate directly through a channel-to-channel attachment, but more commonly attach to a central nonblocking dynamic crosspoint switch. The resulting network topology is similar to a star-wired ring, which provides both efficient bandwidth utilization and reduced cabling requirements. The switching function is provided by an ESCON director, a nonblocking circuit switch. Although ESCON uses 8B/10B encoded data, it is not a packet-switching network; instead, the data frame header includes a request for connection that is established by the director for the duration of the data transfer. An ESCON data frame includes a header, payload of up to 1028 bytes of data, and a trailer. The header consists of a two-character start-of-frame delimiter, two-byte destination address, two-byte source address, and one byte of link control information. The trailer is a two-byte cyclic redundancy check (CRC) for errors, and a three-character end-of-frame delimiter. ESCON uses a DC-balanced 8B/10B coding scheme developed by IBM.

16.3   FDDI

The fiber distributed data interface (FDDI) was among the first open networking standards to specify optical fiber. It was an outgrowth of the ANSI X3T9.5 committee proposal in 1982 for a high-speed token passing ring as a back-end interface for storage devices. While interest in this application waned, FDDI found new applications as the backbone for local area networks (LANs). The FDDI standard was approved in 1992 as ISO standards IS 9314/1-2 and DIS 9314-3; it follows the architectural concepts of IEEE standard 802 (although it is controlled by ANSI, not IEEE, and therefore has a different numbering sequence) and is among the family of standards (including token ring and ethernet) that are compatible with a common IEEE 802.2 interface. FDDI is a family of four specifications, namely the physical layer (PHY), physical media dependent (PMD), media access control (MAC), and station management (SMT). These four specifications correspond to sublayers of the data link and physical layer of the OSI reference model; as before, we will concentrate on the physical layer implementation.

The FDDI network is a 100-Mbit/s token passing ring, with dual counterrotating rings for fault tolerance. The dual rings are independent fiber-optic cables; the primary ring is used for data transmission, and the secondary ring is a backup in case a node or link on the primary ring fails. Bypass switches are also supported to reroute traffic around a damaged area of the

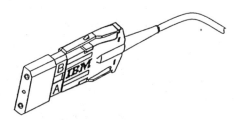

FIGURE 1   ESCON duplex fiber-optic connector.

network and prevent the ring from fragmenting in case of multiple node failures. The actual data rate is 125 Mbit/s, but this is reduced to an effective data rate of 100 Mbit/s by using a 4B/5B coding scheme. This high speed allows FDDI to be used as a backbone to encapsulate lower speed 4, 10, and 16 Mbit/s LAN protocols; existing ethernet, token ring, or other LANs can be linked to an FDDI network via a bridge or router. Although FDDI data flows in a logical ring, a more typical physical layout is a star configuration with all nodes connected to a central hub or concentrator rather than to the backbone itself. There are two types of FDDI nodes, either *dual attach* (connected to both rings) or *single attach;* a network supports up to 500 dual-attached nodes, 1000 single-attached nodes, or an equivalent mix of the two types. FDDI specifies 1300-nm LED transmitters operating over 62.5-micron multimode fiber as the reference media, although the standard also provides for the attachment of 50, 100, 140, and 85 micron fiber. Using 62.5-micron fiber, a maximum distance of 2 km between nodes is supported with an 11-dB link budget; since each node acts like a repeater with its own phase-lock loop to prevent jitter accumulation, the entire FDDI ring can be as large as 100 km. However, an FDDI link can fail due to either excessive attenuation or dispersion; for example, insertion of a bypass switch increases the link length and may cause dispersion errors even if the loss budget is within specifications. For most other applications, this does not occur because the dispersion penalty is included in the link budget calculations or the receiver sensitivity measurements. The physical interface is provided by a special media interface connector (MIC), illustrated in Fig. 2. The connector has a set of three color-coded keys which are interchangable depending on the type of network connection[1]; this is intended to prevent installation errors and assist in cable management.

An FDDI data frame is variable in length and contains up to 4500 8-bit bytes, or octets, including a preamble, start of frame, frame control, destination address, data payload, CRC error check, and frame status/end of frame. Each node has a MAC sublayer that reviews all the data frames looking for its own destination address. When it finds a packet destined for its node, that frame is copied into local memory; a copy bit is turned on in the packet; and it is then sent on to the next node on the ring. When the packet returns to the station that originally sent it, the originator assumes that the packet was received if the copy bit is on; the originator will then delete the packet from the ring. As in the IEEE 802.5 token ring protocol, a special type of packet called a *token* circulates in one direction around the ring, and a node can only transmit data when it holds the token. Each node observes a token retention time limit, and also keeps track of the elapsed time since it last received the token; nodes may be given the token in equal turns, or they can be given priority by receiving it more often or holding it longer after they receive it. This allows devices having different data requirements to be served appropriately.

Because of the flexibility built into the FDDI standard, many changes to the base standard have been proposed to allow interoperability with other standards, reduce costs, or extend FDDI into the MAN or WAN. These include a single-mode PMD layer for channel extensions up to 20 to 50 km. An alternative PMD provides for FDDI transmission over copper wire, either shielded or unshielded twisted pairs; this is known as *copper distributed data inter-*

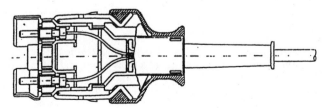

FIGURE 2   FDDI duplex fiber-optic connector.

face, or CDDI. A new PMD is also being developed to adapt FDDI data packets for transfer over a SONET link by stuffing approximately 30 Mbit/s into each frame to make up for the data rate mismatch (we will discuss SONET as an ATM physical layer in a later section). An enhancement called FDDI-II uses time-division multiplexing to divide the bandwidth between voice and data; it would accommodate isochronous, circuit-switched traffic as well as existing packet traffic. Recently, an option known as *low cost* (LC) FDDI has been adopted. This specification uses the more common SC duplex connector instead of expensive MIC connectors, and a lower-cost transceiver with a 9-pin footprint similar to the single-mode ESCON parts.

16.4   *FIBRE CHANNEL STANDARD*

Development of the ANSI Fibre Channel (FC) Standard began in 1988 under the X3T9.3 Working Group, as an outgrowth of the Intelligent Physical Protocol Enhanced Physical Project. The motivation for this work was to develop a scaleable standard for the attachment of both networking and I/O devices, using the same drivers, ports, and adapters over a single channel at the highest speeds currently achievable. The standard applies to both copper and fiber-optic media, and uses the English spelling *fibre* to denote both types of physical layers. In an effort to simplify equipment design, FC provides the means for a large number of existing upper-level protocols (ULPs), such as IP, SCI, and HIPPI, to operate over a variety of physical media. Different ULPs are mapped to FC constructs, encapsulated in FC frames, and transported across a network; this process remains transparent to the attached devices. The standard consists of five hierarchical layers,[4] namely a physical layer, an encode/decode layer which has adopted the DC-balanced 8B/10B code, a framing protocol layer, a common services layer (at this time, no functions have been formally defined for this layer), and a protocol-mapping layer to encapsulate ULPs into FC. The second layer defines the Fibre Channel data frame; frame size depends upon the implementation and is variable up to 2148 bytes long. Each frame consists of a 4-byte start-of-frame delimiter, a 24-byte header, a 2112-byte payload containing from 0 to 64 bytes of optional headers and 0 to 2048 bytes of data, a 4-byte CRC, and a 4-byte end-of-frame delimiter. In October 1994, the Fibre Channel physical and signaling interface standard FC-PH was approved as ANSI standard X3.230-1994.

Logically, Fibre Channel is a bidirectional point-to-point serial data link. Physically, there are many different media options (see Table 1) and three basic network topologies. The simplest, *default topology,* is a point-to-point direct link between two devices, such as a CPU and a device controller. The second, *Fibre Channel Arbitrated Loop* (FC-AL), connects between 2 and 126 devices in a loop configuration. Hubs or switches are not required, and there is no dedicated loop controller; all nodes on the loop share the bandwidth and arbitrate for temporary control of the loop at any given time. Each node has equal opportunity to gain control of the loop and establish a communications path; once the node relinquishes control, a fairness algorithm ensures that the same node cannot win control of the loop again until all other nodes have had a turn. As networks become larger, they may grow into the third topology, an *interconnected switchable network* or *fabric* in which all network management functions are taken over by a switching point, rather than each node. An analogy for a switched fabric is the telephone network; users specify an address (phone number) for a device with which they want to communicate, and the network provides them with an interconnection path. In theory there is no limit to the number of nodes in a fabric; practically, there are only about 16 million unique addresses. Fibre Channel also defines three classes of connection service, which offer options such as guaranteed delivery of messages in the order they were sent and acknowledgment of received messages.

As shown in Table 1, FC provides for both single-mode and multimode fiber-optic data links using longwave (1300-nm) lasers and LEDs as well as short-wave (780 to 850 nm) lasers. The physical connection is provided by an SC duplex connector defined in the standard (see

TABLE 1 Fiber Channel Standard Physical Layer

Media type	Data rate (Mbytes/s)	Maximum distance	Signaling rate (Mbaud)	Transmitter
SMF	100	10 km	1062.5	LW laser
	50	10 km	1062.5	LW laser
	25	10 km	1062.5	LW laser
50-μm multimode fiber	100	500 m	1062.5	SW laser
	50	1 km	531.25	SW laser
	25	2 km	265.625	SW laser
	12.5	10 km	132.8125	LW LED
62.5-μm multimode fiber	100	300 m	1062.5	SW laser
	50	600 m	531.25	SW laser
	25	1 km	265.625	LW LED
	12.5	2 km	132.8125	LW LED
105-Ω type 1 shielded twisted pair electrical	25	50 m	265.125	ECL
	12.5	100 m	132.8125	ECL
75 Ω mini coax	100	10 m	1062.5	ECL
	50	20 m	531.25	ECL
	25	30 m	265.625	ECL
	12.5	40 m	132.8125	ECL
75 Ω video coax	100	25 m	1062.5	ECL
	50	50 m	531.25	ECL
	25	75 m	265.625	ECL
	12.5	100 m	132.8125	ECL
150 Ω twinax or STP	100	30 m	1062.5	ECL
	50	60 m	531.25	ECL
	25	100 m	265.625	ECL

Here LW is long wavelength, SW is short wavelength, and ECL is emitter-coupled logic.

Fig. 3), which is keyed to prevent misplugging of a multimode cable into a single-mode recep-tacle. This connector design has since been adopted by other standards, including ATM, low-cost FDDI, and single-mode ESCON. The requirement for international class 1 laser safety is addressed using open fiber control (OFC) on some types of multimode links with shortwave lasers. This technique automatically senses when a full duplex link is interrupted, and turns off the laser transmitters on both ends to preserve laser safety. The lasers then transmit low-duty cycle optical pulses until the link is reestablished; a handshake sequence then automatically reactivates the transmitters.

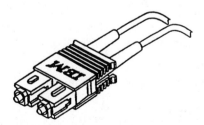

FIGURE 3 Single-mode SC duplex fiber-optic connector, per ANSI FC Standard specifications, with one narrow key and one wide key. Multi-mode SC duplex connectors use two wide keys.

16.5 ATM/SONET

Developed by the ATM Forum, this protocol has promised to provide a common transport media for voice, data, video, and other types of multimedia. ATM is a high-level protocol that can run over many different physical layers including copper; part of ATM's promise to merge voice and data traffic on a single network comes from plans to run ATM over the synchronous optical network (SONET) transmission hierarchy developed for the telecommunications industry. SONET is really a family of standards defined by ANSI T1.105-1988 and T1.106-1988, as well as by several CCITT recommendations.[5-8] Several different data rates are defined as multiples of 51.84 Mbit/s, known as OC-1. The numerical part of the OC-level designation indicates a multiple of this fundamental data rate, thus 155 Mbit/s is called OC-3. The standard provides for 7 incremental data rates, OC-3, OC-9, OC-12, OC-18, OC-24, OC-36, and OC-48 (2.48832 Gbit/s). Both single-mode links with laser sources and multimode links with LED sources are defined for OC-1 through OC-12; only single-mode laser links are defined for OC-18 and beyond. SONET also contains provisions to carry sub-OC-1 data rates, called *virtual tributaries,* which support telecom data rates including DS-1 (1.544 Mbit/s), DS-2 (6.312 Mbit/s), and 3.152 Mbit/s (DS1C). The basic SONET data frame is an array of nine rows with 90 bytes per row, known as a synchronous-transport signal level 1 (STS-1) frame. In an OC-1 system, an STS-1 frame is transmitted once every 125 microseconds (810 bytes per 125 microseconds yields 51.84 Mbit/s). The first three columns provide overhead functions such as identification, framing, error checking, and a pointer which identifies the start of the 87-byte data payload. The payload floats in the STS-1 frame, and may be split across two consecutive frames. Higher speeds can be obtained either by concatenation of *N* frames into an STS-Nc frame (the "c" stands for *concatenated*) or by byte-interleaved multiplexing of *N* frames into an STS-N frame.

ATM technology incorporates elements of both circuit and packet switching. All data is broken down into a 53-byte cell, which may be viewed as a short fixed-length packet. Five bytes make up the header, providing a 48-byte payload. The header information contains routing information (cell addresses) in the form of virtual path and channel identifiers; a field to identify the payload type; an error check on the header information; and other flow control information. Cells are generated asynchronously; as the data source provides enough information to fill a cell, it is placed in the next available cell slot. There is no fixed relationship between the cells and a master clock, as in conventional time-division multiplexing schemes; the flow of cells is driven by the bandwidth needs of the source. ATM provides bandwidth on demand; for example, in a client-server application the data may come in bursts; several data sources could share a common link by multiplexing during the idle intervals. Thus, the ATM adaptation layer allows for both constant and variable bit rate services. The combination of transmission options is sometimes described as a *pleisosynchronous network,* meaning that it combines some features of multiplexing operations without requiring a fully synchronous implementation. Note that the fixed cell length allows the use of synchronous multiplexing and switching techniques, while the generation of cells on demand allows flexible use of the link bandwidth for different types of data, characteristic of packet switching. Higher-level protocols may be required in an ATM network to ensure that multiplexed cells arrive in the correct order, or to check the data payload for errors (given the typical high reliability and low BER of modern fiber-optic technology, it was considered unnecessary overhead to replicate data error checks at each node of an ATM network). If an intermediate node in an ATM network detects an error in the cell header, cells may be discarded without notification to either end user. Although cell loss priority may be defined in the ATM header, for some applications the adoption of unacknowledged transmission may be a concern.

ATM data rates were intended to match SONET rates of 51, 155, and 622 Mbit/s; an FDDI-compliant data rate of 100 Mbit/s was added, in order to facilitate emulation of different types of LAN traffic over ATM. In order to provide a low-cost copper option and compatibility with 16-Mbit/s token ring LANs to the desktop, a 25-Mbit/s speed has also been approved. For premises wiring applications, ATM specifies the SC duplex connector, color coded beige for multimode links and blue for single-mode links. At 155 Mbit/s, multimode ATM links support a maximum distance of 3 km while single-mode links support up to 20 km.

16.6   GIGABIT ETHERNET

Ethernet is a local area network (LAN) communication standard originally developed for copper interconnections on a common data bus; it is an IEEE standard 802.3.[9] The basic principle used in Ethernet is carrier sense multiple access with collision detection (CSMA/CD). Ethernet LANs are typically configured as a bus, often wired radially through a central hub. A device attached to the LAN that intends to transmit data must first sense whether another device is transmitting. If another device is already sending, then it must wait until the LAN is available; thus, the intention is that only one device will be using the LAN to send data at a given time. When one device is sending, all other attached devices receive the data and check to see if it is addressed to them; if it is not, then the data is discarded. If two devices attempt to send data at the same time (for example, both devices may begin transmission at the same time after determining that the LAN is available; there is a gap between when one device starts to send and before another potential sender can detect that the LAN is in use), then a collision occurs. Using CSMA/CD as the media access control protocol, when a collision is detected attached devices will detect the collision and must wait for different lengths of time before attempting retransmission. Since it is not always certain that data will reach its destination without errors or that the sending device will know about lost data, each station on the LAN must operate an end-to-end protocol for error recovery and data integrity. Data frames begin with an 8-byte preamble used for determining start-of-frame and synchronization, and a header consisting of a 6-byte destination address, 6-byte source address, and 2-byte length field. User data may vary from 46 to 1500 bytes, with data shorter than the minimum length padded to fit the frame; the user data is followed by a 2-byte CRC error check. Thus, an Ethernet frame may range from 70 bytes to 1524 bytes.

The original Ethernet standard, known also as 10Base-T (10 Mbit/s over unshielded twisted pair copper wires) was primarily a copper standard, although a specification using 850-nm LEDs was also available. Subsequent standardization efforts increased this data rate to 100 Mbit/s over the same copper media (100Base-T), while once again offering an alternative fiber specification (100Base-FX). Recently, the standard has continued to evolve with the development of Gigabit Ethernet (1000Base-FX), which will operate over fiber as the primary medium; this has the potential to be the first networking standard for which the implementation cost on fiber is lower than on copper media. Under development as IEEE 802.3z, the gigabit Ethernet standard was approved in late 1998. Gigabit Ethernet will include some changes to the MAC layer in addition to a completely new physical layer operating at 1.25 Gbit/s. Switches rather than hubs are expected to predominate, since at higher data rates throughput per end user and total network cost are both optimized by using switched rather than shared media. The minimum frame size has increased to 512 bytes; frames shorter than this are padded with idle characters (carrier extension). The maximum frame size remains unchanged, although devices may now transmit multiple frames in bursts rather than single frames for improved efficiency. The physical layer will use standard 8B/10B data encoding. The standard does not specify a physical connector type for fiber; at this writing there are several proposals, including the SC duplex and various small-form-factor connectors about the size of a standard RJ-45 jack. Transceivers may be packaged as gigabit interface converters, or GBICs, which allows different optical or copper transceivers to be plugged onto the same host card. There is presently a concern with proposals to operate long-wave (1300 nm) laser sources over both single-mode and multimode fiber. When a transmitter is optimized for a single-mode launch condition, it will underfill the multimode fiber; this causes some modes to be excited and propagate at different speeds than others, and the resulting differential mode delay significantly degrades link performance. One proposed solution involves the use of special optical cables known as optical mode conditioners with offset ferrules to simulate an equilibrium mode launch condition into multimode fiber.

16.7 REFERENCES

1. D. Stigliani, "Enterprise Systems Connection Fiber Optic Link," Chapter 13 in *Handbook of Optoelectronics for Fiber Optic Data Communications,* C. DeCusatis, R. Lasky, D. Clement, and E. Mass (eds.) Academic Press, 1997.

2. "ESCON I/O Interface Physical Layer Document" (IBM document number SA23-0394), third edition, IBM Corporation, Mechanicsburg, Pennsylvania, 1995.

3. ANSI Single Byte Command Code Sets CONnection architecture (SBCON), draft ANSI standard X3T11/95-469 (rev. 2.2) 1996.

4. ANSI X3.230-1994 rev. 4.3, Fibre Channel—Physical and Signaling Interface (FC-PH), ANSI X3.272-199x, rev. 4.5, Fibre Channel—Arbitrated Loop (FC-AL), June 1995, ANSI X3.269-199x, rev. 012, Fiber Channel Protocol for SCSI (FCP), May 30, 1995.

5. ANSI T1.105-1988, Digital Hierarchy Optical Rates and Format Specification.

6. CCITT Recommendation G.707, Synchronous Digital Hierarchy Bit Rates.

7. CCITT Recommendation G.708, Network Node Interfaces for the Synchronous Digital Hierarchy.

8. CCITT Recommendation G.709, Synchronous Multiplexing Structure.

9. IEEE 802.3z, Draft Supplement to Carrier Sense Multiple Access with Collision Detection (CSMA/CD) Access Method and Physical Layer Specifications: Media Access Control (MAC) Parameters, Physical Layer, Repeater and Management Parameters for 1000 Mb/s Operation (June 1997).

INDEX

Index notes: The *f* after a page number refers to a figure, the *n* to a note, and the *t* to a table.

ABOUT THE EDITORS

Michael Bass is Professor of Optics, Physics, and Electrical and Computer Engineering in the School of Optics/Center for Research and Education in Optics and Lasers at the University of Central Florida. He received his B.S. in physics from Carnegie-Mellon, and his M.S. and Ph.D. in physics from the University of Michigan.

Eric W. Van Stryland is Professor of Optics, Physics, and Electrical and Computer Engineering in the School of Optics/Center for Research and Education in Optics and Lasers at the University of Central Florida. He received his Ph.D. from the University of Arizona.

The **Optical Society of America** is dedicated to advancing study, teaching, research, and engineering in optics.